ANIMAL HEALTH

Third Edition

ANIMAL

NANCY S. JACKSON, D.V.M

WILLIAM J. GREER, D.V.M.

JAMES K. BAKER, Ph.D.

HEALTH

INTERSTATE PUBLISHERS, INC.

Danville, Illinois

ANIMAL HEALTH

Third Edition

Library of Congress Catalog Card No. 99-75340

ISBN 0-8134-3169-7

3 4 5 6 7 8 9 10   04

Order from

INTERSTATE PUBLISHERS, INC.

510 North Vermilion Street
P.O. Box 50
Danville, IL 61834-0050

Phone: (800) 843-4774
Fax: (217) 446-9706
Email: info-ipp@IPPINC.com

World Wide Web: http://www.IPPINC.com

PREFACE

Animal Health provides basic information on animal health and disease prevention. The responsibility for animal health rests first and foremost in the hands of the owner. The owner or caretaker of farm animals is usually the person who provides the basic husbandry that undergirds animal health and recognizes animal ill health and determines when to call for professional assistance.

Diagnosis and treatment of sick animals is the responsibility of the veterinarian. For the caretaker to compete with the veterinarian in diagnosis and treatment is economically unsound. Instead, the caretaker should cooperate with the veterinarian by providing the present and past history of the patient, herd, or flock. In some cases, the caretaker, under the direction and supervision of the veterinarian, will participate in the treatment and nursing care of convalescents. Thus, animal health is dependent upon a cooperative working relationship between the veterinarian and the owner. The aim of this book is to promote this relationship by preparing the caretaker to play an effective role in the prevention of disease in farm animals.

Animal Health is divided into four sections. Part One contains nine chapters that are basic components of a disease prevention program. It includes fundamental information on the nature of disease, nutrition, sanitation, disinfection immunization, quarantine, housing, hereditary abnormalities, and basic husbandry practices.

Part Two deals with the structure and function of the various systems of the animal body and with the signs of disturbances and diseases that affect the digestive, genitourinary, respiratory circulatory, and nervous systems. The cause, clinical signs, prevention, and treatment are presented for each disease.

Part Three includes the cause, clinical signs, prevention, and treatment for diseases that are not commonly associated with a particular system of the animal body. These include generalized infectious diseases, metabolic and deficiency disorders, localized diseases of the skin and extremities, and disorders due to plant and chemical poisoning.

Part Four is devoted to the understanding of parasitology. It covers the internal and the external parasites that are common to livestock. A description, the damage inflicted on the host, symptoms, control procedures, and the recommended treatment are included for each parasite.

ACKNOWLEDGMENTS

The authors express sincere appreciation to those individuals, companies, and educational institutions that supplied pictures, drawings, and technical material. Special recognition is given to Syntex Agribusiness, Inc., Animal Health Division, for permission to reprint sections of the 12-part Herd Health Series and to the Agricultural Chemical Division, Shell Chemical Company, Elanco Products Company, American Cyanamid Company, and Jensen-Salsbery Laboratories for the many pictures they provided. In addition, the following people provided invaluable technical information: Dr. Jack Howarth, Professor, Veterinary Medicine, University of California-Davis; Rusty Pedersen, Nutritionist, Imperial Premix, El Centro, California; and Dennis Hammett, owner and manager of the Southwest Stockyards Company, El Paso, Texas.

Information sheets, circulars, and bulletins that provided excellent drawings and background data for this publication were supplied by various state universities. These include universities in Arizona, California, Colorado, Florida, Georgia, Illinois, Indiana, Michigan, Minnesota, Missouri, Nebraska, New York, Ohio, Oklahoma, Pennsylvania, and Texas.

The authors also wish to acknowledge the contributions of ranchers, feedlot operators, and agriculture teachers who provided counsel and assistance. A special note of appreciation must be given to Ed Siemens, Hayward, California, for preparing the sketches of the life cycles of parasites, to Joel Torrevillas, Berkeley, California, for preparing the sketches of the various systems of the animal body, Marcove Farm and George Brink Dairy for the numerous photo sessions, and to Mary Carter of Interstate Publishers, Inc., for new computer-generated graphics. A special note of appreciation should also be given to Dr. Jasper S. Lee, Demorest, Georgia, without whose help this edition would not have been published.

CONTENTS

PART 1

Introduction to Disease Prevention

PART 2

Diseases Associated With Systems of the Animal Body

PART 3

Miscellaneous Diseases of Farm Animals

PART 4

Parasitology

PART I

Introduction to

Disease Prevention

CHAPTER I

The Nature of Disease

Animal diseases are one of civilization's oldest and most formidable enemies. History provides dramatic accounts of the influence of animal deaths, due to plagues, on the destiny of civilizations. Many people died from starvation, and, in some cases, they died from the same disease that killed their animals. Their attempts to flee the "wrath of the gods" by migrating to new lands were often disastrous because the diseases from which they hoped to escape were transported in and on the animals they saved and took with them.

Our response to the problems of animal diseases is a fascinating story with five discernible stages in its history: (1) primitive ignorance and superstition, resulting in fear of disease; (2) accumulation of experience in managing disease, culminating in a professional group specially trained to care for sick animals; (3) develop-

Fig. 1-1. Health is vibrant and beautiful, as evidenced by this Polled Hereford bull on the range. (Courtesy, American Polled Hereford Assn., Kansas City, Mo.)

ment of verifiable knowledge of the causes of disease; (4) discovery of sulfonamides, penicillin and other antibiotics, resulting in overreliance on the treatment of sick animals; and (5) current emphasis directed toward preventative rather than curative measures in the control of disease.

Livestock producers, be they commercial producers, FFA members, or 4-H Club members, need to understand the nature of disease—its cause, transmission, and prevention, if they are to realize a profit from their businesses (Fig. 1-1).

DISEASE

Disease may be correctly defined as "not at ease" because the prefix *dis* denotes reversal or separation from the root *ease*. Animal ill health is synonymous with the word *disease*. They both describe a condition that results from any structural defect or functional impairment of the animal body. Some diseases are not easily detected until they are in the terminal stages; however, most diseases are manifested by signs of disturbances called *symptoms*. The caretaker should be fully aware of the appearance, movement, and daily habits of healthy animals so that abnormal behavior can be detected early.

Cause of Disease

Ancient people believed that disease was caused by demons and ill health was a visitation from displeased gods or devils. It was not until the 1400s that scientists began to suspect that some diseases were caused by minute, invisible particles called "germs." Early researchers called these germs "living seeds of disease." Only during the past 100 years or so have the causes of infectious diseases been known.

Bacteria and other microorganisms were first viewed with a microscope in the 1600s. But, the "germ theory" of disease was not proved until the late 1800s. Robert Koch, a German physician, Louis Pasteur, a French chemist, and Delafond, a French veterinarian, experimented with anthrax, an infectious disease of humans and animals. Delafond found that the blood of sheep which had died from anthrax contained microorganisms that would reproduce. Koch showed that animals injected with anthrax germs soon got the disease; whereas, Pasteur developed a vaccine to protect animals from anthrax.

Diseases are often classified into two major categories according to the cause of morbidity: noninfectious and infectious.

Non-infectious Disease

Diseases caused by non-living agents, such as nutritional deficiencies, metabolic disorders, trauma, toxic materials, and congenital defects, are classified as noninfectious. Such diseases are also termed non-contagious since they are not capable of being spread from one animal to another by direct or indirect contact.

NUTRITIONAL IMBALANCES

Faulty nutrition is a direct cause of some diseases and a contributing cause of many others. A lack of vitamins, minerals, fats, and amino acids in the ration or an inadequate quantity of feed may pave the way for or result in disease. Animals which receive too little water often become dehydrated, resulting in digestive and respiratory disturbances. Occasionally, nutritional problems arise from hypernutrition, or feeding animals excessively.

METABOLIC DISORDERS

Diseases of this nature result from a physiological inability of an animal to make proper use of nutri-

ents. The nutrients may be adequately supplied in the ration, but faulty assimilation, excessive nutritive demands for fetal development, or a sudden onset of profuse lactation may reduce the blood levels of the nutrients in the animal's body to a point that causes a disorder. Examples of metabolic diseases are milk fever (hypocalcemia)[1], grass tetany (hypomagnesia), ketosis, and pregnancy disease of ewes which is associated with hypoglycemia.

TRAUMA

Wounds or injuries from electrical sources, mechanical devices, or thermal conditions which damage body tissues or impair their function are referred to as traumatic wounds. Cuts, tears, burns, bruises, abrasions, sprains, excessive pressure, electrical shock, and perforations from sharp objects are examples of traumatic injuries.

TOXIC MATERIALS

Toxic or poisonous substances are common causes of noninfectious diseases of livestock. They are usually classified according to the type of material or organism which produces the poison.

1. *Bacterial toxins.* Poisonous substances produced by bacteria that affect animal health are often classified according to the manner in which the toxic material is released from the organism. Endotoxins are poisonous substances retained within the bacterial cell until the cell disintegrates. Exotoxins, when liberated in the body of an animal, have the ability to stimulate the animal to produce antitoxins. Antitoxins tend to neutralize the toxins produced by invading microorganisms.

2. *Metallic and chemical poisons.* Toxic materials included in this category include lead, arsenic, fluorine, copper, molybdenum, chlorinated napthalene, and salt.

3. *Phytotoxins.* Poisonous substances produced by plants such as cockle-burs, loco-weeds, Sudangrass, johnson grass, nightshades, oleanders, castor-beans, and ferns are called phytotoxins.

---

[1]See Glossary for structural analysis of veterinary terminology.

4. *Zootoxins.* Occasionally livestock are poisoned by toxic substances produced by certain snakes, spiders, and bees.

5. *Mycotoxins.* Toxin-producing fungi or molds may produce toxins when they grow on peanuts, brazil nuts, corn and most other cereals, silage, hay, and grasses. These toxic feedstuffs are then ingested by livestock.

CONGENITAL DEFECTS

Abnormalities that existed at or before birth, such as atresia anus, cryptorchidism, hernias, misplaced limbs, and white-heifer disease are congenital defects classified as noninfectious.

Infectious Disease

An infectious disease is one caused by parasites such as bacteria, viruses, protozoa, or fungi. The word *infection* is derived from the Latin *inficere*, meaning "to put into." Common use of the word *infection*, however, denotes the disturbances resulting from the entrance, growth, and activity of disease-causing organisms in or on the animal's body (Fig. 1-2). Such disturbances are usually manifested by one or more of the following signs or symptoms: (1) abnormal body temperature, (2) inflammation, (3) depression, (4) anorexia (lack of appetite), (5) abnormal breathing, (6) diarrhea, and (7) abortion.

Fig. 1-2. A feedlot steer with an infectious disease of the respiratory tract. Note nasal discharge.

Infectious diseases may or may not be contagious. Those diseases spread from one animal to another are referred to as contagious or communicable. Most infectious diseases are rather easily transmitted by direct contact. Some, however, are not passed from animal to animal. They are termed non-contagious diseases. An example of an infectious, non-contagious disease is tetanus.

BODY REACTION TO INFECTION

Livestock produced under the most ideal management program will eventually be exposed to pathogenic organisms. Animals capable of warding off specific pathogens are said to be "resistant" to that particular disease. Resistance depends on the physical condition and the inherent ability of the animal to fight off the invading organisms. A lack of resistance is called "susceptibility."

Animals in good health have internal and external bodily defenses to aid them in keeping their resistance high. An animal's primary defenses inhibit the entrance of microorganisms into the animal's body. The secondary defenses combat those infectious agents that have penetrated the body tissues.

Primary Defenders

The first line of defense includes the skin which provides a protective armor for the delicate and susceptible parts of the animal's body and the mucous membranes which line the soft tissues of openings to the body. Organisms that are ingested may be destroyed by gastric juices or eliminated in the feces. Those entering the respiratory tract are either expelled by coughing or passed into the digestive tract. Secretions from the tear glands trap and wash out invaders, and organisms present in the urinary tract are often washed out in the urine.

Secondary Defenders

Pathogenic organisms that are not contained by the primary defenses are challenged by antibodies and/or phagocytes.

Antibodies are substances produced by an animal in response to the presence of protein-like substances called *antigens*. Antibodies are specific in their effects on microorganisms, on their toxins, and on other chemical compounds. They are classified according to their response to specific antigens.

1. Agglutinins—cause bacteria to adhere to each other in clumps. This greatly increases the efficiency of phagocytes because the bacteria can be engulfed in clumps instead of individually.
2. Antienzymes—neutralize enzymes.
3. Antitoxins—neutralize toxins.
4. Bacteriolysins—destroy bacteria by dissolving them.
5. Opsonins—render microorganisms more susceptible to destruction by phagocytes.
6. Precipitins—cause extracts of bacterial cells or other soluble antigens to settle down into solid particles.

Phagocytes are cells that ingest microorganisms or other cells. They are divided into two groups. Those that are virtually immobile are called fixed phagocytes. They line the blood vessels in the liver, spleen, and bone marrow and attack organisms that pass their way. Free phagocytes include certain leucocytes, or white blood cells, that are rushed to the site of infection to combat foreign invaders.

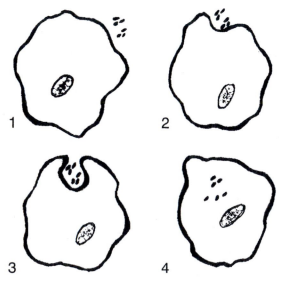

Fig. 1-3.   Phagocytes engulf bacteria.

Phagocytes are more effective in dealing with endotoxins than are antibodies, since they engulf and destroy such cells without releasing the toxic materials (Fig. 1-3). Antibodies combat exotoxins more readily than phagocytes because they tend to neutralize toxins secreted by bacteria. Phagocytes and antibodies make effective secondary defenders for the animal's body, but in situations in which invading organisms are overpowering body defenses, additional assistance in the form of injections of anti-serum or antibiotics will be required to enable the animal to survive.

FATE OF INVADING ORGANISMS

Host Organism Relationships

Several possible fates await disease-causing organisms that enter the animal body. They may be destroyed by the body defense mechanisms or eliminated from the body without causing a disease. When the animal's resistance is low and destruction fails, invading organisms grow and multiply in the body tissues, thus causing an infection. Sometimes, organisms are present in numbers too high for the animal's immune system to handle, and disease will occur. If the infection persists and gradually wears down the animal's resistance, the infection is termed *chronic*. When the pathogenicity of invading organisms rapidly overpowers the animal's resistance, the infection is termed *acute*. In some cases, a situation occurs in which the host animal and invading organism are able to live without serious injury to each other. This is not a voluntary situation on the part of either host or organism. It is a case of the host's inability to eliminate the organism and of the organism's inability to lower the resistance of the host. Quite often, this situation lasts the lifetime of the animal. In other cases, the animal's resistance may be lowered by stress, enabling the infectious agent to become active and cause disease. *Pasteurella* organisms, for example, are commonly found in the respiratory tract of healthy cattle and sheep. Pasteurellosis, or shipping fever, outbreaks do not usually occur unless the animal's resistance is lowered by stress.

When the organism is isolated and surrounded by tissue, the infection is referred to as an arrested case. This does not mean that the animal is entirely cured. In many cases, the organisms become active later, resulting in a flare-up of the disease. Some organisms localize in excretory organs and spread infection to susceptible animals after the host has recovered from the disease. Such animals are called carriers. These animals pose a threat to animal health because detecting a carrier requires commitment by the animal manager, as well as an understanding of the laboratory and veterinary expense that can be incurred in sampling the herd to detect carrier animals.

Elimination by the Host

Some organisms are expelled through secretions or excretions of the infected animal. Milk, for example, is infective from cattle with brucellosis, as is urine in leptospirosis, nasal discharges in respiratory diseases, saliva in rabies, and the discharge of wounds in anthrax. Secretions and excretions from infected animals are common means of spreading diseases of livestock. That is why isolation of sick animals is an important part of a disease prevention program.

Destruction with Carcass

Many organisms perish if the host dies. Proper disposal of the carcass by removal, incineration, or burial, and disinfection of the area in which the animal died are necessary to destroy any remaining organisms and to prevent the spread of disease to healthy animals. Some organisms remain infective within the carcass if it is not composted or incinerated. Many bacteria can cause infections for many years if they are in organic matter that resists disinfection, such as wood and soil. Check with the Environmental Protection Agency for standards of disposal of dead animals in your state.

Means of Transmission

Most infectious organisms are transmitted from one animal to another animal of the same species. Some diseases, however, are spread from one species to another. Certain diseases and parasites, for example, leptospirosis, brucellosis, canine hookworms, and ringworm, are capable of being transmitted from animals to humans. They are called *zoonoses*.

Transmission of infection implies (1) a portal of entry in susceptible animals, (2) a transmitting agent or vector, and (3) the presence of pathogenic organisms.

Portals of Entry

There are certain routes or pathways by which microorganisms enter the body and cause infections. The particular route often depends on the kind of microorganism and, to some extent, on the kind of vector involved. The most important portals of entry are: (1) through cuts, punctures, or abrasions of the skin; and (2) through mucous linings of the respiratory tract, the eyes, the mouth, the navel, the gastrointestinal tract, and the genitourinary tract.

The route by which a microorganism enters the body often determines whether or not a disease will occur. Many disease-producing organisms are able to cause infection only if they enter through a specific portal. Gastroenteric bacteria rubbed into a cut in the skin would probably cause very little trouble, but the same organisms, if ingested, could cause a bad case of scours. On the other hand, tetanus organisms could be ingested and pass through the digestive tract without causing infection, but when these organisms are sealed over in deep cuts or punctures, they produce a severe, often fatal, infection. Some microorganisms can enter through almost any portal and cause infection.

Infectious Agents or Vectors

Transmitting agents, called vectors, may be animate or inanimate. Inanimate vectors or agents include contaminated feed, water, needles, syringes, dehorning equipment, fence posts, troughs, buckets, stalls, bedding, etc., and feces, urine, and saliva from infected animals.

Animal vectors or agents are generally arthropods and mammals, including humans. These include flies, ticks, lice, dogs, rats, wild animals, and farm animals. Transmission of disease by animate vectors usu-

ally involves direct contact of healthy animals with infected animals or carriers.

Characteristics of Microorganisms That Cause Disease

There is a tendency to view all microorganisms as
evil enemies of humanity. This is not true. Most microorganisms are harmless to animals, and the work
they do is beneficial, in fact, necessary to human life.
The basic health and well being of cattle is dependent
on an intense population of microorganisms within
the rumen to break down plant cell walls and convert
nitrogen into microbial protein. Microorganisms are
commonly categorized according to their activities
into two major groups: harmful and beneficial.

The harmful group of microorganisms cause disease and are called *pathogens*. They are parasitic and
motivated for survival. They have no more evil intent
than poisonous plants which, if ingested, can kill an
animal. Most microorganisms cause disease simply because they are able to live on or in the bodies of animals, and this, unfortunately, damages the host.

The second group of microorganisms live in the
outside world and are harmless to animals. They are
called *saprophytes*. These microorganisms, especially
bacterial yeasts and molds, are extremely important to
industries that manufacture cheese, alcohol, butter,
solvents for paints and oils, and other products. They
also play a vital role in maintaining and improving soil
fertility by decomposing organic matter and changing
atmospheric nitrogen into forms which are available
for plant use.

BACTERIA

Bacteria are single-celled microscopic organisms
that vary widely in size, shape, and cultural prefer-
ences (Fig. 1-4). They, like all other true fungi, have no
green coloring material or chlorophyll, as do ordinary
plants. They multiply by binary fission and, therefore,
are assigned to the class Schizomycetes, or fission
fungi.

Shape of bacteria. There are many thousands of
bacteria and probably as many sizes but they can be

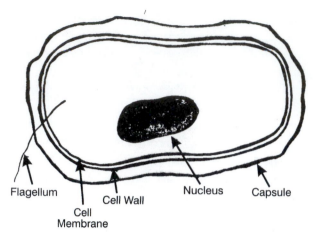

Fig. 1-4.   Bacterial cell structure.

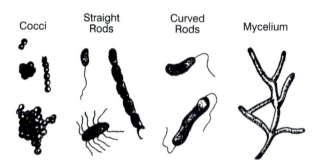

Fig. 1-5.   Bacterial shapes.

grouped into five types on the basis of shape (Fig.
1-5).

1. Spherical- or ellipsoidal-shaped. A bacterium
that is round in shape, or nearly so, is known as a *coccus*
(plural *cocci*). The origin of the term is from the Greek
word *kokkos*, meaning *berry*. Some cocci are slightly
elongated or ellipsoidal in shape. Many cocci exhibit
definite patterns of arrangement (Fig. 1-6). They
form chains, clusters, pairs, and squares which enable
them to be further classified as:

   a. Streptococci—rows of cells in a
chain-like arrangement
   b. Staphylococci—irregular clusters of cells
   c. Diplococci—pairs of cells
   d. Tetrads—four cells arranged as squares

2. Rod-shaped. The rod-shaped bacterium is
known as *bacillus*. Bacillus is derived from the Latin
word *baculum* which means "rod" or "stick." Bacilli are
cylindrical in form and may be long and narrow, short

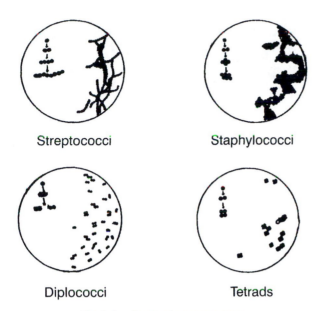

Streptococci   Staphylococci

Diplococci   Tetrads

Fig. 1-6. Bacterial arrangements.

and thick, or almost any combination of these dimensions. The ends may be rounded, flat, or pointed. Very short bacilli with rounded ends resemble cocci cells and are called coccibacilli.

3. Spiral-shaped. A bacterium that is coiled in a spiral form is a spirillium. The cell is corkscrew-shaped. It may have many spirals or only a few, and the coils may be lightly or loosely arranged.

Some species of spiral-shaped organisms do not have rigid cell walls, which is characteristic of most bacteria. They are flexible and resemble the protozoa in this respect. A bacterium of this type is known as a spirochete. The organism that causes leptospirosis in cattle, sheep, and swine belongs to this group.

4. Comma-shaped. Bacteria of this shape are known as vibrios. They are short, cylindrical and slightly curved in the long axis. One of the most important organisms in this group causes vibrionic abortion in cattle and sheep.

5. Filament-shaped. These bacteria are relatively long, slender, and threadlike.

Size of bacteria. The most common unit of measurement of microscopic objects is the micron which is a part of the metric system of linear measure and has a value of .001 millimeter or $\frac{1}{25,400}$ of an inch. There is a great deal of variation in size between the smallest and largest bacteria. Some are so small they are quite

difficult to observe with an optical microscope, while others reach lengths of 100 microns or more.

It is difficult to visualize the smallness of the micron. Cocci organisms range in size from 0.5 micron to 1.25 microns. A coccus magnified 1,000 times would appear no larger than a period on a printed page. Bacilli range from 0.5 to 1 micron in width and from 1.5 to 4 microns in length. It would require 25,000 bacilli 1 micron wide and placed side by side to form a line 1 inch long. Spirella reach lengths up to 10 microns, while vibrios are usually less than 3 microns long. Filamentous bacteria range in length from 20 to 100 microns or more.

Spores. Two genera of bacteria which commonly affect livestock are capable of transforming themselves into highly resistant bodies known as spores. They may also be correctly called endospores because they are produced within the cell. The spore is a stage in the life cycle of rod-shaped bacteria belonging to the genera *Bacillus* and *Clostridium*. Spores cause diseases such as anthrax (*Bacillus anthracis*), enterotoxemia (*Clostridium perfringens*), and tetanus (*Clostridium tetani*).

Spores possess extraordinary resistance to many adverse environmental factors. They are highly resistant to heat, dessication, and chemical disinfectants. Most spores can withstand the temperature of boiling water for short periods of time. Some are unaffected by boiling water for 5 to 6 hours, while a few can be boiled for 15 to 20 hours without being killed. Many spore-forming bacteria live in the soil. They are often referred to as soil-borne organisms.

When spores are provided favorable conditions, within an animal's body for example, they become vegetative and cause disease.

Flagellates. Many bacteria are capable of independent movement in liquid media. Their motility is provided by extremely delicate, thread-like appendages known as flagella. Some bacteria have only one flagellum, but others may have a dozen or more flagella. Many species of bacilli, spirilla, and vibrios are motile; whereas only a very few cocci possess this ability.

Conditions for growth. Bacteria, with the exception of those in the dormant or spore stage, are very sensitive to their environment. They require suitable temperature, moisture, darkness, and an adequate supply of food nutrients to grow and multiply. The majority of bacteria are active in temperatures ranging from 70° to 100°F.

Some bacteria must have free access to the oxygen supply of the atmosphere and cannot live and reproduce if denied this means of respiration. They are said to be strictly *aerobic*. On the other hand, there are many species that are unable to survive in the presence of free oxygen. They are called *anaerobes*. The organism *Clostridium tetani*, which causes tetanus, is an example of an anaerobe. Another large group of bacteria, called facultative organisms, are capable of growing in the presence of oxygen and have the ability to live well in the absence of air.

FUNGI

True fungi are simple plants that do not contain cholorphyll. Since they are without chlorophyll, they are unable to manufacture their own food. They must, therefore, depend upon living or dead plants or animals for their source of food.

Fungi are a versatile group of organisms that grow in many forms. They are tolerant of a wide range of physical and chemical conditions. There are many thousands of species which include molds, yeasts, smuts, blights, mildews, and toadstools (Figs. 1-7 and 1-8). Some scientists include bacteria in this group.

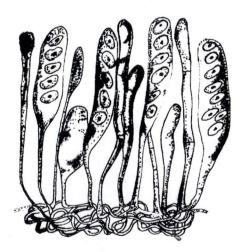

Fig. 1-7. Fungi with spores.

Fig. 1-8. Budding yeast.

Fungi are much larger and more complicated than bacteria. They, like bacteria, perform an indispensable function in the decomposition of organic material in the soil and play a vital role in the maintenance of soil fertility.

In addition, one genus, *Penicillium*, is the source of penicillin, a valuable antibiotic which is relatively nontoxic to livestock. Not all fungi are beneficial. Pathogenic fungi cause many diseases in plants and are responsible for such diseases as lumpy jaw, actinomycosis, and ringworm in animals.

VIRUSES

Viruses are the smallest microorganisms that cause infectious diseases. They are, with a few exceptions, invisible to the most powerful ordinary microscopes and pass through extremely fine clay filters which hold back yeasts, molds, and bacteria. Because of this, they are often called "filterable viruses." A second important characteristic of viruses is their inability to live independently of living cells. They must have living cells to parasitize if they are to grow and reproduce.

Classification of viruses. Viruses may be classified according to their general appearance, the host they normally infect, and the type of tissue they commonly attack.

Size and shape. The electron microscope has enabled scientists to make pictures of viruses. Their size ranges from about 10 to 300 milicrons. The diameter of some of the larger viruses approaches that of a small bacterium. Among the smaller viruses are those that cause foot-and-mouth disease in livestock and poliomyelitis in human beings.

The shape of viruses ranges from spherical to rectangular forms. Some of the round- or cube-shaped organisms have tail-like processes and are often described as "tadpole-shaped." The rectangular forms range from rod-like to blocky, cube-shaped organisms.

Common hosts. Viruses infect a wide variety of animals, plants, and bacteria. Those which infect animals, including humans, are called zoopathogenic viruses. They vary from tiny spherical granules to much larger cube-shaped objects. Viruses which infect and destroy bacteria are known as bacterial viruses or bacteriophagues. These viruses appear to be cube-shaped with short, narrow tails like tadpoles. Viruses which parasitize plants are called phytopathogenic organisms.

Tissue attacked. Viruses are often referred to by the tissue they prefer in their host. Dermotropic viruses cause damage to cells of the skin and mucous membranes. These viruses cause diseases such as cowpox, vesicular stomatitis, foot-and-mouth disease, and warts. Pneumonotropic viruses attack cells of the respiratory tract and cause swine influenza, infectious bovine rhinotracheitis, and equine influenza. Neurotropic viruses invade cells of the nervous system and produce diseases such as rabies, equine encephalomyelitis, and poliomyelitis in human beings.

Nature of viral infection. Upon entry into the host, a virus adheres to a susceptible cell. In some cases the entire virus enters the cell, but in other cases only the nucleic acid core enters the host cell. After the virus or its core enters the cell, a merger of the two occurs, resulting in the loss of identity of the mature virus. The virus, however, dominates the host cell and new viral organisms are produced that closely resemble the virus which originally infected the cell.

PROTOZOA

Protozoa are single-celled animals that are microscopic in size. Often referred to as simple forms of animal life, these little creatures in reality are quite complex. Many grasp and take solid food particles into their single-celled bodies and have sex differentiations that are much more distinct than bacteria, yeasts, or molds (Fig. 1-9).

The amoeba is one of the simplest protozoa. The single cell that makes up its body carries on all the necessary life processes by itself. The cell eats, breathes, and responds to its environment. The amoeba is

Fig. 1-9. Protozoa types.

motile, although it does not have the benefit of cilia, which are sharp, hair-like projections that assist the more complicated protozoans, such as the paramecium, to move about.

The majority of protozoa are free living, but a number of species have become adapted to living as parasites in the bodies of plants and animals. Pathogenic protozoa cause malaria in humans, trichomoniasis in animals (including people), and coccidiosis which commonly affects cattle and sheep.

FACTORS IN DISEASE PREVENTION

Losses from animal diseases and parasites cost producers and the American economy billions of dollars each year. These losses result from lowered production, reduced growth rate, sterility, abortion, the unnecessary cost of feed additives and medicine, and the death of animals. Much of this waste could be reduced by disease prevention programs, carefully designed and executed to eliminate infections and infestations.

In consultation with a veterinarian, the owner should plan and conduct a sound program of disease prevention. The plan should include appropriate practices in animal nutrition, sanitation, environmental management, immunization, and parasite control. In addition, management strategies should be incorporated that will (1) avoid the spread of disease due to contact with infected animals, equipment, etc., (2) reduce stress, (3) increase reproductive performance, and (4) ensure the survival of newborn animals.

Disease Resistance and Challenge

Practical means of disease control are to increase the resistance to a particular disease and to decrease the disease challenge. It is important to build disease resistance. As Fig. 1-10 shows, an animal may be infected but show no signs of sickness as long as the disease resistance is above the disease challenge level. Methods of keeping the disease resistance above the disease challenge level are proper natural and acquired resistance, maternal antibody production, parasite control, and nutrition. Sanitation and the avoidance of exposure by the animal to disease are not always effective, as some disease exposure can occur even in herds practicing excellent biosecurity control. As Fig. 1-11 shows, an increase in the disease challenge or stress level results in clinical disease. Stress through weaning, transportation, overcrowding, parasitism and poor nutrition lowers the immune system of the animal, allowing the resistance to drop. The disease challenge can be elevated in cases where sick animals spew out infectious bacteria and viruses to nearby animals, or the environment is conducive to the rapid growth and transmission of disease organisms. In contrast, Fig. 1-12 shows a spread between disease resistance level and disease challenge level, which becomes larger over time. To get herds to this safe position, the herdsman must work at increasing disease resistance through the proper practices aforementioned. The caretaker must further work at limiting stress and the level of disease challenge.

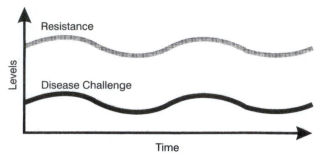

Fig. 1-10. Disease resistance. Herds remain healthy as long as their resistance level (grey line) is maintained above the level of disease challenge (black line).

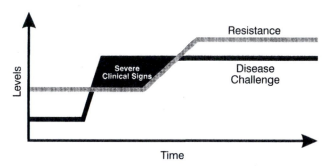

Fig. 1-11. Disease occurrence. Clinical signs occur when disease challenge (black line) surpasses resistance level (grey line).

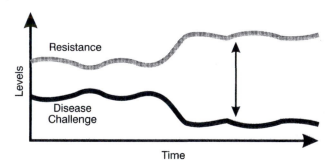

Fig. 1-12. Margin of protection. A comfortable spread between disease resistance (grey line) and disease challenge levels (black line) is desirable if calves are to remain healthy.

CHAPTER 2

Nutrition and Animal Health

An animal health and disease prevention program must be undergirded with proper nutrition. All the life processes of animals are dependent upon an adequate supply of essential nutrients. In addition, proper nutrition is necessary to maintain an animal's defense system against diseases and parasites.

NUTRITION AND RESISTANCE

Well-fed animals in good condition are believed to be more resistant to disease than animals in poor condition. This is especially true of resistance to the effects of parasites and to certain bacterial diseases. Mucous membranes and the skin provide animals with the first line of defense against diseases. Vitamins A, riboflavin, and niacin are necessary to keep these epithelial surfaces in a healthy state. Protein and several of the B complex vitamins are essential for the production of antibodies and phagocytes which serve as secondary defenders to destroy infectious agents that enter the animal's body.

Proper nutrition, in conjunction with good feeding practices, plays a vital role in keeping an animal's resistance at the maximum level provided by nature (Fig. 2-1). It also enables an animal to respond properly to vaccination.

NUTRITION AND DISEASE

The effects of disease, in most cases, greatly increase the nutritive requirements of affected animals. Primary nutritional deficiencies result from the absence of one or more essential nutrients in the ration or from the inability of an animal to absorb or to utilize nutrients that are ingested. A caloric deficiency due to insufficient energy feeds is not at all uncommon. This problem is often compounded by other deficiencies such as protein, vitamin A, and phosphorus deficiencies. Malnutrition of this kind is usually the result of poor management.

Nutritional problems arising from a deficiency of vitamins and minerals are usually quite complicated because they are closely related with various metabolic processes. Vitamin D, for example, exerts a direct effect upon calcium absorption and assimilation.

Specific nutritional deficiencies and metabolic disorders are frequently causes of decreased production and death in livestock. Anemia, rickets, white muscle disease, milk fever, ketosis, and grass tetany are some of the diseases in these categories. They are treated in detail in later chapters.

Fig. 2-1. Proper nutrition is essential for health and for efficient reproductive performance.

Secondary nutritional deficiencies are the result of diseases that interfere with the ingestion, absorption, or utilization of essential nutrients and those that increase the nutritional requirements of affected animals. Anorexia, the loss of appetite, limits the ingestion of feed and often results in complex secondary nutritional deficiencies. Digestive disturbances that cause diarrhea are responsible for the loss of nutrients which are passed through the digestive system before they can be absorbed. Certain diseases and parasites destroy tissues, red blood cells, and vital organs. This increases the nutritional requirements for repair and restoration of damaged cells.

NUTRITION AND THERAPY

The specific role of nutrition in the treatment of disease will, of course, depend upon the nature of the problem itself. Diagnosis of nutritional deficiencies is usually quite difficult and should be left to the care of a veterinarian or nutritionist. Once diagnosis has been made, the caretaker will be expected to provide basic management practices that will aid in the restoration of appetite and in the replacement of nutrients lost during the animal's illness.

Therapeutic nutrition is necessary in the treatment of secondary deficiencies resulting from infectious disease. Animals with hemolytic diseases often respond to iron supplements and additional levels of high quality protein in the ration. Electrolytes, extra protein, and B vitamins may be beneficial in the treatment of diarrheal diseases. Highly digestible carbohydrates are essential in the diet of sick animals to provide a quick source of energy and to release protein for use in the repair of damaged tissues. During illness, an animal's requirement for vitamins, especially the B vitamins, is also greatly increased.

NUTRIENT REQUIREMENTS OF ANIMALS

All animals have basic nutritional requirements that include proteins, carbohydrates, fats, vitamins, minerals, and water. Table 2-1 shows the composition of some feeds commonly fed to ruminants. The amount and quality of nutrients required for good health and efficient production vary according to species, age, sex, and the stress imposed by work, gestation, and lactation (see Tables 2-2 and 2-3).

The greatest variation occurs between ruminants and simple-stomached animals. Ruminants, unlike monogastric animals, utilize microorganisms in the paunch to break down fibrous feeds for energy and to synthesize amino acids from poor quality protein or nonprotein nitrogen compounds such as urea.

Nutrients are often referred to as being essential or nonessential. An essential nutrient cannot be synthesized by the body. On the other hand. nonessential nutrients are those chemical elements or compounds required for growth and maintenance that can be synthesized in the animal's body in sufficient quantities to meet daily requirements.

Proteins

Proteins are complex organic compounds which contain the elements nitrogen and sulfur in addition to carbon, hydrogen, and oxygen. Some proteins also contain phosphorus. Proteins are formed by long chains of amino acids. Amino acids are often called the building blocks of protein because they are assembled in many different combinations to form various kinds of protein in much the same manner that letters of the alphabet are combined to form words.

Plants are capable of forming amino acids from nitrogen, phosphorus, sulfur, water, and carbon dioxide by photosynthesis. Animals depend upon the feed they consume or upon microorganisms within the digestive tract for these nutrients.

Essential Amino Acids

There are 10 amino acids that are classified as essential. The remainder are termed nonessential because they can be synthesized in the body in sufficient quantities for normal growth. It is important to note that the classification for essentiality of amino acids is based upon experiments with rats and, therefore, is related to the needs of monogastric animals. Ruminants, as previously stated, have the unique service of rumen

TABLE 2–1. Composition of Some Feeds Commonly Fed to Ruminants, As-Fed Basis[1]

Feed	TDN	Crude Protein	Digestible Protein	Calcium	Phosphorus
	(%)	(%)	(%)	(%)	(%)
Alfalfa hay, sun-cured, all analyses...............	51	16.0	11.2	1.28	0.24
Alfalfa-grass hay.........................	51	14.5	9.9	1.33	0.23
Alfalfa silage, all analyses	16	4.7	3.4	0.48	0.07
Almond hulls...........................	66	4.1	—	0.19	0.09
Barley grain, all analyses....................	75	11.7	8.8	0.05	0.34
Barley straw	43	4.0	0.6	0.27	0.07
Beet pulp, dehy.........................	8	8.8	4.3	0.10	0.01
Corn, dent yellow No. 2	80	8.9	6.8	0.02	0.29
Cottonseed meal, solv. extd., 41% protein........	72	41.3	35.4	0.16	1.07
Molasses sugarcane (blackstrap)................	60	4.3	0.6	0.74	0.08
Oat grain, all analyses	69	11.9	9.2	0.08	0.34
Oat hay, all analyses.......................	52	8.6	4.5	0.29	0.23
Sorghum grain, all analyses..................	67	11.5	7.1	0.05	0.32
Soybean meal, 44% protein...................	76	44.4	37.8	0.35	0.64
Sudangrass sorghum hay	51	10.9	4.7	0.47	0.28
Urea, 45%............................	—	281.7	—	—	—
Wheat grain, all analyses	77	13.1	10.5	0.05	0.35

[1]From: *Feeds & Nutrition*, second edition, 1990, Table V, Ensminger Publishing Company, 648 West Sierra Ave., Clovis, CA 93612.

TABLE 2–2. Nutrient Requirements in Rations for Growing and Finishing Cattle (Medium-Frame Steer Calves)[1]

Weight	Feed Consumed	1½ lb/day Gain		Feed Consumed	2½ lb/day Gain	
		TDN	Total Protein		TDN	Total Protein
(lb)	(lb)	(%)	(%)	(lb)	(%)	(%)
300	9.7	56.7	11.9	9.9	66.2	15.0
400	12.0	56.7	10.4	12.2	66.2	12.8
500	14.2	56.7	9.5	14.4	66.2	11.3
600	16.3	56.7	8.8	16.6	66.2	10.3
700	18.3	56.7	8.3	18.6	66.2	9.5
800	20.2	56.7	7.9	20.6	66.2	8.8
900	22.1	56.7	7.6	22.4	66.2	8.4
1,000	23.9	56.7	7.3	24.3	66.2	7.9

[1]From: *Feeds & Nutrition*, second edition, 1990, p.704, Table 19–4, Ensminger Publishing Company, 648 West Sierra Ave., Clovis, CA 93612.

TABLE 2–3.   Daily Nutrient Requirements per Horse, 1,100 lb Mature Weight[1]

Animal	Body Weight	Daily Gain	Energy		Protein		Calcium	Phosphorus	Magnesium	Potassium	Vitamin A Activity
			TDN	Digestible Energy	Crude Protein	Lysine					
	(lb)	(lb)	(lb)	(Mcal)	(lb)	(g)	(g)	(g)	(g)	(g)	(1,000 IU)
Mature horses											
Maintenance	1,100		8.2	16.4	1.44	23	20	14	7.5	25.0	15
Stallions (breeding season)	1,100		10.3	20.5	1.79	29	25	18	9.4	31.2	22
Pregnant mares [2]											
9 months	1,100		9.1	18.2	1.75	28	35	26	8.7	29.1	30
10 months	1,100		9.3	18.5	1.78	29	35	27	8.9	29.7	30
11 months	1,100		9.9	19.7	1.89	30	37	28	9.4	31.5	30
Lactating mares											
Foaling to 3 months	1,100		14.2	28.3	3.12	50	56	36	10.9	46.0	30
3 months to weaning	1,100		14.2	24.3	2.29	37	36	22	8.6	33.0	30
Working horses											
Light work[3]	1,100		10.3	20.5	1.79	29	25	18	9.4	31.2	22
Moderate work[4]	1,100		12.3	24.6	2.15	34	30	21	11.3	37.4	22
Intense work[5]	1,100		16.4	32.8	2.87	46	40	29	15.1	49.9	22
Growing horses											
Weanling, 4 months	385	1.87	7.2	14.4	1.58	30	34	19	3.7	11.3	8
Weanling, 6 months											
Moderate growth	473	1.43	7.5	15.0	1.64	32	29	16	4.0	12.7	10
Rapid growth	473	1.87	8.6	17.2	1.88	36	36	20	4.3	13.3	10
Yearling, 12 months											
Moderate growth	715	1.10	9.5	18.9	1.86	36	29	16	5.5	17.8	15
Rapid growth	715	1.43	10.7	21.3	2.09	40	34	19	5.7	18.2	15
Long yearling, 18 months											
Not in training	880	0.77	9.9	19.8	1.95	38	27	15	6.4	21.1	18
In training	880	0.77	13.6	26.5	2.61	50	36	20	8.6	28.2	18
2-year-old, 24 months											
Not in training	990	0.44	9.4	18.8	1.75	32	24	13	7.0	23.1	20
In training	990	0.44	13.2	26.3	2.44	45	34	19	9.8	32.2	20

[1]Adapted from *Nutrient Requirements of Horses*, 5th rev. ed., NRC-National Academy of Sciences, 1989, p. 43. To convert lb to kg, divide by 2.2.

[2]Mares should gain weight during late gestation to compensate for tissue deposition. However, nutrient requirements are based on maintenance body weight.

[3]Examples are horses used in Western and English pleasure, bridle path hack, equitation, etc.

[4]Examples are horses used in ranch work, roping, cutting, barrel racing, jumping, etc.

[5]Examples are horses in race training, polo, etc.

flora which synthesize amino acids in their bodies and provide these essential nutrients to their hosts as they are digested.

The essential amino acids are (1) arginine, (2) histidine, (3) isoleucine, (4) leucine, (5) lysine, (6) methionine, (7) phenylalanine, (8) threonine, (9) tryptophan, and (10) valine.

Functions of Proteins

Proteins are used in the animal body for growth, reproduction, and lactation; for repair of body tissues; for the formation of enzymes, antibodies, and certain hormones; and for energy.

Protein fed in excess cannot be stored in the animal's body. A portion of the excess can be used for energy, while the remaining part is excreted in the urine. Protein is the most costly of the ingredients used in large quantities in a ration and it should be provided in adequate amounts. But too much protein should be avoided for two reasons: (1) excess protein is not an efficient source of energy, and (2) protein is generally more costly than carbohydrates and fats.

Ruminants fed rations deficient in protein are unable to properly utilize other nutrients in the diet properly. This is especially true of cellulose material which comprises a large part of the diet of ruminants. Microorganisms in the rumen or paunch which reduce roughage to digestible forms are dependent upon protein for their existence and effectiveness. When the rumen flora are not properly fed, digestive disturbances are likely to occur that can result in lowered production and impaired breeding performance. Malnutrition, especially protein deficiency, is responsible for some of the sterility problems in cattle and sheep. In addition, growth of the wool fiber in sheep is impaired, and breaks occur in the fleece when nutritional levels are sub-standard.

Inadequate protein in the diet of swine results in slow growth, reduced feed efficiency, and excess fat on the carcass (see Table 2-4, "Recommended Nutrient Allowances for Swine"; Table 2-5, "Feed Ingredients for Swine Rations"; and Table 2-6, "Symptoms of Marked Dietary Deficiencies in Swine").

PROTEINS IN RUMINANT RATIONS

The microbes in the rumen require nitrogen and a carbohydrate for synthesis of microbial protein. The forage diet of ruminants often provides the carbohydrates and protein levels necessary for maintenance and growth, but the high demands of lactation often place the animals in a protein-deficit state. Proteins are broken down in the rumen and are classified as degradable intake protein (DIP) and undegradable intake protein (UIP). DIP plus UIP are equal to the crude protein level in the diet. Degradable proteins are broken down in the rumen to release nitrogen. Undegradable proteins bypass the rumen and are digested in the intestines as amino acid sources. The degradability of the protein source should be matched with the solubility of the carbohydrate source in the rumen for the most efficient and maximal digestion. Rapidly digested starches, like high-moisture corn, match well with rapidly degraded proteins, like alfalfa silage. Slowly digested starch sources, such as mature pasture grasses, would be better matched to a protein source, such as cottonseed meal.

Protein is constantly circulated throughout the ruminant animal by the formation of blood urea nitrogen (BUN), which is circulated to the salivary glands and re-enters the rumen. An excess of nitrogen, and hence BUN, will depress conception rates and milk production levels. The animal must use calories to convert the excess nitrogen and excrete it from the body via the urine, so lowered milk production is the result. A test is available from U.S. Dairy Herd Improvement Association centers in the United States to easily analyze milk urea nitrogen (MUN) levels. MUN should range from 10 to 14 mg/dl in cattle several hours post feeding. Very low levels of MUN can be related to inadequate protein rations. Very high MUN levels can be related to improper feed bunk management, excess dietary protein, or a deficiency of soluble carbohydrates.

PROTEIN REQUIREMENTS FOR CATTLE

Protein requirements vary throughout the production cycle. Dry, gestating cows have the lowest requirements and early lactation cows in peak milk have

TABLE 2–4. Recommended Minimum Nutrient Allowances for Swine (As Fed Basis)[1,2]

Type of Diet Body Weight, lb	Starter 1 <15	Starter 2 15 to 25	Starter 3 25 to 50	Grower 1 50 to 80	Grower II 80 to 120	Finisher 120 to Mkt. Wt.	Gestation	Lactation
Total level, %								
Protein	20 to 22	18 to 20	18	17	16	14	14	15
Lysine	1.50	1.25	1.15	.95	.80	.65	.60	.70
Tryptophan	.21	.18	.17	.15	.13	.11	.13	.14
Threonine	.86	.71	.69	.59	.50	.43	.42	.50
Calcium	.90	.90	.80	.75	.75	.65	.90	.90
Phosphorus	.80	.80	.70	.65	.65	.55	.80	.80
Additions/ton[3] Minerals								
Salt, lb[4]	5	5	7	7	7	7	10	10
Copper, g[5]	15	205	205	15	15	10	15	15
Iodine, g	.27	.27	.27	.27	.27	.18	.27	.27
Iron, g	150	150	150	150	150	100	150	150
Manganese, g	36	36	36	36	36	24	36	36
Selenium, g[6]	.27	.27	.27	.09	.09	.09	.09	.09
Zinc, g	3,000	150	150	150	150	100	150	150
Vitamins								
Vitamin A, million IU	10	10	10	8	8	6	10	10
Vitamin D$_3$, thousand IU	1,500	1,500	1,500	1,200	1,200	900	1,500	1,500
Vitamin E, thousand IU	40	40	40	32	32	24	40	40
Vitamin K, g[7]	4	4	4	3.2	3.2	2.4	4	4
Riboflavin, g	7.5	7.5	7.5	6	6	4.5	7.5	7.5
Niacin, g	45	45	45	36	36	27	45	45
Pantothenic acid, g	26	26	26	20.8	20.8	15.6	26	26
Choline, g	150	150	150	120	120	90	500	500
Biotin, mg							200	200
Folic acid, mg							1,500	1,500
Vitamin B$_{12}$, mg	30	30	30	24	24	18	30	30

[1]From: *Kansas Swine Nutrition Guide*, 1994, p. 38, Cooperative Extension Service, Manhattan, KS.

[2]For corn or milo–soybean meal based diets. All diets (except the gestation diet) are provided ad libitum (full-fed).

[3]For convenience, the mineral and vitamin additions/ton for starter 1 can be included in all diets to market weight.

[4]For pigs <25 lb, added salt levels may be lowered or eliminated if the diet contains >10% dried whey.

[5]To convert from g/ton to ppm, multiply by 1.1.

[6]Maximum legal addition .3 ppm for pigs <50 lb and .1 ppm for pigs >50 lb.

[7]Menadione activity.

TABLE 2–5. Feed Ingredients for Swine Rations[1]

Feed	Digestible Energy	Crude Protein	Crude Fiber	Calcium	Phosphorus	Cystine	Lysine	Methionine	Tryptophan	Maximum % of Feed in Ration 25–50	50–75	75–125	125–
Energy feeds	(kcal/lb)	(%)	(%)	(%)	(%)	(%)	(%)	(%)	(%)	(%)	(%)	(%)	(%)
Barley, all analyses	1,396	11.7	5.0	0.05	0.34	0.20	0.40	0.16	0.15	40	50	70	80
Corn, dent yellow, No. 2 . .	1,586	8.9	2.1	0.02	0.29	0.11	0.19	0.11	0.09	50	60	75	85
Corn, opaque 2, high lysine .	1,643	10.1	3.0	0.03	0.20	0.19	0.42	0.16	0.12	50	60	75	85
Fats, animal-poultry	4,144	—	—	—	—	—	—	—	—	5	5	5	5
Molasses, cane (blackstrap). .	1,135	4.3	0.4	0.74	0.08	—	—	—	—	5	5	5	5
Oats, all analyses	1,278	11.9	10.7	0.08	0.34	0.19	0.40	0.57	0.15	15	15	15	15
Potatoes, tubers, cooked . .	416	2.2	0.7	0.01	0.05	—	—	—	—	25	25	30	40
Rye grain, all analyses	1,474	12.0	2.2	0.06	0.31	0.19	0.42	0.17	0.12	20	25	30	30
Triticale.	1,453	15.4	3.0	0.04	0.30	0.27	0.52	0.22	0.18	40	50	50	50
Wheat grain, all analyses. . .	1,544	13.1	2.6	0.05	0.35	0.22	0.39	0.18	0.15	50	60	75	85
Wheat middlings	1,321	16.4	7.7	0.13	0.89	0.22	0.68	0.19	0.19	15	20	20	20
Protein feeds													
Alfalfa meal, dehy., 17% protein	643	17.4	24.0	1.40	0.23	—	—	—	—	5	5	5	5
Blood meal	1,220	80.5	1.0	0.29	0.25	1.25	6.43	0.94	1.01	3	3	3	3
Canola (rapeseed) meal . . .	1,249	40.5	9.3	0.66	0.93	—	2.15	0.77	0.49	25	20	20	20
Corn gluten meal, 41% . . .	1,637	43.2	4.5	0.15	0.46	0.66	0.78	1.07	0.21	10	10	10	15
Cottonseed meal, 41% . . .	1,209	41.2	12.1	0.17	1.11	0.76	1.69	0.58	0.55	8	8	8	8
Fish meal.	1,317	64.3	0.7	6.63	3.61	0.62	5.26	1.63	0.75	25	20	20	20
Fish meal, menhaden	1,578	61.2	0.9	5.19	2.88	0.58	4.74	1.75	0.65	25	20	20	20
Meat meal.	936	50.7	2.7	8.61	4.58	0.68	3.09	0.73	0.38	25	20	20	20
Meat & bone meal, 50% . .	1,028	50.4	2.4	10.00	4.94	0.46	2.89	0.68	0.28	25	20	20	20
Peanut meal	1,296	49.0	7.7	0.36	0.61	0.54	1.45	0.44	0.48	10	10	10	10
Soybeans, full fat, cooked. .	1,820	38.4	5.4	0.25	0.60	0.42	2.32	0.48	0.56	10	15	20	10
Soybean meal, 44%	1,585	44.4	6.2	0.35	0.64	0.67	2.85	0.59	0.62	25	20	20	20
Soybean meal, 49%	1,591	49.0	3.7	0.25	0.63	0.75	3.08	0.66	0.70	25	20	20	20
Tankage (meat meal) with blood, 60%	1,112	60.5	1.8	11.16	5.41	0.48	3.89	0.75	0.65	25	20	20	20
Whey, dehy.	1,444	13.3	0.2	0.86	0.76	0.30	0.94	0.19	0.20	10	10	10	10

[1]Feed compositions from *Feeds & Nutrition*, second edition, 1990, pp.1270–1506, Table V, Ensminger Publishing Company, 648 West Sierra Ave., Clovis, CA 93612.

TABLE 2–6. Symptoms of Marked Dietary Deficiencies in Swine

Deficiency	Slow or Interrupted Growth	Reduced Feed Intake	Poor Hair and Skin Condition	Lameness and Stiffness	Weakened Bone Structure	Diarrhea	Impaired Breeding or Gestation	Offspring Dead or Weak at Birth	Other (See Code*)
Energy	X						X	X	1
Protein level	X	X	X				X		2
Protein quality (essential amino acids)	X	X	X				X		2
Essential fatty acids	X	X	X						3
Calcium	X	X	X	X	X		X		4
Phosphorus	X	X		X	X		X		5
Potassium	X	X	X						6
Sodium	X	X							7
Magnesium	X	X	X	X	X				8
Manganese				X	X		X	X	9
Iodine	X	X	X				X	X	10
Iron	X	X	X						11, 12
Copper	X		X	X	X				12, 13, 14
Zinc	X	X	X	X	X	X			15
Selenium	X	X							16
Vitamin A	X		X	X	X		X	X	17
Vitamin D	X			X	X				18
Vitamin E	X	X							16
Vitamin K	X								19
Thiamin	X	X					X	X	20
Riboflavin	X	X	X			X	X	X	21
Niacin	X	X	X	X		X	X		22
Pantothenic acid	X	X		X			X	X	23
Vitamin B$_6$	X	X	X	X		X			12, 24
Choline							X	X	25, 26
Vitamin B$_{12}$	X						X		12
Biotin	X	X	X	X					27

*Code

1. Reduced fat deposition.
2. Poor feed efficiency.
3. Loss of hair; scaly, dandruff-like dermatitis, especially of feet and tail.
4. Rickets or osteomalacia; reduced serum calcium and tetany in severe cases.
5. Reduced inorganic blood phosphorus.
6. Decreased feed efficiency; cardiac impairment.
7. Depraved appetite.
8. Hyperirritability and tetany; weak pasterns.
9. Reduced skeletal growth; increased backfat; irregular estrus.
10. Pigs hairless at birth; goiter.
11. High mortality in young pigs; susceptibility to disease; thumps.
12. Anemia.
13. Lack of rigidity of leg joints; hocks excessively flexed; forelegs crooked; use of forelegs impaired.
14. Lowered elastin content of arterial wall.
15. Severe dermatosis; parakeratosis.
16. Liver necrosis; brownish-yellow discoloration of adipose tissue; waxy degeneration of muscle tissue; edema; sudden death.
17. Increased cerebrospinal fluid pressure; incoordination; weakness and paralysis; night blindness followed by total blindness.
18. Rickets; enlarged joints; weak bones.
19. Blood clotting time prolonged; hemorrhage; hyperirritability.
20. Slow pulse; low body temperature; flabby, enlarged heart.
21. Crooked legs; stillborn pigs; incoordination; nerve degeneration.
22. Occasional vomiting; foul-smelling feces; pig pellagra.
23. Incoordinated, wobbly gait (goose stepping); stiff legs.
24. Epilepsy-like fits; convulsions; slowing of growth after first convulsion.
25. Incoordination; improper rigidity of joints; fatty liver; renal glomerular occlusion; tubular epithelial necrosis.
26. May be associated with spraddled-leg condition at birth.
27. Spasticity of hindlegs; loss of hair; dermatosis; cracks in feet.

Courtesy, National Research Council

the highest requirements. A cow in peak milk producing over 70 pounds of milk cannot meet all of her protein requirements through microbial protein, and needs to be fed higher levels of UIP to support the large amount of protein uptaken by the udder.

PROTEINS IN SWINE RATIONS

Swine diets are formulated to provide the proper balance of amino acids, especially lysine. Lysine is called the first limiting amino acid because growth and production is impaired if lysine is not fed at adequate levels, even if other amino acids are fed at excess levels. Lysine needs are highest, as a percentage of the diet, in young, early weaned pigs. The lysine needs of growing pigs decreases, as a percentage of the diet, as the pigs reach market-weight sizes. The gender of the pig also affects the lysine needs, with gilts having higher lysine requirements than barrowst—gilts generally produce leaner carcasses.

Synthetic lysine is added to the nursery-grower-finisher diets to supply the level of lysine needed for optimum performance. Using a corn-soybean meal based diet, the use of synthetic lysine enables a lower amount of soybean meal to be fed, since other amino acids are being fed in excess in order to obtain the needed lysine levels.

The lactating sow requires high levels of lysine to support milk production and large weight gains of the litter. First-parity sows have higher lysine requirements than second- and older-parity sows, as they are still depositing body tissue and growing. For lactating sows, an elevated soybean meal level is preferred over synthetic lysine in order to meet the protein requirements. Top dressing first-parity sow diets in the farrowing barn with one cup of soybean meal, or equivalent, can supply the additional protein needs for growth, lactation, and reproduction.

The swine manager must ensure that the lactation sow diet is formulated and balanced for protein for future reproduction as well as milk production. A deficiency in protein during lactation, even if feed intake is high and the caloric intake is sufficient, will impair subsequent reproductive performance. The inadequate protein intake appears to delay the return to estrus, and suppresses follicular growth and maturation.

Carbohydrates

Carbohydrates are organic compounds that contain carbon, hydrogen, and oxygen. The hydrogen and oxygen are usually present in the same proportions as in water. Carbohydrates are the most important source of energy in livestock rations.

Carbohydrates include sugars, starches, cellulose, and related substances. They are the major organic compounds in plants and usually constitute from 50 to 80 percent of the dry matter in livestock feeds. These compounds are found in the seeds, fruits, roots, stems, and leaves of plants.

Carbohydrates are formed by the process of photosynthesis, which involves the utilization of carbon dioxide and water in the presence of sunlight and through the mechanism of chlorophyll. Glucose is one of the first products of photosynthesis. The following formula explains how a carbohydrate is formed by photosynthesis in the plant leaf:

$$6CO_2 \;+\; 6H_2O \;+\; \Delta \;\rightarrow\; C_6H_{12}O_6 \;+\; 6O_2$$

Carbon Dioxide   Water   Energy (Sunlight)   Glucose   Oxygen

Classification of Carbohydrates

Glucose is called a simple sugar because its molecule cannot be reduced to smaller units that retain the properties of sugar. There are three other common simple sugars: fructose, galactose, and mannose. Glucose is present in most plant and animal cells, fructose is found in most fruits and galactose is a component of lactose, or milk sugar. Mannose is present in some plants, one of which is the ivory nut.

A simple sugar is called a monosaccharide. Carbohydrates containing two molecules of simple sugars are called disaccharides. The most important disaccharides are lactose, sucrose, and maltose. Polysaccharides are carbohydrates containing many molecules of simple sugars. Starch and cellulose are the most important polysaccharides.

Digestibility of Carbohydrates

The digestibility of carbonaceous materials decreases as the complexity of their structure increases. Simple sugars, monosaccharides, are more easily digested, for example, than the more complex polysaccharides. Starch is the principal carbohydrate in swine rations, and it is the major component of most grains. Cellulose, a polysaccharide, is found in the cell walls and woody portions of plants. Feeds high in cellulose, such as hay, straw, and silage, are low in digestibility. They are utilized primarily by ruminants.

Carbohydrates and other energy-producing nutrients must be broken down into simple units capable of passing through cell walls before the energy they contain can be used by animals. This process, known as digestion, takes place in the digestive track.

The digested material is absorbed into the cell and oxidized to produce energy. Although this is a complicated process, the practical aspects of respiration, which is the release of energy in a form usable for animals, is shown by the following equation:

$$C_6H_{12}O_6 + 6O_2 \rightarrow 6CO_2 + 6H_2O + Energy$$

Glucose   Oxygen   Carbon   Water
                              Dioxide

RUMINANT DIGESTION OF CARBOHYDRATES

Energy in the cow's diet is derived from two primary sources: structural carbohydrates (cell-wall fiber found in forages such as hay) and non-structural carbohydrates (sugar and starches, found in feeds such as corn and molasses). A proper balance of fiber and non-structural carbohydrates must be fed to maximize rumen health and feed efficiency. Feeding too much forage reduces total dry-matter intake because of its bulkiness. Feeding too much grain can cause rumen acidosis and reduce fiber digestion in the rumen. Crude fiber, cell contents (acid detergent fiber), lignin, cellulose, crude protein, and cell walls (neutral detergent fiber) are chemically determined components used to predict forage quality, forage intake, and animal performance. Fig. 2-2 shows the breakdown of plant material into the digestible and indigestible components. The digestibility of the plant cell walls decreases with increasing plant maturity, so there are acceptable limits to the NDF in rations. The proper

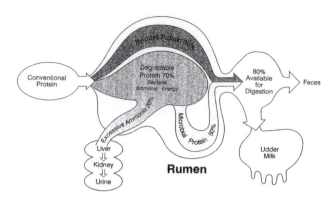

Fig. 2-2. Breakdown of plant material into digestible and indigestible components.

stage to harvest forages is when the protein, nonstructural carbohydrates, and minerals are at a high level, and different forage species have "prime harvesting" characteristics.

The microflora of the ruminant forestomach breakdown the forages and release volatile fatty acids (VFA). The VFAs are absorbed across the rumen wall into the bloodstream where they are utilized by different body tissues and converted to glucose, lactose, and other energy components. The proportion of VFAs released are related to the composition of the diet, regarding soluble carbohydrate levels.

Function of Carbohydrates

The major function of carbohydrates in animal nutrition is to provide energy for growth, muscular activity, reproduction, lactation, and the maintenance of body temperature. Energy is also supplied by the metabolism of fats and amino acids. A lack of energy in the ration can reduce the efficiency of the utilization of protein. In such a case, protein, or a portion of it, is diverted from tissue formation to meet the animal's needs, thereby complicating the original problem of malnutrition.

CARBOHYDRATES IN RUMINANT RATIONS

Ruminants have a requirement for "effective fiber," or long-stemmed forage particles to create a normal rumen enviroment for bacterial fermentaion (Fig. 2-3). Exclusive forage diets for high-producing cows have their drawbacks, as too much fiber limits

Fig. 2-3. Forage fiber is one of the most important nutrients for cattle health.

total dry matter intake and slows the rate of passage of feed through the gastrointestinal tract. Too little fiber will affect milk components, rumen and foot health, and body condition scores in dairy cattle. Provide at least 5 pounds of 1.5-inch long fiber per day to prevent off-feed problems and low milkfat. Guidelines recommend providing 2 percent of the bodyweight from forages, 19 to 21 percent ADF, and 28 to 30 percent NDF levels in the rations. The balance of the expected dry matter intake can be from minerals, proteins, and grains.

Dry Matter Intake

Dry matter intake (DMI) is one of the most critical points in understanding ruminant rations and feed-

ing. Dry matter intake is the amount of dry feed that the animal will consume and digest. For example, a 1,000-pound cow consuming 3 percent of her body weight daily in DMI will be expected to eat 30 pounds of dry matter. In a grazing situation, where the forages are only 25 percent dry matter, the cow will eat 75 pounds of grass to supply 30 pounds of dry matter.

Dry matter intake varies by the stage of lactation for beef and dairy cows (Fig. 2-4). The peak dry matter intake of cows lags behind the peak milk production by about two weeks. For beef and dairy cows, high-quality, palatable forages should be provided to encourage maximum dry matter intake, especially in fresh cows approaching peak milk production (Fig. 2-5). For every 2 pounds of expected milk production, cows should eat at least 1 pound of dry matter. Eating less makes cows prone to metabolic disorders and causes excessive body condition loss. As stated previously, mature forages that are high in NDF and low in protein lead to lower intakes and can be a primary cause of poor milk production and calf weaning weights. For beef cows, providing a protein supplement will enable the rumen bacteria to digest more of the low-quality forages, leading to higher dry matter intakes and thus more feed energy intake.

Carbohydrates in Swine Rations

Feed costs account for almost 70 percent of the costs for raising pigs, so a great deal of attention is

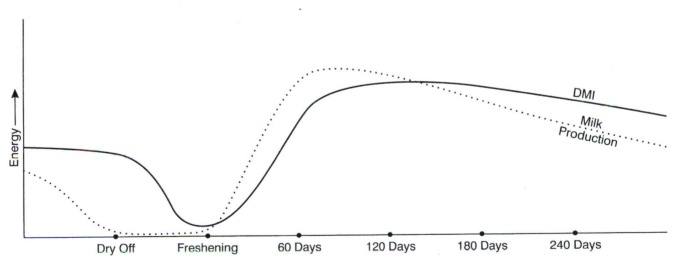

Fig. 2-4. Dry matter intake drops before freshening and peaks two weeks after peak milk yield.

Fig. 2-5. Peak productivity is reached when dry matter intake is maximized by readily accessible, palatable feeds.

paid to formulating the nutritional programs for swine operations. Pigs have the ability to utilize many different feedstuffs, and the digestion and rate of inclusion in the diet vary depending upon the life-cycle stage of the animal. For example, young growing pigs are not fed high levels of fibrous feeds, like oats, because energy is a limiting factor in growth, and oats do not contain the same amount of digestible energy per unit that ground corn contains.

Processing of the carbohydrates will affect the energy available to the pig. Pelleted feeds and finely ground feeds have a large particle surface area, and therefore have higher digestible energy values. Often, some low-energy feeds, like soybean hulls or hay, is offered to sows to increase the "satiety factor" and prevent the waste of animal energy searching for feed, as sows are often fed only 3 to 4 pounds daily of a corn-soybean meal diet.

Fats

Fats and oils comprise one of the three main classes of feedstuffs. The distinction between fats and oils is based primarily upon the differences in melting points. Fats are solids at room temperatures, while oils are liquid. They are lipids—insoluble in water but soluble in ether and similar organic solvents.

Fats like carbohydrates contain carbon, hydrogen, and oxygen. The proportion of carbon and hydrogen to oxygen is much higher in fats than in carbohydrates. On a weight basis, fats produce about $2\frac{1}{4}$ times more energy than carbohydrates.

Fats provide an easily digested, concentrated source of energy for animals. In addition, fats furnish the essential fatty acids required for normal growth and development in young animals. Another function of fats in a ration is to aid in the absorption of the fat-soluble vitamins A, D, E, and K.

Fats in Ruminant Rations

Fats are easy to include in ruminant rations to supply additional calories to high-producing animals, or as a way to supplement pasture cattle rations. Caution must be used in feeding fats to ruminants, as excessive fat levels, or the wrong type of fat in the diet can reduce the bacterial function in the rumen and ultimately result in lower energy intake. A common rule of thumb is to not exceed 6 percent fat of the total ration dry matter. Cheaper sources of energy, such as corn or barley, should be added to the ration before fat. Tallow, animal fats, and rumen-inert fats can be fed successfully. Vegetable oils should be avoided as they contain polyunsaturated fats and are toxic to some rumen organisms. Oilseeds, such as roasted soybeans and cottonseed, are good fat sources and provide additional protein and fiber to the ration.

Fats in Swine Rations

Fats are added to swine rations to increase the caloric density of the feed, control dust in ground rations, and to increase the palatability of some feeds. A swine nutritionist should be consulted to formulate a swine ration with additional fat, paying special attention to the economic pros and cons. Supplemental fat is most advantageous in the hot summer months in which the heat stress causes a decline in feed intake in growing and lactating animals. Supplemental fat will increase the calories per unit of feed compared to a corn-soybean meal ration. The protein level of a fat-supplemented ration must be increased to make sure the animal consumes the required grams of protein per day, because the animal will stop eating when the caloric needs are met, even if amino acid intake is not sufficient.

Fats in Equine Rations

Horses are able to tolerate relatively high-fat diets without impairing fiber digestion as occurs in ruminants. The fiber digestion in the horse occurs in the cecum, which is past the small intestine where fat absorption and digestion takes place. Supplemental fat is used in horses for several reasons, one of which is to increase the caloric intake in performance and growing animals. One important use for supplemental fat in equine diets is for horses affected by polysaccharide storage myopathy, or PSSM. These horses become sore and stiff with exercise, and the removal of starch sources (like corn) and replacing the grain with alfalfa cubes mixed with one cup of corn oil helps prevent PSSM. For good skin and coat appearance in the horse, corn oil provides the most well-balanced, polyunsaturated, fatty-acid profile.

Vitamins

Until early in this century, proteins, fats, carbohydrates, and certain minerals were believed to be the only essential ingredients required for normal body functions. Experimental work in the the early 1900s, however, revealed that other factors were necessary for the maintenance of good health and normal growth. Casimir Funk, a Polish biochemist, isolated one of these "factors" in 1912. He thought it belonged to the class of chemical compounds called amines and that it was vital to life, so he named it vitamine. Although vitamins are not amines, the name has been retained to designate those organic substances that are required in very small amounts to facilitate normal development and to maintain animal life.

Many phenomena of vitamin nutrition are related to solubility—vitamins are soluble in either fat or water. Consequently, it is important that nutritionists be well informed about solubility differences in vitamins, and make use of such differences in programs and practices. Based on solubility, vitamins may be classes as (1) fat soluble, or (2) water soluble.

Fat-Soluble Vitamins

The fat-soluble vitamins are A, D, E, and K. They are soluble in fat and absorbed with the lipids. They are stored in the body in rather large quantities, thereby eliminating the need for daily consumption. Excessive amounts of the fat-soluble vitamins can be toxic.

Vitamin A

The main source of vitamin A in the natural state is animal oils, fish liver oils in particular. Plants do not contain vitamin A, but high quality alfalfa hay, young, green pasture grasses, and legumes contain carotene which is a precursor of vitamin A. Carotene is easily destroyed by oxidation.

Vitamin A is the only vitamin that is commonly deficient in the diet of cattle grazing on rangeland. Green forage provides a precursor to vitamin A, which can be stored in the body. Body stores of vitamin A can be depleted in 90 to 120 days, and late winter illnesses in cattle receiving unsupplemented rations is not uncommon.

Vitamin A deficiency is probably the most common vitamin deficiency disease of cattle (Fig. 2-6). Beef cattle are produced in range areas and semi-arid regions where summer weather and extended periods of drought destroy the carotene content of pasture grasses very rapidly. Feedlot cattle on high concen-

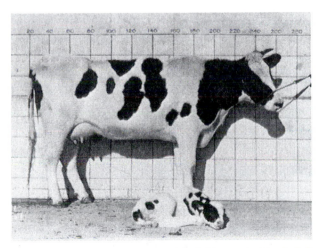

Fig. 2-6. A weak calf born blind due to vitamin A deficiency in the dam. (Courtesy, Michigan State University-East Lansing)

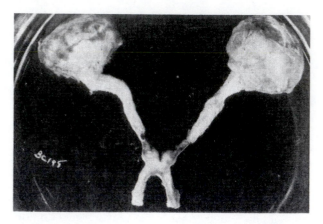

Fig. 2-7. Constriction of optic nerves in a calf due to a lack of vitamin A. (Courtesy, Dr. L. A. Moore, Michigan State University-East Lansing)

trate rations are also likely to suffer from vitamin A deficiency.

Sheep appear to be more economical in converting carotene to vitamin A than cattle. They can tolerate a low carotene diet for longer periods of time than cattle.

A deficiency of vitamin A results in disturbances of the epithelial tissues of the eye and in the digestive, reproductive, nervous, respiratory, and genitourinary systems. Therefore, it is apparent that a wide variety of clinical signs may result from a deficiency of vitamin A.

Young animals are more severely affected by a lack of vitamin A than older animals. They are born with low blood levels of vitamin A and depend upon colostrum until carotene from other sources can be obtained. Young animals born of dams fed rations low in vitamin A may be born dead or so weak they die shortly after birth. They may also be born blind due to malformation of the optic nerves (Fig. 2-7). Scours and respiratory problems often attack young animals suffering from a deficiency of vitamin A.

The first signs of vitamin A deficiency in cattle are watery eyes, rough hair coat, reduced growth rate, and night blindness.

Incoordination followed by posterior paralysis is a common sign of vitamin A deficiency in growing pigs (Table 2-6). Impairment of reproductive performance may be expected when gilts and sows are fed rations low in vitamin A.

VITAMIN D

Vitamin D, the anti-ricketic vitamin, is essential for proper utilization of calcium and phosphorus in bone development. It also serves an important function in the absorption of calcium and phosphorus from the small intestine and helps to maintain normal blood levels of these minerals.

The vitamin D requirement of animals is provided by exposure to sunlight. The sun's rays supply the necessary energy to convert 7 dehydrocholesterol, an animal sterol located under the skin, to vitamin D_3. The sun's rays also irradiate ergosterol, a plant sterol, to form vitamin D_2. Animals consuming sun-cured feed utilize vitamin D_2 produced in plants and the D_3 form produced in their bodies to satisfy the requirements for vitamin D.

A lack of vitamin D causes rickets in younger animals and osteomalacia in mature animals (Fig. 2-8).

Fig. 2-8. Rickets due to a ration deficient in vitamin D and no exposure to direct sunlight. (Courtesy, USDA)

VITAMIN E

Vitamin E, *Alpha-d-tocopherol*, is the reproductive vitamin contained in lettuce leaves, vegetable oils, whole oats and wheat, egg yolks, beef liver, and in unusual potency in the wheat embryos. It is not present in milk.

Vitamin E occurs as a viscous, nearly colorless oil. It is stable to heat, but it is destroyed by association with rancid fats. Vitamin E is important in the body as an antioxidant and as an important part in the immune system through the relationship with selenium. Serum levels of vitamin E drop before calving, and supplementation of vitamin E by injection may help prevent cases of mastitis in the peri-parturient period. In sheep, swine, and rats, a deficiency of vitamin E results in reproductive failure, though goats are not similarly affected. In suckling pigs, defective muscular nutrition results when this vitamin is withheld from the sows. In female rats, a deficiency of vitamin E results in embryonic mortality, and, in males, degeneration of the testes. In cattle, added vitamin E may be necessary under certain conditions because of its relationship to vitamin A utilization and the prevention of white muscle disease.

An absence of vitamin E, which is stored only to a slight extent in the body, is especially noticed in animals during the late winter months when the previously stored vitamin E is exhausted, and the animals have not yet had an opportunity to replenish their supply by the consumption of green growing feed. Adequate amounts of vitamin E and the presence of adequate levels of selenium prevent and even correct "white muscle disease" in calves and lambs.

Vitamin K

In order for coagulation of blood to take place whenever bleeding is profuse, vitamin K is essential for the synthesis in the liver of four blood-clotting proteins: factor II, prothrombin; factor VII, proconvertin; factor IX, Christmas factor; and factor X, Stuart power. Therefore, in the absence of vitamin K, the chain of events in normal clotting is not complete, and the resulting bleeding may progress to a fatal termination.

"Warfarin," the substance used to destroy rats and other rodents, produces internal bleeding; in this instance, dicoumarol, which at times is present in sweet clover, is the agent in "warfarin." Vitamin K is used as the treatment of "sweet clover disease," and in accidental "warfarin" poisoning.

Water-Soluble Vitamins

The water-soluble vitamins include vitamin C (ascorbic acid), which is not a problem in the nutrition of farm animals, and the B-complex vitamins which are required by monogastric animals (pigs, poultry, fish, and mink). In the past, it has been commonly assumed that ruminants do not need dietary supplementation with the B-complex vitamins because the rumen microflora synthesize these vitamins in sufficient quantities to fulfill the host's requirement. However, there is now experimental evidence that supplemental sources of three of the B vitamins—thiamin, niacin, and choline—need to be provided to ruminants when (1) they are receiving high-concentrate and low-forage rations (i.e., feedlot cattle and sheep, and high-producing dairy cattle), or (2) they are stressed, such as newly weaned or shipped calves or lambs.

The B vitamins may be classified into two groups, based upon their major function in the animal's body. The first group includes thiamin, riboflavin, niacin, and pantothenic acid which are involved in the release of energy from feed nutrients. The second group includes folic acid and vitamin B_{12} which play a vital role in the formation of red blood cells and in the prevention of anemia. Pyridoxine is an exception, however. It functions in both categories.

Thiamin

All animals must have a dietary source of thiamin, unless there is rumen synthesis, as in cattle and sheep. Thiamin is widely distributed in feeds such as cereal grains, brewer's yeast, and animal by-products.

Thiamin is essential for the proper metabolism of carbohydrates. A deficiency of the vitamin will result in incomplete carbohydrate metabolism, and pyruvic acid will accumulate in the tissues. Without thiamin, there could be no energy.

RIBOFLAVIN

Riboflavin (vitamin B_2) is a constituent of several enzymes and functions in the respiration process whereby energy is released for use in the cell by the oxidation of carbon-containing compounds.

A lack of riboflavin in the ration of young pigs causes slow growth, vomiting, and rough hair coats. In addition, pigs may have cataracts and suffer from skin eruptions and ulcerations that terminate in the loss of hair. Riboflavin is important for proper reproduction in swine, and it is added at increased levels for sows compared to growing pigs. Sows deficient in riboflavin may fail to cycle, resorb their fetuses, or farrow pre-term.

Horses occasionally suffer from a deficiency of riboflavin. This occurs when green feeds (pasture, hay, silage) are not available. The common signs of riboflavin deficiency in horses are periodic ophthalmia (or moonblindness), characterized by conjunctivitis in one or both eyes, a watery discharge, and discomfort in the presence of light.

Green, leafy legumes and grasses, milk, milk by-products, brewer's yeast, and some fermentation and distillery by-products are good sources of riboflavin. The vitamin in crystalline form is widely used in B vitamin premixes. Riboflavin is stored in small amounts in the liver and kidneys, but it must be supplied on a daily basis to satisfy the dietary requirements of animals.

NIACIN

Niacin (nicotinic acid) is a component of several enzymes involved in the metabolism of carbohydrates. Pigs can convert a portion of the amino acid tryptophan to niacin, but this is undesirable from an economical standpoint since synthetic niacin can be supplied more inexpensively than tryptophan.

Niacin is indispensable for swine. A deficiency of nicotinic acid in swine results in loss of appetite and decreased gain, followed by diarrhea, occasional vomiting, dermatitis, and loss of hair. Niacin-deficient chicks show poor feathering, scaly dermatitis, and sometimes a spectacled eye. Dogs show a thickening of the tongue (black tongue) and mouth lesions.

A deficiency of niacin in humans causes a disease called pellagra, characterized by a bright red tongue, mouth lesions, loss of appetite, and nausea.

Good sources of niacin include rice bran, meat scraps, tankage, fish meal, wheat bran, wheat shorts, milk, brewer's yeast, and distiller's solubles. A crystalline form is commonly used in B vitamin premixes.

PANTOTHENIC ACID

Pantothenic acid is essential for the metabolism and synthesis of fats, carbohydrates, and many other compounds. Signs of a deficiency of pantothenic acid in all species include reduced growth, loss of hair, and enteritis. Pantothenic acid deficient calves are characterized by loss of appetite, rough coat, dermatitis, and loss of hair around the eyes. Pigs walk with a goose-stepping gait. Chicks show dermatitis and embryonic death.

Feeds high in pantothenic acid include milk products, brewer's yeast, wheat bran, green, leafy alfalfa hay, and fish solubles. The crystalline form of pantothenic acid is widely used in B vitamin premixes.

CHOLINE

Choline does not meet the exact requirements for the definition of a vitamin since it is synthesized in considerable amounts in the body, and it is utilized in much larger quantities than other B vitamins.

Choline has several important functions, it is vital for the prevention of fatty livers, the transmitting of nerve impulses, and the metabolism of fat.

A deficiency of choline is unlikely when pigs are fed good rations. Signs of deficiency include unthriftiness, lack of coordination, spraddled hind legs at birth, fatty liver, poor reproduction, poor lactation, and decreased survival of the young. Choline is added to gestating sow diets as it has been proven to increase the number of pigs born live and litter sizes.

Good sources of choline include tankage, meat scraps, fish meal, and soybean meal. Crystalline choline is commonly used in vitamin premixes.

CYANOCOBALAMIN

Cyanocobalamin (vitamin B_{12}) was first called the "animal protein factor" because early studies indicated that additions of animal protein to swine rations were essential for optimum growth, reproduction, and lactation. It was later called vitamin B_{12}. When the chemical structure was determined, the chemical name cyanocobalamin was applied.

Research on the value of cyanocobalamin in swine rations is indefinite. Hogs finished on good rations containing corn, soybean meal, and alfalfa meal can perform quite efficiently without the addition of vitamin B_{12}. Vitamin B_{12} should be included in pig starters for early weaned pigs, creep rations for pigs up to 30 pounds, and grower rations for pigs up to 100 pounds. It is also wise to include adequate amounts of vitamin B_{12} in the rations of brood sows and bred gilts.

Vitamin B_{12} is required by young pigs for growth and for the formation of red blood cells. Good sources of B_{12} include liver meal, tankage, meat scraps, fish meal, and milk by-products.

Minerals

The minerals most commonly required in animal nutrition are calcium, phosphorus, sodium chloride (salt), magnesium, potassium, sulfur, iodine, iron, copper, cobalt, manganese, zinc, selenium, and molybdenum. They are required for skeletal development; for use as constituents of the proteins and fats that make up muscles, blood cells, and internal organs; and for use in many enzyme systems of the body. Some minerals are responsible for maintaining osmotic pressure and for establishing acid-base relationships, and they exert characteristic effects on the irritability of muscles and nerves.

Nutritional research indicates that a highly complex relationship exists among the minerals. Calcium, iron, and copper are known to interfere with the metabolism of other minerals and nutrients. It is, therefore, not only wasteful to fortify animal rations with excessive minerals but also dangerous to the animal's health. Table 2-7 presents toxic levels and symptoms of dietary excesses of certain required minerals.

Calcium and Phosphorus

Calcium and phosphorus make up over 70 percent of the minerals in the body and are particularly important for skeletal growth and bone strength. Approximately 99 percent of the calcium and 80 percent of the phosphorus of the body are present in the bones and teeth. The 20 percent of body phosphorus which occurs in the soft tissues is widely distributed and is an important part of many enzyme systems that have essential functions in body metabolism.

Calcium functions in blood clotting, skeletal formation, cardiac rhythm, and enzyme activation. Phosphorus is a component of bone, red blood cells, muscle and nerve tissue.

A large excess of either calcium or phosphorus interferes with the absorption of the other. Thus, it is important to have a suitable ratio between the two minerals. The most favorable calcium-phosphorus ratio for swine ranges from 1:1 to 1.5:1. A 2:1 ratio is preferred in the total diet for cattle. Cattle on pasture need to receive supplemental minerals, and the mineral balance will depend upon the type and maturity of the forage being consumed. Proper calcium: phosphorus ratios are important to prevent milk fever problems in dairy cattle.

A deficiency of calcium or phosphorus in swine can result in reduced growth rate, inefficient feed conversion, rickets, broken bones, and posterior paralysis.

A phosphorus deficiency in cattle can result in lack of appetite, lowered growth rate, inefficient feed utilization, and a decrease in milk production. A de-

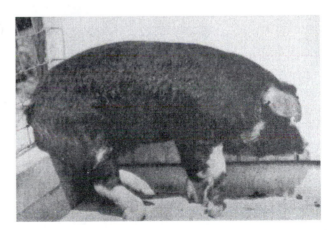

Fig. 2-9. Rickets caused by a ration deficient in calcium. (Courtesy, The Fertilizer Institute)

TABLE 2–7. Toxic Levels and Symptoms of Dietary Excesses of Certain Required Minerals[1, 2]

Mineral	Toxic Level	Species/Age Most Affected	Symptoms
Copper	Horse, 800 ppm; chicken, 300 ppm; swine, 250 ppm; cattle, 100 ppm; and sheep, 25 ppm.	Copper poisoning commonly occurs only in cattle and sheep, with sheep being more susceptible than cattle. Monogastric animals are more resistant to copper poisoning than ruminants.	Animals that develop signs of copper poisoning die in 1 to 2 days. Poisoning is characterized by severe gastroenteritis, marked by abdominal pain, nausea, salivation, and diarrhea. Chronic copper poisoning is marked by loss of appetite, thirst, jaundice, dark brown urine, and depression.
Iodine	50 ppm.	All species/all ages.	Specie difference. Horses, big head; in mammals, abortions may occur; in poultry, reduced egg production and poor hatchability.
Iron	1,000 ppm.	All species/all ages.	Free (unbound) iron is very toxic. Symptoms: decreased feed intake and rate of gain.
Selenium	2 ppm.[3]	All species/young animals most affected.	Blind staggers, or alkali disease, characterized by emaciation, loss of hair, soreness and sloughing of hoofs, lameness, anemia, excess salivation, grinding of the teeth, blindness, paralysis, and death. In poultry and other birds, egg production and hatchability are reduced and such deformities as lack of eyes and wings are common.
Sodium and chlorine (salt)	10%, provided water is available.	All farm animals, but sheep and swine are most affected.	Sudden death, 1 to 2 hours after eating salt. Extreme nervousness; muscle twitching and fine tremors; much weaving, wobbling, staggering, and circling; blindness; normal temperature, rapid but weak pulse, and very rapid and shallow breathing; and diarrhea.
Zinc	500 ppm.	All species, but young swine seem to be most susceptible.	Anemia, depressed growth, stiffness, hemorrhages in bone joints, bone resorption, depraved appetite, and in severe cases death.

[1]From *Feeds & Nutrition*, second edition (from throughout the book), Ensminger Publishing Company, 648 West Sierra Ave., Clovis, CA 93612.
[2]To convert ppm to mg/lb, multiply by 0.4536. To convert ppm to percent (%), move decimal 4 places to left.
[3]Suggested maximum tolerable level for all species. There is a species difference; beef cattle can tolerate the most selenium, poultry the least.

praved appetite often will follow, which leads to the consumption of sticks, rocks, bones, and soil. This may create secondary complications, such as traumatic gastritis and pericarditis. A phosphorus deficiency also has an adverse effect upon reproductive performance.

A deficiency of calcium or phosphorus may impair proper bone formation, causing rickets in young animals and osteomalacia in older animals (Fig. 2-9).

High-quality roughages, especially legumes, are good sources of calcium. They are often low in phosphorus, though. Wheat bran, cereal grains, and most high-protein feeds are good to fair sources of phosphorus. Steamed bone meal, dicalcium phosphate, and defluorinated rock phosphate are excellent sources of phosphorus. Ground limestone or steamed bone meal are commonly used to supplement rations low in calcium.

Sodium and Chlorine

Hydrochloric acid, a substance rich in chlorine, which is obtained from salt, is very essential in the digestive processes. Furthermore, some liquids must pass through some of the membranes in the animal's body. This takes place only when the amount of salt in solution in the liquids on one side of the membrane exceeds the amount of salt in solution on the opposite side. Under these conditions, the liquids will pass from one side of the membrane to the opposite side until there is an exact balance between the salt content of the liquids on both sides of the membrane. This process is known as *osmosis*. Salt is also a constituent of many of the tissues of the body, such as the blood, tears, and other secretions.

Salt must always be supplied to animals in addition to the amounts contained in the usual well-balanced ration. Swine, because of their rapid growth, require 0.25 to 0.5 percent salt added to their ration.

Good cattle management calls for free access to salt at all times. Whether on drylot roughage rations or on grass, salt may mean the difference between a profit and a loss. In a winter feeding test, steer calves on a roughage ration, with free access to salt, gained an average of 40 pounds more per head in 138 days than similar steers which had no salt. The "salted" calves, which consumed larger amounts of feed, made more efficient gains (Fig. 2-10).

Steers on a full feed of grain with salt free choice for 210 days averaged 2.21 pounds of gain per head daily. Steers on the same ration without salt gained 2.15 pounds per head per day. This test showed little difference in feed consumption or economy of gain when steers were on full feed of grain. It was observed that steers wintered on dry grass consumed as much or more salt than those on summer pasture.

At the Kansas Agricultural Experiment Station, steers consumed 1½ pounds of salt per month when on pasture; 3 pounds per month when on alfalfa hay; and 9 pounds per month when on corn ensilage. Generally, at least 20 pounds of salt per head per year should be provided.

Sheep and lambs have a very high salt requirement. Tests prove that for the fastest gains lambs should receive ½ pound of salt per month and ewes 1 pound. When salt is added to mixed feeds, it is customary to add 0.5 percent to the complete ration or 1 percent to the concentrate portion.

Horses lose 30 grains of salt in every pound of perspiration and therefore require additional salt in their grain rations.

Iodine

The iodine requirements of the animal body, in comparatively infinitesimal amounts, are very specific. In most parts of the United States, with possibly the exception of those in the Great Lakes region and in the far Northwest, there is enough iodine in the soil to impart to the plants grown therein enough iodine to meet the needs of people and animals. In those sections of the country where the soil is deficient in iodine so that people and animals do not receive the necessary maintenance amounts, it is a legal requirement that a small amount of iodine regularly be incorporated in table salt, which is marketed as iodized salt. The same iodized salt is usually offered for sale in states in which there is no need for extra iodine, but the amount is so small that its consumption is free from known harm, and the addition of iodine has not increased the cost of the salt to the consumer.

The National Research Council recommends that cows in lactation receive a dietary iodine concentration of 0.6 parts per million (p.p.m.). When stabilized iodine is used, a level of 0.0076 percent iodine in salt is adequate. The iodine requirements of beef cattle and swine are met by adding 0.0076 stable iodine in the ration.

Fig. 2-10. Cattle should always have free choice access to salt.

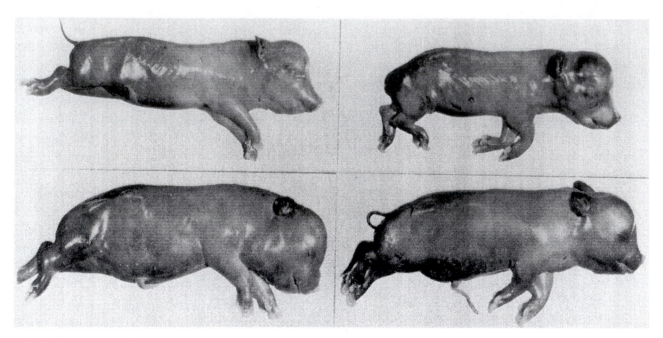

Fig. 2-11.   Hairlessness and big neck in pigs whose dam was fed an iodine-deficient ration. (Courtesy, Kansas State University)

Iodine deficiency in sheep, cattle, swine and horses is usually manifested by the birth of young animals that are weak, goitrous, or dead. Young pigs are often born hairless (Fig. 2-11). Signs of disturbances are not commonly observed in mature animals.

Iron and Copper

Iron and copper are necessary for the formation of hemoglobin in the red blood cells and for the prevention of nutritional anemia. Hemoglobin serves as a carrier of oxygen throughout the body.

In newborn calves, lambs, and foals, there is usually a sufficient amount of iron and copper stored in the liver to tide them over until they start to eat feeds other than milk. This is not true, however, in the case of baby pigs. There is not enough iron and copper stored in the pigs' bodies to ensure proper growth and development for more than 10 days to 2 weeks. A single intramuscular injection of 100 to 200 mg of iron from iron dextran into baby pigs at 2 to 3 days of age is usually sufficient to prevent anemia in baby pigs.

Milk is very low in both iron and copper. Research has not uncovered any way of increasing the content of these minerals in milk. Once young ani-

mals start to consume feed materials other than milk, the danger of nutritional anemia is greatly reduced.

Anemic pigs lose their appetite and become weak and inactive. In the latter stages, the pigs' breathing becomes heavy, causing a condition commonly referred to as "thumps." In this condition, the pigs are more susceptible to other diseases. Death may occur in severe cases.

A copper deficiency in cattle usually causes scours, loss of appetite, anemia, and impaired breeding performance. Copper is responsible for normal pigmentation of the hair. A deficiency often results in a rough, bleached, or graying hair coat. A part of the syndrome of molybdenum toxicity, including change of the hair coat, is caused by interference with copper metabolism. Feeding additional copper will generally alleviate the symptoms.

Magnesium

Magnesium is closely allied to calcium and phosphorus in the body, and about 65 percent is stored in the bones and teeth. This reserve is not easily mobilized for the animal's use. Consequently, an abrupt change from a normal diet to a magnesium-deficient diet can result in hypomagnesia.

Although lactating cows grazing on lush green pastures during the winter and early spring are the most likely victims, hypomagnesia can also occur in other stock. Signs of hypomagnesia are usually loss of appetite, restlessness, nervousness, tetany, convulsions, and eventually death. A deficiency of magnesium may cause grass tetany in cattle (and sometimes in sheep).

Zinc

Zinc functions as a catalyst and as part of hundreds of enzyme systems in the body. Deficiencies of zinc impair reproduction through abortion, fetal mummification, lowered birth weights, and impaired uterine contractility. It influences the rate of absorption of carbohydrates and proteins from the gastrointestinal tract. A deficiency of zinc results in impaired growth and in the depression of certain enzyme actions. In swine, a lack of zinc is the cause of a skin condition known as *parakeratosis.*

Which ion aggravates the zinc deficiency is not known, but it might be presumed that calcium phosphate blocks the absorption of the zinc ion rather than competing with zinc in the body. Factors which may combine to produce parakeratosis are: (1) a relatively low zinc content in the feed, (2) increased quantities of calcium or phosphate in the ration, and (3) relatively higher requirements of the pig for zinc.

The National Research Council (NRC) recommended nutrient requirements of zinc for beef cattle is 30 p.p.m. of ration dry matter, with a range of 20 to 40 p.p.m. For all classes and ages of dairy animals, the NRC recommends rations with a content of 40 p.p.m. of zinc. Deficiency symptoms include decreased performance and listlessness, followed by the development of swollen feet and a dermatitis that is most severe on the neck, head, and legs. Deficiencies may also result in vision impairment, excessive salivation, decreased rumen volatile fatty acid production, failure of wounds to heal normally, and impaired reproduction in both bulls and cows.

Supplementation of the diet with the proper levels of zinc carbonate or zinc sulfate will correct deficiency symptoms in calves within three to four weeks. Changing a supplemented diet to a deficient diet will produce deficiency symptoms within three weeks. Animals adjust quickly to levels of dietary zinc when the percentage of zinc absorbed is increased or decreased, as necessary. Ruminants fed a zinc-deficient diet show a rapid and large increase in percentage of dietary zinc absorbed and a reduction in endogeneous fecal losses. Absorption tends to decrease with age and as growth rate decreases.

The NRC lists the maximum tolerable level of zinc at 500 p.p.m., but the NRC also reports that steers have been fed rations containing 1,000 p.p.m. zinc for 13 to 18 months without marked reduction in performance. The toxicity symptoms are listed in Table 2-7.

Manganese

The NRC recommended nutrient requirements of manganese for cattle are: 40 p.p.m. for mature cows and bulls, and 20 p.p.m. for growing-finishing cattle. Rations containing high levels of calcium and phosphorus increase the requirements for manganese. Most roughages contain high levels of manganese. The manganese levels in pastures, grains, and forages are variable because of variations in plant species, soil types, soil pH, and fertilization practices. Nevertheless, there is no need to provide supplemental manganese for ruminants except in the northwestern United States, where manganese deficiency has been shown to cause *crooked calves.*

In swine, a deficiency of manganese is manifested by lameness, weakened bone structure, irregular estrus, offspring born dead or weak, and increased backfat. Fortunately, manganese is usually present in adequate amounts in most swine rations, but it may not be adequate for the optimum reproductive performance of sows. So, it is recommended that the ration of sows be supplemented with manganese at the rate of 10 p.p.m.

Cobalt

The only known function of cobalt is that of an integral part of vitamin B_{12}, an essential factor in the formation of red blood cells.

When cobalt is given to animals which utilize microbial fermentation, the element is used for the synthesis of vitamin B_{12}; hence, ruminants do not need supplemental sources of the vitamin provided cobalt is supplied. However, most nonruminants and very young ruminants do not have a need for cobalt as such, but do require dietary sources of vitamin B_{12}.

Vitamin B_{12} is needed for the normal formation and metabolism of the red constituent (hemoglobin) of the blood. In animals with a single stomach, such as the horse, hog, dog, cat, etc., there does not appear to be a direct relationship between cobalt and hemoglobin formation. In ruminants, however, one of the symptoms of cobalt deficiency is a profound paleness of the visible mucous membranes, indicating anemia.

Many of the symptoms observed in a deficiency of vitamin B_{12}, such as loss of appetite and general starvation, are also noticeable in cobalt deficiency. At times it is not clinically possible to distinguish between phosphorus and cobalt deficiencies. Under such circumstances, veterinarians usually adopt the attitude that if the symptoms are observed in calves only, or in both cows and calves, the owner should suspect cobalt deficiency, but if the symptoms are noticed in cows only, a phosphorus deficiency should be suspected.

Since cobalt is a component of vitamin B_{12}, it must be supplied in ruminant rations to enable the microorganisms to synthesize the vitamin. For nonruminants, cobalt is not added to rations because vitamin B_{12} is generally supplied in association with animal protein, or in the vitamin supplement. A common cobalt supplement for ruminants is made by adding cobalt to salt at the rate of 0.2 ounces/100 pounds of salt, using cobalt carbonate, cobalt chloride, cobalt oxide, or cobalt sulfate. Also, several good cobalt-containing commercial minerals are on the market. Other cobalt-containing supplements are brewers' yeast and rice polishings. Grazing animals may be given pellets composed of cobalt oxide and iron administered orally with a balling gun. The pellets lodge in the rumen and are gradually dissolved over a period of months.

Selenium

Selenium has an important interrelationship with vitamin E. Both these nutrients are needed by animals, and both have a nutritional role in the body. Both also have an antioxidant effect. In some cases, selenium will substitute for vitamin E to a certain extent, and vitamin E will substitute for selenium in varying degrees. However, neither one can replace the other entirely. The use of selenium in animal rations will reduce the need for vitamin E; however, vitamin E will still be needed.

The amino acids, cystine and methionine, are also of some benefit in protecting the animal against a deficiency of selenium and vitamin E. This may be due to an antioxidant effect of these two sulfur-containing amino acids.

Selenium deficiencies occur throughout the United States. Some selenium-deficient areas have been shown in 40 states. But, the most deficient areas occur in the eastern part of the United States, in the Pacific Northwest, and in parts of the Midwest. It is difficult to be free from selenium deficiencies in the United States, since feeds produced in a deficient area can be shipped to almost anywhere in the country.

Selenium deficiencies affect calves, lambs, and foals. In calves, it is commonly called white muscle disease. In lambs, it is referred to as stiff-lamb disease. In calves, white muscle disease is characterized by lameness or inability to stand, and heart failure; and it most commonly affects calves two to four months of age. In lambs, the symptoms and signs are: a stiff, stilted way of moving, a humped or "roached" back, and stunted growth.

Selenium is essential, but only in minute amounts. In 1987, the Food and Drug Administration (FDA) approved a maximum allowance of 0.3 p.p.m. selenium in the complete feed of all farm animals and poultry. Since excess selenium is highly toxic, regulations and guidelines should be carefully followed.

Selenium poisoning of farm animals is usually caused by the ingestion of plants grown on seleniferous soils. These soils are found in many areas of the Western States, particularly in the Central and Northern Great Plains.

The maximum tolerable level of selenium varies according to species, with beef cattle tolerating the most selenium and poultry the least. The maximum toxic level in the feed for each species is as follows: beef cattle, 10-30 p.p.m.; dairy cattle, 3-5 p.p.m.; sheep, 3-20 p.p.m.; swine, 5-10 p.p.m.; chickens, 2 p.p.m. The suggested maximum tolerable level for all species is 2 p.p.m.

The chronic form of selenium poisoning, commonly called *alkali disease*, is characterized by a lack of interest in feed or water, dehydration, emaciation, a rough hair coat, and the loss of hair from the mane and tail of horses, the tail of cattle, and a general loss of hair in swine. The hoof often separates or breaks at the coronary band, resulting in such lameness that animals resort to walking on their knees. Death may occur due to starvation.

Acute selenium poisoning, commonly called *blind staggers*, occurs when animals ingest large amounts of toxic materials in a short period of time. Sudden death, especially in sheep, may occur. Clinical signs for poisoned animals that survive a short period of time include impaired vision, a lack of coordination, labored breathing, abdominal pain, and prostration, followed by death.

There is no effective treatment for acute selenium poisoning. A high protein ration, arsenic-salt mixes and arsenic acid added to the ration may reduce the incidence of chronic selenium poisoning of cattle on seleniferous pastures. Arsenic, however, is a very toxic substance and should never be used without prescription from a veterinarian. Pasture rotation, the addition of supplemental feeds from nonseleniferous areas and removal of animals from pastures in selenium belts are practical means to control chronic selenium poisoning.

Water

Uses of Water

Life is impossible without a certain amount of water. Seventy to eighty percent of the weight of newborn calves, lambs, and pigs is water, though when the animal is three months of age or older, the water content is reduced to slightly above two-thirds of its weight. Whenever there is a loss of 10 percent of the total water content, the animal will be seriously distressed, and a loss of 20 percent will cause death.

Water is essential to dissolve all food before it may be utilized by the body. The body, lymph, gastric juice, joint water, and spinal fluid are largely water.

Fig. 2-12.   All animals should have access to clean water.

The very essential part of the microscopic cells of which all living matter is composed, known as protoplasm, is largely water. The waste product of the body is mostly removed in solution in water such as in the urine, in perspiration, and in tears. The refuse of the digestive tract cannot be removed until it has been softened with water. The normal temperature of the body will not be maintained without an adequate supply of water (Fig. 2-12).

Intake and Outflow

Whenever there is a large outflow of water from the body because of excessively high outside temperatures, or for other reasons, more fluid must be taken in than the amount that the animal may make in its body or that is taken in with the food. In fact, the need of the animal body for water is largely a matter of balancing the intake with the outflow. If this principle is disregarded, as is so often the case with working animals during days of hard work and high temperature, heat stroke can occur; or, if there is insufficient intake to moisten fibrous foods, such as straw, cornstalks, and similar materials in the digestive tract, choking, and constipation, with its attendant evils, follow.

Water Requirements

The exact water requirement of an animal cannot be definitely determined, since it varies so widely with the amount and nature of the ration, with the stage in the life cycle, and with environmental conditions. To say that an animal needs a definite amount of water is meaningless unless all the conditions are specified. For example, it is a matter of common observation that cows and horses on alfalfa hay drink more water and urinate more freely than those on timothy or other non-legume roughage. This is due in part to the larger amount of waste matter excreted through the kidneys. Swine on corn and on cob meal are reported to drink a third more water than on cornmeal, due perhaps to the greater amount of undigested material excreted through the feces.

A cow in heavy milk production may drink up to four or five times as much water as when dry. Particularly striking is the extra water needed for regulation of body temperature in hot weather. Experiments at the Missouri and California stations show that high-producing cows cannot vaporize water fast enough in hot weather to maintain normal body temperature. The critical continuous surrounding temperature for a high-producing Holstein is around 70° to 75°F. Above this temperature, she will eat less feed and thus produce less body heat to be removed. At 90°F milk production may cease entirely. The critical continuous temperature of Jerseys is around 85°F, and for Indian cattle around 90°F or higher. The difference in ability of the breeds to dissipate heat, although not clearly understood, is perhaps related to differences in hair coat and skin texture.

Frequency of Watering

Frequency of watering is an important factor in assuring an adequate water intake. Swine drink small amounts at a time and subsist largely on rations naturally low in moisture; therefore, they require a constant water supply in close proximity to the feed, with proper allowance of drinking space per animal.

Less frequent watering is required by ruminants on green pasture since they drink large amounts at at time and since the grass is naturally high in moisture. If the grass is lush, they may get enough from this source alone to supply their needs, four or more pounds of water for each pound of dry matter in the grass. On such pasture, one watering per day is ample. On dry pasture, more frequent watering is advisable.

Dairy cattle will drink readily after milking. To maximize milk yield, an adequate water suppy must be available when cattle exit the parlor. As a guide, provide at least 2 feet of tank perimeter for every 15 cows, with at least two waterer locations per group of cows. Extra waterer space may be needed if heifers are grouped with older cows, or if cows are observed waiting for the waterer to fill or to gain access to the waterer.

CHAPTER 3

Sanitation and Disease Control

Approved practices in the care of livestock imply that these animals should be provided with clean feed and water, that they should be kept reasonably clean, and that they should be given comfortable, healthful surroundings (Fig. 3-1). Under "native" conditions, natural processes, together with the animal's instinct, tended to preserve comfort and health; however, modern production schemes have disturbed the "native" conditions and the natural processes.

FEED

Importance of Clean Feed

It is very important for feed to be offered in a reasonably clean manner (Fig. 3-2). The practice of throwing feed, especially grains, on the ground, so that it is impossible for animals to get these without, at

Fig. 3-2. Feed should be fed in clean, sturdy, easily accessible feed bunks.

Fig. 3-1. Sanitation is a prerequisite for health in confinement operations. These individual hutches allow individual care and prevent disease transmission.

the same time, getting a liberal amount of foreign matter, is to be condemned. Unfortunately, since the feeding area is usually one where animals generally congregate, the ground becomes badly contaminated with worm eggs and with other dangerous forms of infection.

Thus, various types of self-feeders have been developed. These self-feeders usually are designed to keep the animals out of them, to prevent animals from depositing feces in them, to be easily cleaned, and to be readily accessible (Fig. 3-3). They are usually designed with smooth corners, so that decomposing old food cannot be harbored. Hay feeders are often constructed of wood, because the average individual can work with this material easily, because it is readily available, and because the racks and self-feeders made out of it can easily be moved to some other part of

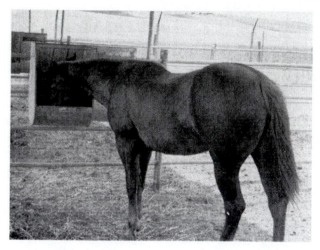

Fig. 3-3. A feeder that keeps feed clean and enables the horse to eat in comfort.

the farm. The chief disadvantage of wood is that it absorbs disease-producing agents and it is difficult to disinfect. Whenever feasible, metal or concrete should be used for building feed racks and troughs, self-feeders, feed platforms, and similar devices. Concrete is durable and should be formed to be smooth on the eating surface. Reducing the amount of surface pitting reduces mold formation and feed spoilage. Many college agricultural experiment stations have conducted valuable research regarding the most efficient and sanitary methods of feeding and watering livestock and poultry. Their publications are available upon request.

Clean Storage of Feed

Lofts and bins for the storage of feed should keep feed dry and clean. If feed is repeatedly moistened, with no compensating exposure to the sun's action, the growth of bacteria, molds, and fungi is unhampered, and decomposition soon takes place. Bins for the storage of grains should be insect, rodent, and bird proof; otherwise, such intruders will eat and contaminate the feed, rendering it undesirable for feeding purposes. In addition, several diseases can be transmitted to animals from feed contaminated by pests, such as equine protozoal myelitis in horses.

Sudden Changes in Feed

No single factor in feeding practice causes as much trouble in animals as a sudden change from an accustomed feed to a new and unaccustomed feed. Usually, the new feed is highly relished by the animals. Their bodies are suddenly called upon, without gradual preliminary preparation, to handle this, and gorging, with its attendant evils, also takes place. In ruminants, microorganisms and feed are in close relationship with each other; no single organism species is responsible for the complete digestion of a particular feed. Therefore, if the proper balance is upset by a sudden change in feed, resulting in either an inadequate feed intake or gorging, the function of the rumen is impaired. Many illnesses in dairy cattle, around the time of calving, are attributed to the rapid change in diet. Current recommendations are to have a far-off, dry cow group on a lower-protein, lower-energy ration; a close-up group where the protein and carbohydrates are increased; and a fresh cow group where a higher-fiber ration (higher than the main milking herd receives) is fed.

When animals are to be changed to a new ration, and, especially if the ration is quite different from that which they have been accustomed to over a period of time, the change should be made gradually. Rumen microorganisms require two weeks of adaptation time to a new ration to adjust the microbial population for efficient digestion to occur. On the first day of the new feeding, a good practice is to let the animals have their usual accustomed ration, and then offer them some of the new feed in addition. Gradually, from day to day the old ration may be reduced in amount and the new ration correspondingly increased. An abrupt change in diet is most likely to be harmful when a dry ration has been fed and then is changed to a green, succulent one. The change from old prairie hay to new prairie hay or alfalfa, or from old seasoned corn to new or green corn is frequently a cause of trouble. Sometimes animals are on a large grain ration because heavy work is demanded of them, and under these circumstances, they are able to utilize this feed. Sudden discontinuance of work for a day, as over a Sunday or a holiday, without a corresponding decrease in the grain ration, is likely to cause horses to contract the serious so-called "Monday morning sickness" or "holiday disease." Veterinarians speak of it as azoturia or hemoglobinuria because of the presence of blood in the urine.

Another example of the evils attending a sudden change to a new diet, as well as gorging, is the development of inflammation in the feet known as "founder" or laminitis. There seems to be a sympathetic relationship between inflammatory processes in many of the bodily organs and the deeper sensitive portion of the feet, so that the latter frequently become secondarily inflamed to produce "founder." This reaction in the feet is most likely to occur as a sequence to inflammatory processes in the digestive tract, as induced by sudden changes and gorging, though there may be other accessory causes and conditions.

WATER

Providing Clean Water

Water is so important to the health of animals and to their efficient production that it should always be accessible and free from adulteration (Fig. 3-4).

It is well known that the germs of some animal diseases will live in water for a short time, and, under these conditions, water may be the carrier of diseases such as hog cholera, foot-and-mouth disease, strangles, anthrax, leptospirosis, salmonellosis, and others; but, unless there is frequent and repeated pollution of the water by these germs, flowing water will be cleansed by natural processes in the course of time. Although there is a real danger of transmitting germs by water, with the possibility of spreading diseases, it does not follow that waters once polluted will remain permanently harmful if the source of germ pollution is removed.

There is doubtless greater danger to animals from drinking water that has been contaminated with intestinal discharges, because the latter usually have a rich intermingling of worm eggs which live for a comparatively long period in water. It is not implied that water is the natural habitat for many varieties of worm eggs, nor that prolonged soaking in water with attendant changes in temperature will destroy the eggs. But

Fig. 3-4. Clean, fresh water is indispensable for animal health and efficient production.

livestock producers should be aware that the drinking of water obtained from sources where it is subject to contamination by intestinal discharges is responsible, in a large measure, for the almost universal infestation of intestinal parasites in animals.

Danger of Flowing Water

The influence of flowing streams may not be disregarded in the spread of animal diseases. Many flowing streams are badly polluted with human and animal fecal material. Pathogenic bacteria and viruses may be carried for some distance by flowing streams, as witnessed when an upstream outbreak is followed by successive outbreaks on bordering downstream farms. The spores of anthrax and blackleg germs will live for some time in the slime of flowing streams, concentrate at locations that overflow and cause rampant disease outbreaks. There are sections in the Southwest where the raising of calves without protective vaccination against blackleg is practically impossible because flowing streams have deposited the highly resistant spores of the blackleg organism on soil and vegetation accessible to cattle. The eggs of many animal parasites may be similarly transported.

Contaminated flowing water undergoes a certain degree of self-purification because it is exposed to the sun's rays and to the consuming and destroying action of harmless bacteria and other forms of vegetable life as well as being exposed to fish and other forms of aquatic life. Furthermore, some of the harmful organisms settle to the bottom of the stream, and pure water flowing into the main stream provides a beneficial diluting action. The time required for limited self-purification to take place depends upon several factors, such as the rapidity of flow of the water in the polluted stream, the number of brooks and springs flowing into it below the point of pollution. There is a limit to the waste loads waters can carry without exhausting their supply of dissolved oxygen. All states have environmental protection agencies regulating the discharge of sewage or other wastes into streams.

Fencing Off River Water

Generally, river water does not become fit for human consumption through self-purification. So long as the water flows only through wooded and sparsely animal-populated areas, it is reasonably safe for animal consumption. However, when it flows through extensive livestock-producing areas, it should be fenced off.

Flowing surface water exposed for some time to full sunlight and air is frequently freer from contamination than waters that are from deep sources, though a carefully conducted investigation of the latter may disclose that there are surface waters rather deeply deposited in the soil. A decided disadvantage of many deep waters is that in their subterranean travels they dissolve much material from rocks, chemicals, and iron in the soil. It is these substances that make waters "hard" or "alkali."

Damage Due to Salt Water

The degree of hardness in some sections of the United States may vary from 36,000 to 200,000 parts of minerals to 1 million parts of water. In other words, 3.6 percent to 20 percent of the weight of the water, may consist of these dissolved solids. One of the solids frequently present is common salt, and when it is present to the extent of 3.6 percent, there is approximately between 4 and 5 ounces of it in each gallon of water. The disposal of "salt" water is one of the serious problems in "oil country," where it occurs in large quantities from some deep oil wells. If pumped onto the surface, it destroys vegetation. If put into streams, it may make the latter unpalatable for

consumption by livestock. There are many claims for damages against oil producers by livestock owners for alleged damage to, or death of, animals that had consumed these waters.

Carefully conducted research projects have established that, in general, when water contains more than 15,000 parts of total solids (total solids may be salt alone or a mixture of various inorganic chemicals) in a million parts of water, and if animals have no other source of supply, it becomes dangerous to their well-being.

Sheep seem to tolerate these waters better than other animals, perhaps because they drink less water. They have been known to get along fairly well, even though the water consumed by them may contain as much as 2.5 percent or 25,000 parts of solids per million parts of water.

Cattle not producing milk and during the cooler months of the year, when a minimum of water is consumed, have apparently suffered no harmful effects from the daily consumption of water containing 2 percent of total solids.

That animals can adjust themselves to those high levels seems astonishing. Of the animal functions, lactation and reproduction are generally the first to be disturbed by a continuous intake of high saline water. Milk production may be greatly reduced or even terminated.

Effects of Freezing Water

Water may become harmful to animals when the inclemencies of the weather have altered it physically. The harmful effects of sudden chilling of the animal, following the unaccustomed consumption of quantities of ice water, are well known. In part, at least, very cold water will inhibit the important bacterial action in the rumen, which is essential in the preliminary digestion processes. Thus, a decrease in, or even complete stoppage of, the milk flow and the premature birth of young are ascribed to this factor. Strange as it may seem, animals accustomed to the regular daily consumption of very cold water are seldom or never harmed by it.

Confinement swine on full feed are susceptible to salt poisoning, which can occur if water lines freeze and the water supply is interrupted. Piglets from lactating sows may starve when the water supply is frozen or the water flow rate is hampered.

Effects of Chilled Water

Confined cattle having continuous free access to water from automatic fountains can consume it whenever it is not frozen without harm. Even water at 45°F may be harmful to heated work animals. If there is any doubt about the matter, the animals first should be permitted to quench their thirst only in part, then given some feed, and, after they have cooled off, given access to more water.

Extensive research has been performed using chilled water for heat-stressed dairy cattle. The research has shown an increase in milk production from cattle drinking chilled water compared to the control animals. The expense of chilling water can be significant, especially when considering that a high-producing cow can drink up to 50 gallons of water daily.

Water Analyses

There may be situations when knowing exactly what materials are in water obtained from a definite source may be desirable. Thus, analysis becomes necessary. Generally, such analyses may be classified as bacteriological and chemical, the former to determine the absence or presence of harmful organisms, and the latter to determine the amount of chemicals and inorganic and organic substances in the water. Occasionally, the examination is to determine whether or not pollution exists within the higher forms of life, such as animal parasites and their eggs. When such analyses are desired, the water should be submitted to a public or private laboratory, along with a statement indicating why it is to be examined.

Preparing the Sample Correctly

A good method of collecting and shipping the specimen of water is to gather at least ½ gallon in a glass container that has previously been sterilized in boiling water. The specimen should be an average sample—that is, it should be collected from wells after a few bucketfuls have been discarded, so that contamination from the pump itself may be largely eliminated. In very warm weather, the water and its container should be surrounded with ice, and, in cold weather, it should be protected from freezing during shipment to the laboratory. It should reach the laboratory within 24 hours from the time of its collection. Sometimes several samples collected at different times must be analyzed before definite conclusions can be drawn from the results.

From a sanitary standpoint, harm to animals may be minimized if watering troughs and other containers are kept clean, if water is placed in containers where it will not be subject to contamination by animal droppings, and if animals are kept away from sources of water supply other than those supplied for them in a reasonably sanitary manner.

GROOMING ANIMALS

Necessity for Grooming

All farm animals, if given an opportunity, prefer clean to filthy surroundings. For those animals that have access to pasture, heavy rains and sunshine help keep them clean. Rubbing against posts and trees and rolling on the ground helps remove loose hair and other skin debris, as well as many parasites. However, when animals are denied access to pasture and are stabled, caretakers must assume the task of regularly grooming them if they want their animals to maintain health and vigor and physical comfort. Grooming signifies care of the skin and hair coat; care of the feet; and care of the genitals and external orifices of the body.

Resistance to disease may be lowered considerably by a filthy body surface, this being especially true in relation to parasitic skin diseases. The skin of all farm animals becomes filthy as a result of accumulated exfoliated parts of its superficial layer, the secretion of fatty material, and sweating. When these natural processes are combined with a deposit of dust on the skin, consisting of the animals' own intestinal discharges, these conditions are not wholesome.

Sweat Glands in Animals

When the skin is sealed with dirt, one of its functions is impaired. The skin is one of the important organs of the body in the elimination of waste products through the sweat glands. This is especially true in horses. Sweat glands in the horse are present in all the skin of the general body surface, and, to a less extent, this is also true for sheep. In other animals, the sweat glands are confined to certain skin areas, such as the snout in pigs and mainly in the muzzle of cattle. If the openings of the sweat glands become clogged with dirt, either from the skin itself or from deposits thereon, they cannot function properly. Some animals, such as the goat, rabbit, and fowl do not sweat.

Grooming a Horse

Grooming involves not only cleaning the horse but also improving its health. It follows that a properly groomed horse will look better, eat better, feel better, and, as a result, act and perform more efficiently.

Vigorous grooming massages the underlying body muscles and improves fitness. It cleans the hair and stimulates the pores to produce natural oils that bring a shine to the horse's coat. Artificial oils are unnecessary when horses are groomed regularly, properly, and thoroughly. Grooming provides an opportunity to examine the entire body of the horse. In doing so, parasite eggs, lice, mange, or skin disorders may be detected and controlled.

Proper Equipment

Grooming equipment should include:

1. A rubber or plastic currycomb.

2. A dandy brush—stiff-fiber brush used for the mane and tail and the extreme lower portion of the legs.

3. A body or finishing brush—a soft-fiber brush.

4. A grooming cloth, such as Irish linen, salt sack, or even burlap, which will shake out readily.

5. A sponge.

6. A hoof pick to clean the horse's feet.

Grooming Procedures

The groomer should:

1. Take the horse out of the stall and cross-tie it securely. If this is not possible, tie the horse before the grooming process begins. This is an important safety procedure. There are four accepted ways to secure a horse for grooming: direct tie, cross-tie, rope-in-hand, and hobble.

2. Remove the stable sheet or blanket properly. This is another important safety factor. Release the back strap first, then the middle strap, and finally the front strap. Pull the blanket off the horse along the lay of the hair. Turn it inside out, shake it, and hang it where it will air while the horse is being groomed. When the blanket is placed back on the horse, the front strap should be fastened first.

3. Start grooming on the left side. Take the currycomb in the right hand and the body brush in the left. Start with the currycomb and follow along with the body brush. Begin on the neck immediately behind the head. Then work the chest, the withers, the shoulders, and the foreleg down to the knee. Next, work the back, the side, the belly, the rump, and the hindleg down to the hock. The motion for the currycomb should be irregular, back and forth, or in a small circular motion. Never use it about the head or on the legs from the knees and hocks down. Clean the currycomb by tapping it against the heel of the boot. Clean the brush every few strokes with the currycomb.

4. Next groom the right side in the same order as the left, changing the brush to the right hand and currycomb to the left. Finally, brush the head and face, using the body brush.

5. Take the dandy brush and brush the mane and tail. Start at the bottom or ends and work gradually toward the roots. Free any entanglements with the fin-

gers. Wash the tail occasionally with warm water and soap. Be sure to rinse thoroughly.

6. Use the grooming cloth or rub rag to wipe the ears, the face, the eyes, the nostrils, the lips, the sheath, and the dock, and to give a final polish to the coat.

7. Soak the sponge in water and press out as much water as possible. The sponge should be damp, but not dripping. Stroke the eyes with an outward circular motion, the ears with an upward motion. Clean the nostrils and the lips with the damp sponge. Then go to the rear of the horse, raise the tail and clean the portion of the dock where there is no hair. This is where sweat and dirt collect, often causing a sore tail. Give extra care to the knees and the elbows, or hocks. These places are abused when the horse lies down and gets up. Use a damp sponge to straighten out the hair.

The groomer should not abuse the horse by severe grooming. Care should be exercised with thin-skinned or short-haired horses to avoid excessive pressure that irritates the skin. To do the best job, stand erect at arm's length from the horse so that when the arms are extended forward the groomer will just touch the horse with the palms of the hands. Then, lean forward toward the horse and provide sufficient pressure to create movement between the skin and the underlying muscles. This massaging action is most effective.

Care of the Feet

Few things in grooming and caring for the horse are as important as properly cleaning the feet (Fig. 3-5).

Near forefoot: Slide your left hand down the cannon to the fetlock. Lean with your left shoulder against the horse's shoulder. Reverse for picking up the off forefoot. When the horse shifts weight and relaxes on the foot, pick it up.

For a quick cleaning, hold the hoof in your free hand. When shoeing, or for a long cleaning job, place the horse's foreleg between your legs. Hold your knees together to help support the weight of the horse's leg.

Near Hindfoot: Stand forward of the hindquarter and stroke with your right hand from the point of the hip down the hip and leg to the middle of the cannon. As you move the right hand down, place the left hand on the hip and press to force the horse's weight to the opposite leg. Grasp the back of the cannon, just above the fetlock and lift the foot forward.

When the horse is settled, move to the rear, keeping the leg straight, and swing your left leg underneath the fetlock to help support the horse's leg. Never pull the foot to the side—your horse will resist. Reverse sides for picking up the off leg.

Fig. 3-5.   Recommended procedures for cleaning a horse's feet. (Courtesy, University of Illinois)

Lack of proper grooming of the horse's feet, particularly failure to clean thoroughly the depths of the commisures and cleft of the frog, can cause thrush. Contributing factors include a lack of frog pressure, insufficient exercise, filthy stables, dry feet, and cuts or tears on the horny frog; but failure to clean a horse's feet properly is most frequently the cause of thrush.

If the horse is shod, shoes should be replaced or reset every 6 to 8 weeks; unshod horses, broodmares, and colts need their feet trimmed every 8 to 10 weeks.

Hoof dressing and oils improve the appearance of the horse's feet for show, but most of them do little to soften hard and brittle hooves. Standing the horse in mud for a few hours will do this best.

A fault in the grooming of the feet of horses or other hoofed animals is to rasp the outer surface and thus remove a "varnish" which nature has placed there to prevent evaporation of the hoof's moisture. If repeatedly done, dryness sets in, followed by contraction and pain, as manifested by lameness.

The use of high-energy, low-fiber rations, as well as confinement on concrete, has led to lameness being the second most common disease in dairy cattle. Regular trimming of the hooves by an experienced hoof trimmer will lead to less production losses, as well as a longer productive life for the cow.

Care of the Stallion's Genitals

Washing the sheath and genitals of stallions and geldings is to be recommended for the removal of an accumulated, dried, and hardened fatty secretion of those parts known as smegma. If this material is not removed, the hardened mass, colloquially known as a "bean," may so obstruct the urinary passageway that straining and severe pain will occur. In addition, severe tail rubbing of geldings is frequently the result of filth in the sheath and may sometimes be overcome by a thorough washing of this organ. In male hogs, a collection of urine and even concrements (stones) have been found in the sack in the dorsal wall of the sheath near its opening. Inanimate foreign bodies, such as sticks, splinters, and hard vegetable burrs, occasionally find lodgement within some of the natural bodily orifices.

Advantages of Clipping

In all animals, removal of the skin covering, whether hair, wool, fur, or bristles, is a distinct aid in the treatment of skin diseases and in the removal of skin parasites. The greatest advantage of clipping is that it aids in keeping the animal clean, and that, after perspiring, the animal will dry more quickly. Although it is primarily done to sheep for the purpose of obtaining the fleece, other animals are often clipped to treat parasites and skin infections. As a general rule, cattle are not subjected to clipping when their skin is in good condition. However, in many dairies, it is a regular practice to clip the hair of the udder, as well as adjoining parts of the body, to assist in the sanitary production of milk.

MANURE DISPOSAL

The movement of food-producing animals into confinement operations, as well as the extreme size of some operations, has led to the topic of manure disposal being very expensive and controversial. Many people object to the odor and potential water pollution from animal waste. The declining number of farm families has led to fewer and fewer people involved in animal agriculture; therefore, there is great legislative pressure to write manure storage and handling laws for each state. Extensive research has been conducted to decrease animal manure odor, as well as develop manure processing and utilizing equipment.

Waste Management Systems

Many devices and strategies have been developed to separate confined animals from their excrement. Elevated calf pens, slatted floors over a holding tank, lagoons, oxidation ditches, and manure drying equipment are only a few of the measures livestock producers are incorporating into their management systems to control disease and to meet Environmental Protection Agency requirements to prevent air and water pollution from animal waste (Fig. 3-6).

Fig. 3-6. Manure and urine pass through slots to keep feedlot cattle clean and dry.

Farm animal waste management may be placed in five categories as follows: collecting, processing, storing, transporting, and utilizing.

Collecting

Among the places where manure is collected are:

1. The gutter behind a row of stalls.
2. The alley in a free-stall barn.
3. The feedlot or exercise lot.
4. The manure pack in a loafing shed.
5. The tank under slotted floors, at the end of a free-stall alley, or at the edge of a lot.
6. The floor under poultry cages.

The first three collection places listed above require frequent removal of the waste to a processing or storage facility or directly to a field for utilization. The last three also serve for storage.

Processing

Processing involves changing the characteristics of livestock waste so as to improve the storing, transporting or utilizing aspects of a waste management system. Processing includes separating solids from liquid, composting, drying, biological degradation, chemical treatment, and protein production. The following devices are used to process livestock waste:

1. Gutter drains for separation of liquid from solids.
2. Settling tanks for milking parlor waste.

3. Settling basins to settle out solids that are carried in feedlot runoff.

4. Oxidation equipment for aerobic degradation and possible microbial protein production. This might be an oxidation ditch or a mechanically aerated lagoon.

5. Lagoons for biological degradation. Since it is usually impractical to make lagoons naturally aerobic, some odor may be expected.

6. Composting facilities. These are seldom used by farm producers.

7. Drying equipment. Experimental work has been mostly with poultry manure.

Storing

Long-term storage facilities are often required because bad weather and field conditions may prevent the transportation of manure, the odor released when manure is spread may force delays, or the stage of development of crops may require a producer to continue storage. Under certain weather conditions, application of manure to land may also create potential water pollution. For these reasons, the following long-term storage facilities should be considered: manure stacking structure; manure pack in a loafing shed;

Table 3-1.   Approximate Daily Manure Production, Without Bedding[1]

Animal	Cu Ft/Day Solids and Liquids[2]	Gallons/Day[3]
1,000-lb cow.........	$1\frac{1}{2}$	11
1,000-lb cow.........	1	$7\frac{1}{2}$
10 head of sheep......	$\frac{1}{2}$	4
10 head of hogs:		
50 lb.............	$\frac{2}{3}$	5
100 lb.............	$1\frac{1}{3}$	10
150 lb.............	$2\frac{1}{4}$	17
200 lb.............	$2\frac{3}{4}$	$20\frac{1}{2}$
250 lb.............	$3\frac{1}{2}$	26
1,000-lb horse........	$\frac{3}{4}$	$5\frac{1}{2}$

[1]Adapted from *Michigan State University Circ.* 231.
[2]There are about 34 cu ft in a ton of manure.
[3]One cubic foot = $7\frac{1}{2}$ gallons.

holding pond for feedlot runoff; and tanks for liquid manure, including tanks under slotted floors. The necessary manure storage capacity can be calculated from Table 3-1, which shows the approximate daily manure production, without bedding, of farm animals.

Transporting

There are times when manure and waste water have to be moved from storage to the utilization area. Manure must be removed from sheds and stacking areas, manure tanks must be pumped, holding ponds must be emptied, and lagoon levels must be lowered.

The types of equipment used for transporting wastes are the conventional manure spreader, liquid manure spreader, irrigation equipment, and low-capacity pump and pipe with gravity discharge

Utilizing

The vast majority of livestock waste is utilized on the farm on which it is produced by applying it to fields where crops are grown. There are two good reasons for doing this. First, manure adds nutrients, improves soil tilth, enlarges water-holding capacity, lessens wind and water erosion, increases aeration, and

Table 3-2.   Quantity, Composition, and Value of Fresh Manure (Free of Bedding) Excreted By 1,000 Pounds Liveweight of Various Kinds of Farm Animals

(1)	(2)	Composition and Value of Manure on a Tonnage Basis[2]						
		(3)	(4)	(5)	(6)	(7)	(8)	(9)
Animal	**Tons Excreted/ Year/1,000 Lb Liveweight**[1]	**Excrement**	**Lb/Ton**[3]	**Water**	**N**	**P_2O_5**[4]	**K_2O**[4]	**Value/Ton**[4]
			(lb)	(%)	(lb)	(lb)	(lb)	($)
Cow (beef or dairy)	12	Liquid	600	79	11.2	4.6	12.0	5.99
		Solid	1,400					
		Total	2,000					
Steer (finishing cattle)	8.5	Liquid	600	80	14.0	9.2	10.8	7.64
		Solid	1,400					
		Total	2,000					
Sheep	6	Liquid	660	65	28.0	9.6	24.0	13.48
		Solid	1,340					
		Total	2,000					
Swine.	16	Liquid	800	75	10.0	6.4	9.1	5.65
		Solid	1,200					
		Total	2,000					
Horse	8	Liquid	400	60	13.8	4.6	14.4	7.05
		Solid	1,600					
		Total	2,000					

[1]*Manure is Worth Money—It Deserves Good Care*, University of Illinois Circ. 595, 1953, p. 4.

[2]Columns 5, 6, 7, and 8 from *Farm Manures*, University of Kentucky Circ. 593, 1964, p. 5, Table 2.

[3]From *Reference Material for 1951 Saddle and Sirloin Essay Contest*, compiled from *Fertilizers and Crop Production* by Van Slyke, published by Orange Judd Publishing Co.

[4]P_2O_5 can be converted to phosphorus (P) by dividing the figure given above by 2.29, and K_2O can be converted to potassium (K) by dividing by 1.2.

[5]Calculated on the assumption that nitrogen (N) retails at 25¢, and K_2O at 17¢ per pound in commercial fertilizers.

Fig. 3-7. Spreading manure on land where a crop will be grown. (Courtesy, Ford-New Holland, New Holland, Penn.)

promotes the growth of beneficial soil organisms. Second, there is no economic system that can degrade livestock waste to the extent that the effluent is suitable for discharge into a stream or into a municipal waste treatment facility. The amount, composition, and fertilizer value of manure excreted by farm animals is given in Table 3-2.

Application to the land must be at a time and in such a manner that the waste is not carried off by rains or melting snows. Spreading waste on frozen, snow-covered, or saturated soils should be avoided, particularly if the land is rolling and extensive runoff is probable (Figs. 3-7 and 3-8).

Soils have varying capacities to accept manure and to serve as a treatment process without polluting underground water or surface runoff. Installing terraces, establishing buffer strips, or using selected vegetation to provide an acceptable manure-utilization area may be necessary.

A lagoon could serve as a final disposal in some operations if the lagoon does not overflow. It is likely that a liquid build-up will exceed evaporation in most areas, requiring removal of some liquids to the land at certain times (Fig. 3-9).

Another alternative to land application is to sell dried or composted fertilizer. This method has limited application because of technical and marketing problems.

Recycling manure as a livestock feed is the most promising of the nonfertilizer uses. Various processing methods are being employed, and some manure is being fed without processing. Poultry manure is the most collectable and the most nutritious of all animal wastes.

Fig. 3-9. A lagoon provides for waste handling of the swine finishing unit (right) and farrowing unit (left). Both swine units have slotted floors. (Courtesy, *National Hog Farmer*, Minneapolis, Minn.)

Systems to Reduce Water Pollution

Suggested livestock waste management systems are described in Figs. 3-10, 3-11, 3-12, and 3-13. Components of one suggested system may be combined with another, and all the components indicated for each of the drawings are not always required. For example, in a liquid manure system the irrigation equipment may not be needed, depending upon the desired type of management. The system most suited for a given operation must be based on a thorough evaluation of all the physical and management factors.

Fig. 3-8. "Honey wagon" spreading liquid manure. (Courtesy USDA, Soil Conservation Service, Washington, D.C.)

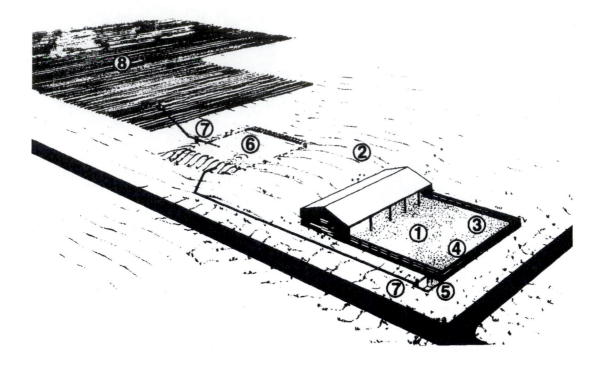

Many feedlots are boxed in by buildings, roads, lanes, or waterways so that runoff water may have to be pumped to another area. Or, the livestock operation may have to be relocated.

① **Feedlot.** Animal wastes will accumulate on the feedlot and in the open shelter. Most of these wastes will be handled as a solid with conventional equipment and applied to the utilization area, but runoff from the lot will transport some solid and liquid waste that must be intercepted.

② **Clean water diversions.** To minimize the amount of runoff that can transport waste, clean water should be diverted from entering areas where wastes are deposited or stored. Diversions may be needed above and adjacent to the feedlot, and buildings should be guttered. The feedlot size may be reduced to the minimum recommended area needed for good animal growth and management.

③ **Runoff collection.** The runoff from the feedlot must be collected and directed to a central storage area. This may be accomplished by the natural feedlot slope, or by diversions, gutters, curbing, or pipes.

④ **Setting basin.** A portion of the existing lot or a narrow concrete channel along the outside of the lot may serve as a settling basin to slow the velocity of runoff so that most of the solids will settle from the runoff water. The basin should be large

enough to accumulate the solids that are carried in the runoff for a six-month period. Other solids should not be scraped into the basin. The basin should be shaped to permit cleanout with available equipment.

⑤ **Sump pit.** A sump pit at the end of the settling basin collects runoff water that is pumped to a place on the farm where space is available to build a holding pond. The sump and pump should be sized to collect and pump water away about as fast as it runs off the most intense storm.

⑥ **Holding pond.** A holding pond receives the liquid from the settling basin. This pond must be large enough to store the runoff from average precipitation during the period from November through April. It must be emptied after major runoff periods and before winter to provide storage for subsequent runoff.

⑦ **Waste water transport.** A system must be included to transport the stored runoff to its utilization area.

⑧ **Utilization.** Liquids should be applied to the utilization area at times and at rates that will not cause runoff, excessive odors, or damage to crops.

Fig. 3-10. Feedlot runoff control (restricted space). (Courtesy, University of Illinois, Urbana)

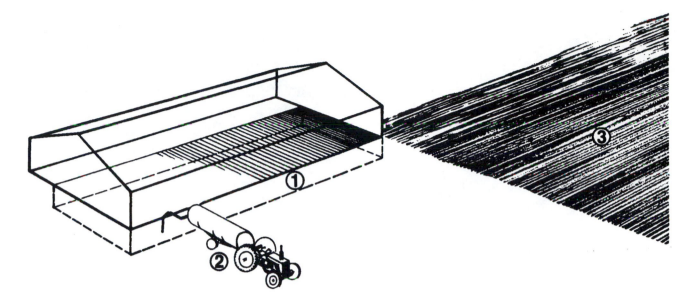

In this liquid manure system, wastes area collected in tanks, and the tanks are emptied periodically by spreading the contents on a utilization area.

① **Collection.** The collection tank may be under a part or all of the building floor, or it may be outside the building. Tanks under buildings will usually have slotted floors over them, although some dairy manure tanks may have a solid cover with wastes scraped into protected openings. Tanks built outside buildings should be covered to prevent collection of rainwater, odor problems, fly breeding, and safety hazards. The tank should be large enough to hold a 90- to 180-day accumulation of manure and waste water. The most suitable storage period depends upon the type of soil, crops, topography of the utilization area, and the availability of labor.

② **Pumping and hauling.** A pump or vacuum system is used to remove wastes from the collection tank. A liquid manure wagon is needed to transport the wastes to a utilization area. The wastes may be spread on the surface of the land, but direct injection into the soil reduces odor and the chances of surface runoff.

③ **Utilization.** Land must be available on which to apply wastes and utilize the nutrients. The wastes should be applied at times and at rates that will not cause runoff, excessive odors, or damage to crops.

Fig. 3-11. Liquid manure system (hauling). (Courtesy, University of Illinois, Urbana)

The size and shape of each component for an actual system will depend upon the calculations made for that system. The following information will help the owner in making estimates for preliminary plans. A holding pond should have the capacity to receive 15 inches of runoff from a concrete lot and 12 inches of runoff from a dirt lot. There are about 34 cubic feet in 1 ton of liquid manure, and 1 inch of runoff per acre equals 3,600 cubic feet or 27,000 gallons. The following formula should be used to calculate the volume of a rectangular pond with 3:1 side slopes:

Volume = (area of bottom × depth) + (1.5 × depth × depth × distance around bottom)

Example: Volume = (50 feet long × 40 feet wide × 10 feet deep) + (1.5 × 10 feet deep × 10 feet deep × 180 feet around bottom) = 20,000 + 27,000 = 47,000 cubic feet

PASTURE ROTATION

Pasture rotation is a practical and an important method of disease control, not only for the preven-

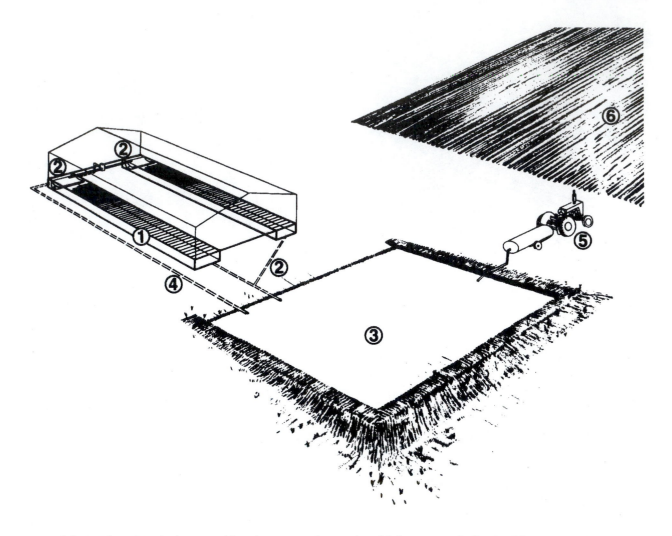

A type of system being used in a few cases in one in which manure is flushed from a confinement building to a lagoon or tank. This system is best adapted to swine farrowing houses or to nursery buildings. In some free-stall dairy buildings, however, waste collected in alleys can be flushed with water collected from cow washing.

① **Collection.** Wastes are deposited in gutters or alleys that are sloped for drainage.

② **Flushing.** A tank is filled with water and dumped periodically to flush waste from the building.

③ **Storage and treatment.** Waste may go to an aerated lagoon or, in some small operations, to a holding tank that can be pumped and the waste applied to cropland.

④ **Water reuse.** Water from the lagoon may be used to refill the flushing tanks.

⑤ **Protection against lagoon overflow.** To keep the lagoon from overflowing, equipment may be needed to haul or pump some excess liquid to a utilization area. A pump and pipeline or a tank wagon may be used for this purpose. Recycling some water for flushing the gutters or alleys in the building should minimize the problem of lagoon overflow. For this reason, equipment can be relatively small and perhaps not permanently installed.

⑥ **Utilization.** The utilization area for excess liquids may be a relatively small area of cropland set aside for this purpose. The waste water should be applied at times and at rates that will not cause runoff, excessive odors, or damage to crops.

Fig. 3-12.   Gutter finishing system in a confinement building. (Courtesy, University of Illinois, Urbana)

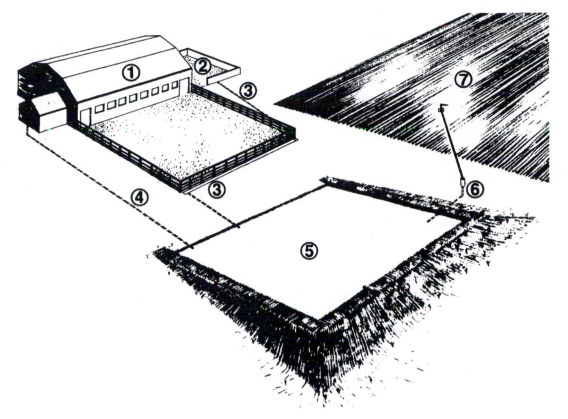

Waste from a stanchion dairy barn may be stored in a manure-stacking facility, and waste in a loose-housing barn may be allowed to accumulate in a manure pack. Liquids that drain from these wastes must be collected, and the milking parlor and milkhouse wastes must also be disposed of properly.

① **Collection.** The barn may be equipped with a gutter cleaner, or waste may be scraped, using small tractors. Bedding is used to absorb urine and keep the livestock comfortable and clean.

② **Stocking.** When daily spreading of manure is impractical, the waste can be stacked in a properly planned area. Adequate storage space should be provided for the anticipated volume of waste that may accumulate in a 180-day period. This area may be covered to minimize the liquid-handling problem and fly breeding.

③ **Lot runoff.** A system must be provided to collect the seepage and runoff from the manure stack and the cow lot next to the barn.

④ **Milking parlor wastes.** When a milking parlor is a part of the system, large volumes of waste and water from this facility may need to be included in the liquid-handling system. Whenever possible, these liquids should be combined with those from the stocking operation and any contaminated runoff from outside feedlots, and delivered to a lagoon sized for the anticipated amount of waste.

⑤ **Lagooning.** A tank or lagoon may be used to store the liquid that drains from the stacking area, exercise lot, and milking parlor. This storage facility must be large enough to contain the anticipated liquid that may be delivered during a 180-day period.

⑥ **Protection against lagoon overflow.** Lagoons must not be allowed to overflow. Sufficient freeboard should be provided to take care of some storm water, and the lagoon must be pumped down as necessary. A pump and pipeline or a tank wagon may be used for this purpose.

⑦ **Utilization.** Land must be available on which to apply waste water at times and at rates that will not cause runoff, excessive odors, or damage to crops.

Fig. 3-13. Solid and liquid system for dairy facilities. (Courtesy, University of Illinois, Urbana)

tion of some of the communicable diseases but also for the prevention of parasitic infestations.

Some disease-producing organisms go into a resting or seed stage during which they are designated as "spores." These are much more resistant to destruction by the elements or by man-made and applied agents than are the parent germs, so that yards and pastures once contaminated by them will remain so for an indefinite number of years. Anthrax and blackleg are examples of diseases caused by spore-forming bacteria. Frequently, protective vaccination is the control method adopted. This protects the vaccinated animal, though the pasture retains its infective potentialities for a long time. In fact, pastures on lower ground may become dangerous because of drainage. Plowing such land is questionable as it does not prevent earthworms and other forms of subterranean life from conveying the highly destruction-resistant spores to the surface.

Protection against parasites is an important element in animal husbandry. To the average person it is an astounding revelation that animals are so generally parasitized. Theoretically, parasites can be controlled by sanitation, and all efforts devoted to this cause will repay many fold in better health and vigor of the hosts.

Damage to Host

A parasite may be defined as an organism that lives in or on another organism known as the host and usually with some degree of harm to the latter. The parasite abstracts nourishment from the tissues of the host, or it may appropriate for its own use food intended for the host and at the same time be deriving shelter and warmth. Some parasites produce injuries by bites; not infrequently, they inject irritant poisons at the time of the bite. Parasites may mechanically obstruct certain organs; for example, roundworms of swine invading and lodging in the bile duct, or stomach bots of horses filling the stomach. When crushed within the tissues of the host, the parasite may set up certain reactions (anaphylaxis), resulting in serious injury to the host and possibly causing death.

The impact of internal parasites is very important economically. Internal parasites are often present in subclinical infections—those that do not cause visible disease. Parasites cause millions of dollars in losses annually due to impaired growth and milk production and death in severe cases.

Internal Parasites

Those that live within the host are most difficult to contend with successfully. Sanitary preventative methods are of the greatest importance. This is dramatically exemplified in the little or total absence of infestation in humans living in sanitary conditions. In animals, however, the problem usually resolves itself in the economic question of whether it is more profitable to control the parasites or to permit infestation. (The answer probably lies midway between the two.) There can be no question, however, that prevention is of greater importance than cure, though the average livestock owner is inclined to favor the latter, forgetting that after an animal has once become heavily infested, the parasite may have inflicted such severe damage upon its tissues that elimination or cure will not restore the animal to normal.

Young Animals Most Susceptible

A general rule in relation to parasites that live within the host is that young animals are much more susceptible to parasite invasion than are mature ones.

Most intestinal worm parasites need moisture either to hatch their eggs, or to enable the immature worm (larva) to reach a favorable location on vegetation prior to animal consumption. Therefore, if infested pastures must be grazed by susceptible hosts—especially the highly susceptible young ones—they should be kept out of such pastures until the wetness of dew and rains has evaporated.

Deep plowing of infected and parasite-infested land, as well as draining all low places, is a very helpful control practice. Most parasite eggs and larvae rarely reach the surface again after having been plowed under several inches of soil. If such land can be applied for a season to the growing of crops, it will be practically free from parasite contaminants at the end of this period.

Parasitic infestations of livestock may very generally be controlled by a system of pasture rotation. Burning off pastures has some merit, but it does not possess the efficiency with which it is usually credited, since those parasites and their eggs that are in or close to the ground may escape destruction. In general, permanent pastures used regularly by recognized host species are to be regarded as highly dangerous for profitable livestock production, and a system of changing land areas from pasturage to crop production is advisable.

Host-Specific Parasites

Fortunately, some parasites as well as some germs, will only live on certain hosts. Although, in general, sheep, goats, and cattle are susceptible to the same parasites, the stomach worms of horses, and the roundworms (strongyles) of horses rarely infest cattle, sheep, and swine. Thus, pastures may be safe for some species, though unsafe for others.

Pastures that have been freed of parasites are soon contaminated again when the mature animals, which are almost invariably heavily parasitized, are again placed on them. If the surrounding daily temperature reaches 95°F, and when moisture conditions are favorable, the invading stage of the parasite will be reached in 3 to 4 days; at 70°F, it requires 6 to 10 days; at 50°F, 3 to 4 weeks. In those portions of the United States where hard freezes are common, both parasitized and non-parasitized sheep may safely be placed in a common pasture for the period extending from late October to early March and then transferred to a clean pasture where they may be safely kept for another month.

Such a procedure is safe for all parasitic diseases in which the eggs hatch outside the host, such as stomach worms of ruminants, but not for those in which the eggs hatch inside the host, such as roundworms of swine. When warmer weather arrives, parasite development gains momentum; therefore, hosts comparatively free of intestinal parasites must be moved to clean pastures ev-

ery two weeks. As the cooler weather sets in, the time permitted on such pastures may again be lengthened.

The consumption by animals of forage or roughages grown on parasite-infested land is not an entirely safe method because such vegetation may have on it the eggs or larvae.

To control livestock diseases by pasture rotation, the caretaker should do the following:

1. Select well-drained pastures.
2. Avoid overstocking pastures.
3. Separate the young from the old as soon as possible, and raise the former on a dry pasture or on rotated pastures.
4. Use clean water and feed in raised troughs or racks.

DISPOSAL OF CARCASSES

Destroying diseased animals having a low value, or those chronically or incurably ill, is a sanitary precaution worthy of the widest application. However, with larger and more valuable animals, the sanitary steps may have to be limited to a quarantine of the ailing.

Reservoirs of Infection

Even after the death of animals from contagious diseases, their carcasses remain as disease-producing reservoirs. During the cold winter, viruses may remain virulent for as long as a month; however, in mid-summer direct sunlight will kill unprotected viruses in less than 24 hours. The viruses will remain virulent for up to 11 weeks during the winter months in a body buried in the fall, although it will reach the non-infective stage sooner if it is subjected to the higher above-ground temperature. Thus, cold makes infective material dormant, while heat hastens its metabolic processes and earlier death in unfavorable surroundings, usually outside the body.

Putrefaction in dead bodies also results in the early destruction of viruses. Statements regarding the infectivity of viruses in carcasses apply to anthrax, swine erysipelas, and, in general, to most communicable diseases. Whenever there is any doubt as to the cause of death, the carcass should be treated as a potential source of danger and disposed of in a proper manner. This is an important step in animal disease prevention.

Suggested Precautions

Many animal diseases are transmissible to humans. These shared diseases are known as *zoonoses*. They include African sleeping sickness, anthrax, brucellosis, leptospirosis, Q fever, rabies, trichinosis, tuberculosis, and tularemia.

The caretaker should not:

1. Dispose of a carcass by depositing it in or near a stream of flowing water, because this will carry infections to points down-stream.

2. Use carcasses for animal feeds. There is too much danger that an infection is responsible for the death of an animal, and using it for animal food may infect other animals. Infected parts may be carried to distant places, thus setting up new centers of infection. Furthermore, the consuming animal may contract serious indigestion from gorging.

3. Permit the carcass of an animal dead of a contagious disease to remain so that biting insects can reach it, since it is usually loaded with infection. It is best to cover the dead animal with a heavy sheet of plastic tarpaulin or screen until disposal.

4. Open a carcass for an autopsy or any other purpose unless approved by a licensed veterinarian, because of the danger of transmitting certain animal diseases to humans. Also, the blood and tissues of animals that have died of diseases such as anthrax, blackleg, and others are rich in the causative germ. Because these are spore formers, the opening of the carcass would likely result in a more or less permanent infection of the premises. Therefore, unless approved by a veterinarian, and then only in an easily disinfected place, it is not safe to open carcasses of animals that have died as the result of disease.

Carcass Disposal Methods

Disposal of dead animals is of concern not only for disease control and eradication but also for envi-

ronmental protection. All states have environmental laws that prescribe in detail the manner in which animal carcasses are to be handled. Livestock producers should be familiar with environmental protection standards for dead animal disposal in their state.

Most states require the owner to dispose of the carcass of any dead animal within 24 hours by burning, burying, or releasing it to a person operating a truck under permit to transport livestock carcasses.

Burning

Standards of the Environmental Protection Agency usually prohibit the burning of animal carcasses, except in incinerators that meet specifications of the agency. An approved incinerator for small carcasses can cost up to several thousand dollars.

Burying

Some state environmental protection agencies permit the burial of animal carcasses provided that:

1. Burial is in a location where runoff will not contaminate water supplies.

2. Burial is at a shallow depth so as to avoid contamination of subterranean water.

3. Lime or other chemicals that would delay natural bacterial decomposition are not applied to the carcass (the abdominal cavity should be punctured to permit escape of putrefactive gases).

4. Precautions are taken to keep scavenger animals from disturbing the carcass after burial (that is, the burial site must be fenced or covered).

Composting

Composting has been proven to be an economical and efficient way to dispose of carcasses, especially for the swine and poultry industry. The basis of composting is to allow microbial and chemical destruction of the carcass. A concrete pad with side walls approximately 10 feet wide and a top is sufficient for a composting site. Hay or straw is used as a carbon source, with the animals being added in layers as a nitrogen source. By always placing hay on the top layer, odor and flies are at a very minimum. The temperature in a compost pile becomes very elevated, destroying all pathogenic organisms, such as *Salmonella* species.

Disposal Pit

A disposal pit may be used for disposing of small carcasses (poultry or baby pigs), provided that the same precautions are taken as required for disposal by burial and that each carcass is covered with soil at the time of disposal.

Dead Animal Collection Service

The dead animal disposal agent is a valuable resource with whom the livestock owner should be personally acquainted, if possible. The livestock owner should keep the name, address and telephone number of the dead animal collection service handy and should be familiar with the fees. When the owner calls the disposal agent, the exact location of the carcass to be collected should be communicated.

Again, in the handling of carcasses, the owner should assume that the condition is the result of infection and should adopt the proper sanitary precautions to prevent disease dissemination. Later, if the diagnosis or the autopsy findings do not demonstrate the presence of an infection, only a little labor will have been lost. If findings indicate a disease of an infectious nature, the health of the remaining animals will have been safe-guarded. Violation of or disregard for the foregoing general principle of sanitation, or its adoption after it is too late, has been an expensive experience for many livestock producers.

CHAPTER 4

Disinfection and Disinfectants

Intensive confinement livestock operations are continually threatened with a condition referred to as disease build-up. Farrowing houses, nursery units, and dairy calf-rearing facilities are particularly vulnerable. As disease-producing microorganisms and parasite eggs accumulate, health problems become severe and may be transmitted to each succeeding group of animals raised on the same premises. Under these circumstances, disinfection becomes extremely important in breaking the disease cycle.

REMOVING ORGANIC MATERIALS

The first requirement in disinfection is to start with clean surroundings and surfaces. There is no substitute for this. No disinfectant applied to a dirty surface will destroy all germs. Organic matter, such as is found in dirt and manure, protects the infective organism or germ against the germ-killing activity of the disinfectant. Even if the disinfectant could penetrate the organic matter, the amount might be far too small to kill many organisms. The organic matter reacts with the disinfectant to form an insoluble compound that is inactive.

Cleaning can be done with a shovel and brush or speeded up by the use of high pressure pumps and detergents. A high velocity stream of water does an excellent job of removing dirt, as do the more expensive steam cleaners. Steam is an effective cleaner, but when used alone, it is only effective in killing disease organisms when it is applied directly through a nozzle and at close range, with the nozzle held not more than 6 to 8 inches away.

When there is an excessive amount of manure or dirt present, sanitation can be made more effective by the use of a detergent. Detergents hasten dirt removal by increasing the wetting speed and breaking the organic material into small particles. They act similarly on fats, proteins, and carbohydrates present in the environment by emulsifying and suspending them. Thus, they are easily removed, permitting the disinfectant to penetrate and kill the infective agents.

The selection of a disinfectant is important. For buildings, disinfectants should work well in the presence of organic matter, be compatible with detergents and soaps, be harmless to building materials, and be relatively nontoxic.

Phenol Coefficient

The term *phenol coefficient* of a disinfectant indicates its killing strength compared to phenol. Since phenol (its common name is carbolic acid) is a well-known disinfectant, its value is taken as unity, and the value of any other disinfectant similarly used can be expressed by a number called its "coefficient." This indicates how many times more, or in some instances less, the disinfectant can be diluted than phenol and still retain an equal disinfecting value.

The phenol coefficient is usually determined by comparing the disinfectant to be tested with phenol in its action on the typhoid fever germ, thus determining the dilutions of the samples that are as efficient as given dilutions of phenol. A low phenol coefficient definitely indicates inferiority of a product, but it cannot be said that a high phenol coefficient always indicates superiority because the test is made under carefully controlled conditions in a laboratory which are not duplicated in the actual practical application of the disinfectant. The phenol coefficient of a disinfectant is nevertheless a useful base to evaluate disinfectants and should be stated on the label of all disinfectants offered for sale to the public.

Germicidal Action

Modes of action of disinfectants vary. Generally, they exert their action:

1. By oxidation, which means that the oxygen of the disinfectant or the oxygen derived from other sources, because of the action of the disinfectant, combines chemically with the germ or actually oxidizes or burns it.

2. By the removal of water, or at least the elements that form water. Some disinfectants have a very great affinity for water and withdraw this from all sources, including germs and the media in which they live, so that death results, or the surroundings become untenable.

3. By coagulation, in which the fluid state of a substance is changed to a soft, jelly-like solid, with the resulting inhibition of life processes.

4. By chemical reactions with germs so as to form new complex compounds which are free from harmful qualities.

The nature of the solvent used is of major importance in the action of the disinfecting solution. Chemical disinfectants can most readily gain access to the bacterial cell via the aqueous (water) phase natural to the organism. Therefore, a disinfectant is most active in water solution. Any solvent which reduces the concentration of the disinfectant in the aqueous phase has the consequent effect of reducing the activity of the disinfectant. Thus, the use of oils and oil-related compounds and alcohols has this effect on phenol, thus decreasing its activity with these solvents.

Desirable Characteristics of Disinfectants

The ideal disinfectant has not yet been discovered and probably never will be, because of the vary-

ing conditions under which disinfectants are asked to operate. Some of the desirable features of a disinfectant or an antiseptic in livestock sanitation are as follows:

1. It must be reasonably low priced; otherwise, the large quantities needed make it economically unprofitable.

2. It must be free from strong or objectionable odors, especially in dairies; otherwise, products will become tainted.

3. It must not be excessively destructive to materials or tissues other than the agent being disinfected. As an extreme example, flame will destroy all germs, but if the germs happen to be on combustible material, the flame will also burn the material. Undiluted carbolic acid will destroy many infections, but, at the same time, it will also destroy the tissues in which the germs are located.

4. It must not remain strongly poisonous after its application because of the danger to the life of those animals that graze or drink in areas that have been disinfected or those that lick one another.

5. To remain active, it must not combine chemically with materials, utensils, blood, and the like the moment it establishes contact with them.

6. It must be neither excessively irritating nor poisonous when inhaled. Some of the gaseous disinfectants, such as chlorine, formaldehyde, and others, are unbearably irritating to the mucous membranes lining the air passages, and, in sufficiently high concentration, they may even destroy the life of those animals inhaling them.

7. The disinfectant used must be effective at ordinary temperatures, and it must not lose an appreciable amount of this effectiveness when the general surrounding temperature drops considerably below that of the usual "mild" day.

8. It must be effective after it has been diluted with water; in other words, it must not be effective only when concentrated.

9. It must mix readily and uniformly with water. As an example, droplets of practically pure carbolic acid, when added to water in certain dilutions will remain on the surface of the water, and if used on living tissues, may result in severe injury.

10. It must be in such concentration and in such form that it may be readily and economically transported from its place of manufacture to its place of ultimate utilization.

Chemical Disinfectants

The lethal action of a disinfectant is due to its capacity to react with the protein and essential enzymes of microorganisms. Any agent which will coagulate, precipitate, or otherwise denature proteins will act as a general disinfectant.

There are four primary chemical groups that possess this ability: phenols, halogens, quaternary ammonium compounds, and formols. Other chemical substances, such as dyes, heavy metals, soaps, and alcohols, also possess the ability to act as disinfectants.

A common characteristic of all disinfectants is that at progressively weaker concentrations their lethal activities become slower and less effective until they give way to inhibitive action and finally cease to exert any influence at all.

The ultimate effect of any disinfectant, whether it be bactericidal or bacteriostatic, is determined by a number of factors, particularly the (1) concentration of the disinfectant, (2) contact time between disinfectant and microorganism, and (3) temperature.

Phenols

This group of chemical disinfectants includes many derivatives of coal tar oils, including carbolic acid, cresol, and pine oil.

Carbolic Acid

Carbolic acid or phenol is derived from coal tar oils. In its pure form, carbolic acid occurs at ordinary temperatures as closely welded needle-like crystals. It is kept liquid in a salable condition in drug stores, first by liquefying it through exposure to heat and then by adding 10 percent of either water or glycerin.

Carbolic acid is one of the oldest known disinfectants. Lister, the pioneer surgeon, who demonstrated its value in keeping wounds free from infections, lauded it very highly.

Carbolic acid has certain decided advantages as a disinfectant. It is a very stable compound, in that it appears to suffer no deterioration from exposure to air or light. It does not injure fabrics immersed in a sufficiently strong solution of it for a period of time necessary for their disinfection. It does not corrode metals. It is reasonably effective against germs in the presence of organic matter, such as stable and barnyard refuse. It is readily available in most all drug stores. It is effective as a disinfectant against all germs, and if permitted to act long enough (at least 24 hours), against the highly resistant spores.

There are also certain disadvantages to using carbolic acid. It is extremely poisonous and, therefore, is unsafe to have on the premises. It has a penetrating odor which is readily absorbed by milk, meat, and other food products. It is of very low efficiency as a germ destroyer when it is in a cold solution, or when used on a cold surface; however, dissolved in hot water, it is a very effective disinfectant. It is not effective against the viruses of diseases such as foot-and-mouth disease, fowlpox, and others. It is not soluble in water above 7 percent, though in many instances the use of a stronger dilution is very desirable. When its dilutions approach 7 percent, or when the diluting water is cold, droplets of the carbolic acid may float on the surface, and these are dangerously destructive to animal or human tissues.

Of the domesticated animals, cats are notoriously susceptible to the poisonous action of carbolic acid, even in its external application, as well as to that of other disinfecting agencies derived from coal tar oil. Carbolic acid even in the weaker dilutions is numbing to the hands of operators and interferes seriously with their touch perception. For the purpose of disinfecting stables and barnyards, the quantity of a disinfectant required is so large that it make carbolic acid prohibitively expensive; therefore, its place for this purpose has largely been taken by cheaper agents derived from coal tar oil.

Because of its anaesthetic or numbing action, it is frequently employed as a douche in ½ percent strength to overcome not only infection in the genitalia of animals incidental to parturition, but the subsequent straining as well.

CRESOL

The basic portion of this compound is derived from coal tar. It is one of hundreds of coal tar disinfectants.

Cresol is effective against many kinds of bacteria and some viruses.

The advantages of cresol are:

1. Metals such as those of surgical instruments are not corroded by it and when immersed in it will keep a sharp edge.

2. It is almost non-irritant to the skin when used in the usual recommended strength; it is soapy, which aids in its penetration.

3. It is less affected than most other disinfectants by factors which tend to inhibit the action of germicides.

4. It is less poisonous and more germicidal than carbolic acid because its phenol coefficient is approximately 2.

5. As a deodorant, it destroys putrefactive germs, and it replaces bad odors.

6. Its cost, based on the unit of efficiency, is low.

The principal disadvantage of saponated cresol solution is that its odor will taint all foods in the same room with it.

For the disinfection of barns, fences, and other animal enclosures, and for the discharges of diseased animals, it is used by mixing it with soft or lye-softened water to the extent of from 2 to 3 percent—usually about 4 ounces of the compound solution of cresol in a gallon of water.

One and one-half pounds of freshly water-slaked lime may be added to each gallon to make it clearly visible where the cresol compound has been applied to woodwork.

PINE OIL

Pine oil disinfectant is widely used in animal sanitary procedures as a substitute for the cresol preparations because of its pleasing aromatic odor and its relatively good germicidal efficiency. The usual emulsified preparation consists of 70 percent steam-distilled pine oil, 21 percent rosin soap, and not over 9 percent of water and other extraneous inert substances. As a matter of general interest, many aromatic oils and spices have some germicidal value. These include agents such as thymol, derived from oil of thyme, oil of cinnamon, and oil of cloves.

Halogens

This group consists of four substances: iodine, chlorine, bromine, and fluorine. Iodine and chlorine compounds are used extensively as disinfectants.

IODINE

Iodine is a solid, occurring usually as blackish-gray scales. The so-called resublimed iodine, obtained from Chili saltpeter and from seaweeds, is readily dissolved in alcohol, chloroform, and other liquids, and in water that has previously had some iodide of potash dissolved in it.

Resublimed iodine has a phenol coefficient of from 170 to 235, though this is very much reduced when it is in the form of tincture of iodine or Lugol's solution. Iodine in the form of any of its preparations, though especially the tincture, continues to be the standard skin disinfectant. It is one of the few agents that is almost immediately destructive to the highly resistant bacterial spores. If it is used too often, it also acts as an irritant so that it finally lowers the natural resistance of the skin to infections.

Iodine is also very efficient as a local or deep application against mold diseases, such as ringworm and lumpy jaw in cattle. For either surface or deep wounds, it is the agent of choice to destroy the germ causing tetanus.

Tincture of iodine is the form of iodine recognized by most people. It is made by dissolving resublimed iodine scales to the extent of 7 percent in a 5 percent alcoholic solution of iodide of potash. It is the most extensively used of the iodine preparations. There is also a 2 percent tincture which is, in general, as effective as the "stronger" 7 percent, though less irritating.

Lugol's solution of iodine is a disinfectant with which many are familiar. It is made by dissolving resublimed iodine scales to the extent of 5 percent in a 10 percent water solution of iodide of potash. It is cheaper than tincture of iodine. It has approximately the same uses as the tincture of iodine, though it lacks the additional value conferred upon the latter by the alcoholic solvent.

Another compound of iodine is iodoform which is frequently used as a wound dressing. It liberates its iodine slowly when in contact with wounds, and this is probably the reason that it has received such extensive and favorable recognition by clinicians as one of the best agents to use for the control of wound infections. It is non-irritating, and it is not soluble in water. It may be held in suspension, however, in a heavy mineral oil, and in this manner can be used for disinfecting the female genitals. Its disadvantages are its almost prohibitive price and its exceedingly pungent, penetrating odor. It imparts this odor to everything with which it comes in contact. If used over a prolonged period on fat subjects, it may result in iodine poisoning, as the fat will dissolve it and absorb it into the animal tissue.

Povidone-iodine at a 7 ½ percent concentration is an excellent solution for cleaning wounds. It is also commonly used as a surgical scrub.

CHLORINE

Chlorinated lime (also known as bleaching powder, or simply as "bleach," and sometimes incorrectly designated as chloride of lime or calcium chloride) is a powerful disinfectant because of the large amount of chlorine gas incorporated in it. It is also an effective deodorant.

Chlorinated lime is made by passing chlorine gas into hydrate of lime to the extent of from 24 to 35 percent. Commercial chlorinated lime usually contains at least 24 percent of available chlorine. The chlorine is in loose combination with the lime, so that in the presence of organic material, such as germs and stable refuse, it is liberated as a gas to combine with the organic material. It is because of this feature that large amounts of chlorinated lime must be used as a disinfectant, so that there will be an excess for action against infection. This is always done in the treatment of public water supplies with chlorine gas and accounts for the frequently noticeable taste of chlorine in it. Also, because of the loose combination, it should be stored in tightly closed containers to prevent the

escape of the chlorine gas; otherwise it may lose as much as 1 percent a month. "Bleach" is one of the few substances that spontaneously evolves an efficient disinfecting vapor in the form of chlorine gas which will kill germs if there are at least 5 parts per 1,000 in dry air. Or ⅛ of this amount is equally effective in moist air.

The disadvantages of chlorinated lime for stable disinfection are (1) its expense, due to the comparatively large amounts that must be used, (2) its powerful bleaching characteristic, which removes colors from fabrics and leather, (3) its corrosion of metal, (4) its irritation to the mucous membranes of the nose and throat, (5) its disagreeable odor, which it imparts to foods, such as milk and meat, and (6) its instability (the exact chlorine content is practically impossible to estimate).

As a disinfectant, it has a phenol coefficient of 21 when there is 10 percent of available chlorine, but it quickly loses this in the presence of organic matter. There is evidence to indicate that it is a destroyer of viruses, such as that of foot-and-mouth disease, and other diseases, and is quickly lethal to most germs, except the "acid-fast" germs, including those of tuberculosis, which are so resistant to its action that it is practically valueless against them.

For disinfecting purposes, chlorinated lime may be used in powder form or dissolved in water. Six ounces of it in 1 gallon of water is recommended for general household and farm use in the disinfection of drains and sewers. For the disinfection of teats of dairy cows and of dairy utensils, a solution containing 200 parts of chlorine gas in each 1 million parts of water may be used.

Sodium hypochlorite has appeared on the market under various trade names. It is usually sold in solution or liquid form. The chlorine content varies from 2 percent to 20 percent. The original container indicates the percentage of available chlorine. The sodium hypochlorite alone in solution, or in combination with calcium hypochlorite, has considerable value as a disinfectant for dairy equipment and as a disinfectant for households. However, like all chlorine preparations, its value as a disinfectant is greatly reduced in the presence of organic material, such as manure, milk, blood, etc. In other words, it is a clean-surface, though germ-contaminated, disinfectant. Also, because the chlorine odor is penetrating, it should not be used in places where human food is stored.

Chlorines are corrosive to metals, and not effective against TB organisms and spores.

Quaternaries

The quaternary ammonium group of compounds has emerged into prominence and use because of the many claims made for their performance. They constitute a large group of compounds, cationic by nature, with high surface activity. Hence, they are known as surface-active cationic germicides.

The bactericidal action of the quaternaries has been attributed to inactivation of enzymes, denaturation of essential cell proteins, or disruption of the cell membrane. Evidence indicates that bacterial cells treated undergo changes which result in leakage of the cell constituents into the suspending fluid.

The quaternaries are more active against Gram-positive organisms than against the Gram-negative types, although the difference sometimes may not be very great. Gram's method of staining germs to that the organism may be identified is the same for all bacteria. However, when the stain can be washed away with alcohol, the germ is said to be Gram-negative, while those that do not decolorize are Gram-positive. While effective on bacteria, quaternaries do not always kill bacterial spores. Generally, they are effective against fungi. There is evidence that some quaternaries will kill Newcastle disease virus; however, viruses are more resistant to destruction by quaternaries than are bacteria fungi.

As with most other disinfectants, the antibacterial activities of quaternaries are suppressed in the presence of organic material. Quaternaries are incompatible with certain compounds. Because of their chemical make-up, they should never be used with soaps. Aluminum is detrimental to their effectiveness. Acidity decreases their efficiency and temperature affects these compounds, as it does other disinfectants.

Formols

Formaldehyde is a gas that is soluble in water. The water solution of this gas is known as Formalin, or as "solution of formaldehyde."

Formaldehyde is an effective disinfectant, in the gaseous state and in a water solution. It has the particular advantage of being lethal not only to viruses, bacteria, and fungi, but also to spores.

It is highly reactive with amino acids and proteins. Thus, they have substantial antibacterial and anti-viral properties. All bacteria, including spores, are readily killed by formaldehyde gas or solutions.

Humidity is an important factor to be considered in relationship to speed of disinfection by formaldehyde gas. As relative humidity is increased, the activity of formaldehyde gas increases. However, above 60 percent relative humidity, there is little additional activity.

Formaldehyde is more active at higher temperatures, both in solution and in the gaseous state. There is little need to use temperatures above 65°F. If temperatures and humidities are low when formaldehyde gas or liquid is used, equipment and animal quarters should merely be exposed for a longer period of time to obtain the lethal results needed to destroy viruses, bacteria, fungi, and spores.

FUMIGATING BUILDINGS

Formaldegen is a commercial preparation containing 91 percent formaldehyde gas. It is a white powder that is stable at room temperature and can be stored for long periods without losing potency or effectiveness.

Fumigation by this method requires only an electric outlet, a Formaldegen frying pan for small areas, a Formaldegen generator for large areas, and a supply of Formaldegen. When an electric switch outside the room is not available, an automatic timer may be used to turn off the unit when fumigation is completed. The fumigator should take these steps:

1. For ideal fumigation, make sure the temperature of the room is above 65°F and the humidity is 60 percent or more.

2. Calculate the number of cubic feet to be fumigated (length × width × height).

3. For room and equipment fumigation use 1 pound (454 grams) for each 5,000 cubic feet of space.

4. Make the areas to be fumigated as airtight as possible and free of all animal life.

5. Post warning signs at all points of entry and lock all doors to prevent accidental entry during fumigation and neutralization of the formaldehyde gas.

6. Dump Formaldegen into either the frying pan or generator when it is cold. Turn the switch on and allow sufficient time for full release of the gas.

7. Open doors and windows from the outside to allow the gas to escape.

8. For quick neutralization of the formaldehyde gas and odor, put twice the amount of Formaldegen neutralizer as Formaldegen used into the frying pan or generator and keep the doors closed for an additional few minutes.

Other Chemical Substances

SODIUM ORTHOPHENYLPHENATE

Sodium orthophenylphenate is a compound that has been combined satisfactorily with detergents to provide a highly effective one-step sanitation program. It has no objectionable odor, is readily soluble in water, has high germicidal activity, and is active in the presence of detergents. It performs well in the presence of moderate amounts of organic matter. Although it is irritating to the eyes and mucous membranes, simple precautions can be observed during use. It is sold under various trade names.

Sodium orthophenylphenate is effective against microorganisms causing brucellosis, tuberculosis, vesicular exanthema, and vesicular stomatitis (Table 4-1). It is approved for use in federal-state animal disease eradication programs and as the official disinfectant for livestock shows.

LYE

Chemically, lye is hydroxide of soda, hydroxide of potash, or a mixture of the two. It is usually used in the form of household lye. It is highly rated as a disinfectant against all viruses, including foot-and-mouth disease of cattle, fowlpox, and dog distemper, and against some germs, such as anthrax and its highly resistant spore, and the germ causing brucellosis (Bang's disease) in cattle (Table 4-1).

Table 4-1. Chemical Disinfectants Recommended for Control of Infectious Agents Causing Diseases in Livestock

Disinfectant	Percent Solution	Mixtures	Disease
Cresol[1]	4	4 ounces to 1 gallon water	Brucellosis Shipping fever Swine erysipelas Tuberculosis Vesicular exanthema Vesicular stomatitis
Sodium carbonate (soda ash)	4	1 pound to 3 gallons water	Foot-and-mouth disease
Sodium Hydroxide (lye) Caustic soda	2	13½-ounce can to 5 gallons water	Foot-and-mouth disease
Sodium orthophenylphenate (USDA approved)	1 2	1 pound to 12 gallons water 2 pounds to 12 gallons water	Brucellosis Tuberculosis Vesicular exanthema Vesicular stomatitis
Sodium hydroxide (lye)	5	5 (13½-ounce) cans to 10 gallons water	Anthrax Blackleg

[1]A list of cresylic disinfectants permitted for use for official disinfection is published periodically by the USDA, Agricultural Research Service, Animal Health Division. This list serves as a guide for use of specific products.

Lye is of no practical disinfecting value against the group of "acid-fast" germs, which includes the germ of tuberculosis and that of Johne's disease, nor against that group known collectively as "Gram-positives." A very important point in its action as a germ destroyer is that it is more effective when its solution is used at room temperature rather than hot, and this obviates the necessity of heating water for dissolving it. However, this action of a cold solution is only against germs and must not be interpreted to mean that a cold lye solution is to be used in cleaning hog houses infested with worm eggs. Lye mixed in scalding hot water should be used to clean hog houses. The hot water is destructive to worm eggs and the lye dissolves the dirt.

HYDROGEN DIOXIDE

Hydrogen dioxide, the basic substance, is a syrupy, colorless liquid, heavier than water, and so chemically unstable that it will keep only at a temperature several degrees below the freezing point.

The hydrogen dioxide described in the preceding paragraph is diluted with water so that the dilution represents approximately a 3 percent solution of it. It is this 3 percent solution that is sold in drug stores as solution of hydrogen dioxide, solution of hydrogen peroxide, or as peroxide.

The solution of hydrogen dioxide is still an unstable preparation in spite of its dilution. It should therefore be kept in a cool place and in amber-colored, well-corked bottles. If there is evidence of gas pressure when the cork is withdrawn, it means that the contents have lost strength.

Under laboratory conditions, the solution of hydrogen dioxide is a good disinfectant, but unfortunately clinical experience does not substantiate the laboratory findings in many instances. In the presence of organic matter, it releases oxygen in a nascent (highly, active, newborn) state. This oxygen is supposed to oxidize or "burn up" the germs susceptible to its action. In this process there is very marked effervescence, so that when it is used on a wound one often hears the expression, "The poison is boiling out." Its action is always transitory and comparatively feeble, though it does assist in the removal of dead tissue from wounds. It is not irritating to tissues, though it must never be used in wound cavities where its complete effervescence is prevented.

Theoretically, it would seem that the solution of hydrogen dioxide should be very effective against those germs collectively designed as anaerobes (obligates), which thrive best in surroundings free from oxygen. Actual practice, however, indicates that no more than ordinary germicidal power is exhibited. In other words, the solution of hydrogen dioxide is not a specific against anaerobic infections, such as blackleg, gas gangrene, and others. Except as an agent for occasional use in the treatment of wounds and some protozoan infections, hydrogen dioxide may be disregarded in the field of animal sanitation.

NOLVASAN SOLUTION

Chlorhexidine diacetate, the active ingredient in Nolvasan, is effective against bacteria and viruses and is not very irritating to the tissues.

COPPER SULFATE

Copper sulfate, or blue vitriol, has at various times been lauded for its alleged disinfectant properties, but careful research has failed to substantiate these claims. For example, a solution of almost 16 percent strength of copper sulfate failed to kill anthrax spores after an exposure of 10 days.

For the destruction of algae, fresh water plants causing green scum on stagnant water, it is the agent of choice. Concentrations of 1 part of sulfate of copper in 500,000 parts of water will destroy most forms of algae. This strength does not make the water unfit for animal consumption, and it does rid ponds, water reservoirs, and tanks of these noxious algae. The most practical method of use is to place a quantity of the copper sulfate in a small cloth bag and to stir this through the water so as to give the water a very slight bluish tint. This copper sulfate treatment should not be used if there are fish in the reservoir because copper sulfate is detrimental to many forms of aquatic life.

Vinegar

Vinegar of a good grade should have in it from 5 to 6 percent of acetic acid as its active ingredient. In general, it compares favorably with the official diluted acetic acid, though the latter is preferable because it is colorless, of greater purity, and of uniform strength, which is exactly 6 percent of acetic acid in distilled water. However, diluted acetic acid is not always available in the home, though vinegar is.

Vinegar, on the basis of 6 percent of acetic acid, decidedly has value against many germs and has been used for many years in preserving foods such as pickles. As an application to wounds, it will free them from certain pus-producing germs. During the earlier stages of so-called foot rot in animals, some veterinarians consider its action to be almost specific in destroying the causative germs of this malady.

Sulfonamides

The sulfonamides are chemotherapeutic agents that stop the growth of infecting germs in concentrations that are not poisonous to the host. They are appropriately classified as being bacteriostatic in their action, since they prevent or arrest the growth of bacteria. Sulfonamides, under proper conditions, may also be classified as antiseptics which prevent the growth or multiplication of bacteria in the body without being injurious to the host.

Under improper conditions, they may be injurious to the host by reducing the white blood cell count, by forming crystals in the urinary tract, and by inhibiting thyroid function by combining with its iodine.

The important basic principles to determine in the use of sulfas are:

1. What blood levels do they produce?

2. How persistent is their action?

3. How active are they against the organisms causing the ailment?

In a measure, answers to the foregoing questions depend on the rate of absorption by the tissues of the body and the rate of their excretion. The following are generally accepted:

1. Sulfanilamide—good absorption and rapid excretion.

2. Sulfapyridine—moderate absorption and moderate excretion.

3. Sulfathiazole—poor absorption and rapid excretion.

4. Sulfamerazine—good absorption and slow excretion.

5. Sulfamethazine (formerly known as sulfamezathine)—good absorption and slow excretion.

The bacteriostatic action of sulfonamides is not the sole influence of these chemicals in controlling infections. During the time that the sulfonamides are checking the growth of the invading germs, the host has time to develop antibodies, the specific bodies that make people or animals immune to diseases. Furthermore, certain blood cells phagocytize, which means that the cells ingest invading germs and other substances, the disease-producing germs.

Potentiated Sulfonamides

Potentiated sulfonamides contain a sulfonamide in combination with a diaminopyrimidine. Trimethoprim is a diaminopyrimidine that is combined with sulfadiazone.

Nitrofuran Derivatives

These are substitution products of furan, which is a colorless liquid made from wood tar. There are many of them now made chemically from wheat husks, corn cobs, oat hulls, etc. The nitrofurans are both bacteriostatic and batericidial. Developing bacterial strains that are resistant to nitrofurans is difficult in the test tube, though this resistance does occur with some of the sulfas, penicillin, streptomycin, etc. The nitrofurans most frequently reported on are:

1. Nitrofurazone (N.F.Z.), which has been used in the prevention of some animal ailments, including some forms of swine enteritis.

2. Furazolidone nf. 180 (Furaxone), which seems to be most effective in the handling and prevention of infections in the organs of digestion.

3. Furadantin, which is used as a urinary disinfectant.

Nitrofurans are also used as feed additives; and veterinarians prescribe them as ointments for the relief of infectious eye and ear ailments.

Non-chemical Disinfectants

At least two non-chemical methods are recognized as being aids to disinfection or as having disinfecting action of their own. These are mechanical agencies and natural agencies. Frequently, the latter closely approach the chemical disinfectants in their action.

Mechanical Agencies

Taken as a whole, their value depends upon the removal of superficial dirt and infectious material, so that the remaining germs may be made more accessible to the action of chemical antiseptics. The mechanical agencies include processes such as sweeping, brushing, scraping, scrubbing, and using water under pressure.

Probably the best method is to dampen all material that may be removed by mechanical means and transport this to some area where it may be safely rendered harmless by burning, burying, or spreading on sun-exposed arable land. The preliminary dampening prevents germs from being transported with dust from infected to clean areas. Dampening is also better than resorting to the use of water under pressure during the mechanical cleaning, because with the latter the great volume of water used is likely to flush germs to clean areas.

Since the very best that may be anticipated following the use of mechanical agencies is the removal of dirt and some of the germs, many other germs remain upon the completion of these processes. Therefore, it must be followed by more potent methods, those due to natural forces or those by chemical means.

Natural Agencies

Various natural agencies are known to have some influence in eradicating germ life. These include sunlight, heat, sedimentation, time, electricity, bodily juices, germ antagonisms, and antibodies.

Sunlight

Sunlight can kill germs and their spores if it can reach them. However, its ultra-violet rays, which destroy germs, cannot pass through ordinary window glass and can pass to a very limited extent through a smoky atmosphere or haze. Thus, they are practically ineffective during the early morning and late afternoon hours.

The germ-destroying rays of the sun are effective only against surface infections, since they do not penetrate material behind which germs may be screened, such as mucus, stable manure, dirt, and comparable material. Since virtually all disease-producing germs multiply and grow only in the bodies of their host, exposure to sunlight can have no effect on the control of this aspect of disease.

Sunlight as a disinfectant is too slow to serve the purpose of controlling or stamping out any specific infectious disease, though it serves as a surface disinfectant when the action of its rays is neither impeded nor obstructed by glass or foreign material. Exposure to bright sunlight is to be encouraged on the basis of its lethal action against germs.

Heat

If applied in sufficient concentration, such as that produced by fire, heat is deadly to all forms of life, including that of animal parasites and germs. It is also highly effective as steam under pressure and as scalding water. The heat generated in heating packed manure piles is surprisingly effective. As heat decreases, so does its effectiveness, such as the heat of the sun which is expressed through its ultra-violet rays.

Because of its germ-killing powers, heat is utilized in the process of pasteurization, in the burning of carcasses and barnyard refuse, and occasionally, in the destruction of extremely disease-contaminated frame buildings of low value.

At least two standard types of pasteurization are recognized: the holding method, and the high-temperature–short-time method. The holding method is in most general use. It consists of heating the product for the duration of 30 minutes to a temperature of between 140° and 148°F, or an average temperature of 143.5°F.

The high-temperature–short-time method, also known as the "flash" method, is a process in which milk is quickly heated to a high temperature and held there under perfect control with sufficient pasteurization and without scorching. It is accomplished by passing a high-frequency alternating electrical current of 2,500 to 6,000 volts through milk, for approximately 10 seconds, the final temperature reached being 161°F, or by holding the temperature at 161°F for 15 seconds.

SEDIMENTATION

Sedimentation is one of nature's methods of purifying water that is not in motion. Animal disease germs contaminating water under this condition settle to the bottom where, in the course of time, they die. This process must not be depended upon in small pools or ponds, which are repeatedly reinfected or reinfested from surface contamination, and are frequently agitated by various animate or inanimate forces.

TIME

All living things, including germs and parasite eggs, die in time. The rate at which dissolution takes place is extremely variable and may, in some instances, require several years. It is always considerably retarded by cold, when life's processes are slow. The germ responsible for brucellosis will live in pastures for several months during freezing weather, though only for a comparatively brief period during the warmer months. In general, this also is true for other germs. A combination of time, bright sunshine, and low optimum conditions for the evolution of life's processes will free pastures of specific infections in due course.

Extreme dryness accounts for the death of many more germs.

ELECTRICITY

Electricity, by its heat-generating ability, or by its powers to free elements from chemical combinations, such as chlorine, which is strongly disinfectant, from common salt (chloride of soda), or possibly from the direct action of the electrical current, is to some degree, destructive to germ life.

BODILY JUICES

The bodily juices considered here are blood serum, white blood cells (phagocytes) and red blood cells, and gastric juice. Blood serum can kill germs, and this ability may be increased by injecting limited, though gradually increasing, doses of germs or their poisons into the bloodstream of a living animal so as to produce in that animal a hyperimmune serum. This serum when withdrawn, and later injected into another animal, confers upon the latter a degree of resistance against a specific infection.

The white blood cells actually "eat" or ingest germs that get into the bloodstream, by gradually growing around them and finally completely disposing of them. Technically, this process is known as *phagocytosis*. If the white cells win the victory, that animal is said to have resisted an infection; if the germs gain the upper hand, disease sets in.

Another of the bodily juices having germ-destroying ability is that secreted by the normal stomach, the gastric juice. Being an acid, it is destructive to germ life.

GERM ANTAGONISMS

It is an observed bacteriological phenomenon that many germs are antagonistic to each other. For example, many of the specific disease-producing germs get out of their natural elements when they leave the animal body to go into the soil and the germs that have the normal abode in the soil may destroy them in the course of time. In other instances, they cannot compete successfully for a food supply with the natural inhabitants of the soil. In the laboratory, this antagonism is shown at times in the growth of germs in their artificial food supplies. Adding antagonistic organisms to the surroundings of germs causing certain infectious diseases will help to destroy them. The antagonism between bacteria is known as *antibiosis*.

ANTIBIOTICS

Antibiotics are anti-bacterial substances of biologic origin derived from (1) bacteria, including pyocyanese, tyrothricin, and bactracin; (2) actinomyces, including actinomycin and streptomycin; (3) molds and fungi, including penicillin and flavicin; and (4) laboratory synthesis, including ampicillin and neomycin.

Antibiotics can be obtained from natural substances other than bacteria. For example, allicin can be obtained from common garlic; canavalin from soybean flour; chlorellin from fresh water seaweed; lysozyme from saliva, tears, egg whites, and many animal fluids.

Probably the best known antibiotic is penicillin, derived from the mold *Penicillium notatum*. Commercially, it is usually in chemical combination as either sodium or calcium salt. It is used with a high degree of efficiency for infections due to staphylococci, streptococci, pneumococci, and Gram-positives. With animals, it is extensively used for mastitis.

Amoxicillin and Ampicillin are members of the penicillin family which are derived semisynthetically and are active against large numbers of both Gram-positive and Gram-negative bacteria. However, they may be destroyed by penicillinases produced by numerous bacteria.

Aminoglycosides are a group of antibiotics that have similar chemical and antimicrobial properties. Narrow spectrum aminoglycosides are streptomycin and dihydro-streptomycin.

Streptomycin is derived from *Actinomyces (streptomyces) riseus*. It is most highly effective against conditions resulting from Gram-negative organisms. However, it is not as effective as penicillin in the handling of Gram-positive organisms. Dihydrostrep-

tomyin is also mainly active against Gram-negative bacteria.

Broad spectrum aminoglycosides include neomycin, gentamicin, and spectinomycin. These drugs are effective against a much broader spectrum of bacteria. However, most of them are in the Gram-negative group.

Oxytetracycline is a very broad spectrum antibiotic; that is, it is effective against large numbers of Gram-positive bacteria as well as Gram-negative bacteria.

Macrolides are a group of antibiotics that have a large lactons ring in their structure. They are much more effective against Gram-positive bacteria than against Gram-negative bacteria. In addition, macrolides are effective against mycoplasma and some rickettsias. Tylosin and erythromycin are the drugs commonly used in veterinary medicine.

Another antibiotic is aureomycin derived from a strain of *Actinomyces aureofaciens*. In humans, it is extremely effective in the handling of Rocky Mountain spotted fever. It is also in frequent use as a feed additive for animals.

Bacitracin is an antibiotic substance produced by the growth of "Tracy 1" strain of *Bacillus subtillis*. It is chiefly effective against Gram-positive organisms. It is more effective than penicillin for local (topical) application, and its effectiveness is apparently not interfered with by wound exudates. As a dry powder, it is commercially handled under the name of bacitracin topical.

Nitrofurazone is a synthetic antibiotic which is effective against both Gram-positive and Gram-negative bacteria. It is used topically and is available as a spray powder or salve.

Most of the forces and agencies included under the general heading "Natural Agencies" certainly exert an immense collective action in holding disease under control. With the exception of heat, the antibiotics, and specific immunizing blood serums, their influence in attempts to immediately or quickly control outbreaks of specific infectious diseases is altogether too slow for practical application.

CHAPTER 5

Vaccination and Immunization

It is important for the livestock producer to realize that vaccination and immunization are *not* synonymous terms. Vaccination is the mechanical act of administering a vaccine for the purpose of developing immunity in an animal. The amount of immunity developed from vaccination is dependent upon many factors, including the animal's state of health, the condition of the vaccine, and the manner in which it is infected.

Vaccination is an important part in the control of many animal diseases, but, as stated above, immunity is a relative thing. It may be overcome by massive exposure to germs. Moderate exposure to a highly virulent strain of the infecting agent or stress, such as poor environment or nutritional conditions, may overcome the response to the vaccines. Vaccination should not be considered as the final answer in disease control. It should always be combined with proper nutrition, management practices to reduce stress, and sanitary measures designed to prevent the introduction and spread of infection.

ANTIGEN-ANTIBODY REACTION

When a vaccine is injected, the tissues react to form immune bodies against the specific disease. This reaction is sometimes accompanied by signs of distress, especially if the animal is in poor health. If a live vaccine is used in an unhealthy animal, it may produce the disease. Effective active immunization can be accomplished only when vaccines are high in specific antigen content or, in the case of passive immunization with serum, rich in antibodies. Some degree of protection may be obtained by injecting artificial preparations of dead or attenuated (weakened) organisms into the animal.

When a mild modified virus vaccine is introduced into the body, the virus immediately invades the cells and begins to multiply. As the new virus particles are released from the invaded host cells, they begin to invade other cells. Because the vaccine virus has been altered and, therefore, lacks disease-producing power, the animal does not have a severe reaction, as it would to a fully virulent virus. The mild virus merely causes many of the cells to become resistant to further invasion by a similar type virus under natural circumstances.

Antibody build-up is a normal body reaction to infection. The germ contains a protein or related compound that causes the tissues to form a specific antibody to which the antigen is attracted. The antibody then combines with the antigen, neutralizing the disease-causing organism so that it can be disposed of by the body. This keeps germs from pursuing their normal infective course and produces immunity in the host. Supplemental or booster vaccinations are sometimes used to maintain antibody production by the animal body.

TYPES OF IMMUNITY

Immunity or resistance to disease can be divided into two groups: natural and acquired (Fig. 5-1). Natural immunity refers to the protection an animal has when it is born and is not associated with antibodies in the body fluids. Acquired immunity, on the other

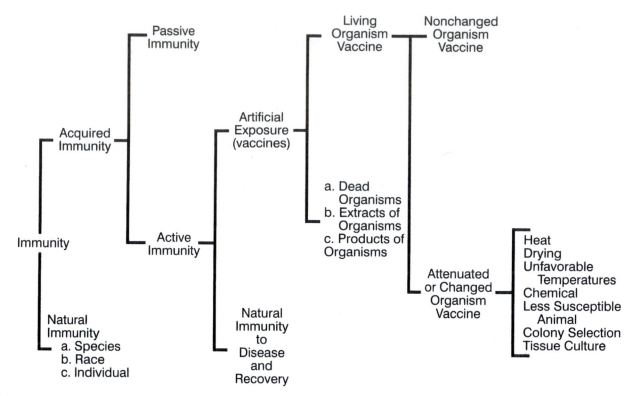

Fig. 5-1. Immunity and vaccines.

hand, is associated with the presence of antibodies from another immune animal or from exposure to the disease.

Natural Immunity

Natural immunity is inherited. It exists in the following three forms:

1. *Species resistance*. Some disease-producing organisms are specific to certain species of animals. Hog cholera, for example, is a disease that affects swine. Scrapie is specific to sheep.

2. *Breed resistance*. Variation in degree of resistance to certain diseases exists among breeds within a particular species. Scrapie, for example, is more common in Suffolk than Dorset sheep. The occurrence of milk fever (parturient hypocalcemia), is less frequent in Holstein than Jersey cattle. Dark pigmented breeds of sheep and cattle have more resistance to photosensitization than the light-colored breeds.

3. *Individual resistance*. In outbreaks of contagious diseases, certain individuals within a herd or flock of-

ten show little or no sign of the malady, while other animals on the same premises become sick and die. Their physical condition and body defenses enable them to resist the disease-producing organism.

Acquired Immunity

The second way livestock and dairy cattle are protected from disease is through acquired immunity. It is the result of exposure to pathogenic organisms or of immunization after birth. There are two distinct types of acquired immunity:

1. *Active immunity*. It is acquired through direct contact with specific disease-causing organisms, with the body developing specific antibodies to counteract the invasion. Animals develop active immunity after being diseased or artificially immunized by a vaccine or by other biologicals. Active immunity is relatively long-lived and often lifelong.

2. *Passive immunity*. It is acquired by transferring antibodies from an immunized animal to a susceptible one. Colostral immunity is critical to the health and

well being of newborn livestock and horses. Newborn animals have the ability to absorb the antibodies from colostrum for only 24 hours, so the timing of feeding dairy calves and insuring that other neonates have nursed is critical. Commercial tests are available to test the amount of immunoglobulins absorbed into the bloodstream of the neonatal animal. Many cases of scours, joint ill, pneumonia, and neonatal mortality can be attributed to the failure of passive transfer. An animal that is stressed from a difficult delivery or a cold environment is not as effective in absorbing colostral antibodies; and the farm manager needs to pay additional attention to caring for these animals. Two common ways of providing this immunity are by feeding colostrum to a newborn animal or by injecting serums produced by another animal.

BIOLOGIC AGENTS

Immunizing agents, or biologicals, as they are often called, are simply antigens or antibodies of diseases whose causative organisms have been identified. The organisms have been prepared in such a way so that they may be safely given to livestock. The livestock may then carry with them protection from diseases to which they have not been previously exposed. Having accomplished this, they can ward off severe exposure to diseases which they might not otherwise survive.

Living Viruses

Living virus vaccines will not cause healthy animals to contract the disease. They are used to stimulate antibody productions against a specific virus.

Attenuated or Modified Viruses

Growth and subsequent change of an organism in another animal or medium, so that the organisms will stimulate antibody formation in the vaccinated animal without producing disease symptoms, is attributed to attenuated or modified viruses. Modified or attenuated virus vaccines are widely used to protect cattle from bovine viral diarrhea, infectious bovine

rhinotracheitis, and parainfluenza$_3$, and to protect horses from rhinopneumonitis.

Killed Viruses

Physical or chemical agents are used to destroy a pathogenic organism in the preparation of killed virus vaccines. A suspension of the inert viral material is then administered to a susceptible animal to stimulate active immunity. Killed viruses are widely used to protect horses from equine encephalomyelitis and equine influenza.

Vaccines

A vaccine is any preparation of killed, living, or attenuated microorganisms introduced into the body by oral or parenteral route to produce immunity to a disease.

Bacterins

Although the words *vaccine* and *bacterin* are used rather freely, and sometimes interchangeably, bacterins are suspensions of killed pathogenic bacteria used to stimulate immunity.

Toxoids

A toxoid is a physically or chemically inactivated toxin which is obtained from toxin-producing organisms grown in the laboratory. Toxoids usually stimulate strong active immunity in the infected animal.

IMMUNIZING AGENTS[1]

Modern science has developed numerous immunizing agents: anti-serums, antitoxins, bacterins, toxoids, virus vaccines, and others. They are manufactured under exacting conditions which provide safe and effective means of protecting livestock against a multitude of invaders. These agents generally fall into two groups: those which protect against diseases

---

[1]Courtesy, Myzon Laboratories, Inc., Syntex Division.

caused by viruses, and those which protect against diseases caused by bacteria. Their manufacture is quite different.

Viral Vaccines

The virus antigen is unique in that it must be propagated in living cells. This may be done in three ways: in the host animal, in a non-host animal, or in a tissue culture. When it is grown in the host animal, a susceptible animal is exposed to and allowed to develop a specific viral disease. Animal tissues are prepared in a suspension then attenuated (weakened) or inactivated (killed) so that the vaccine is safe yet effective. This was the first method developed for the manufacture of viral vaccines.

A second step in development was propagated agents in a non-host animal. They are produced by inoculating growing chick embryos with the virus, allowing them to develop, then separating the virus fluids from the embryos and preparing them for use either as a modified live virus or as an inactivated vaccine.

The most significant recent breakthrough in biological production has been the technique of propagating virus antigens in living tissue cultures, such as kidney, lung, or possibly lymphocyte tissue. This method allows the development of standardized antigen with superior immunizing characteristics which can be produced in large-scale economic units. Many systems of tissue culture production have been developed. Properly executed, they all produce safe, effective vaccines. The researcher selects an antigen (virus) that can provide a strong antibody response and that can be attenuated for safe usage. A system of propagation is then developed which will reproduce this predictable seed antigen repeatedly over a long period of time.

The propagation system may vary, but whether tissue cultures from organs of the host animal or from a non-host animal are used, the end product should produce uniform results and provide effective immunity in susceptible animals.

The freeze-drying (desiccation) process added another significant step in making modified live virus

vaccine practical. Once the antigen is successfully reproduced and stabilized, with certain proteins and chemicals, it is freeze dried to remove the moisture. This process lengthens storage life greatly and makes the vaccine commercially feasible. A diluent is added at the time of use to convert the vaccine into an injectable suspension.

Bacterial Vaccines

Bacterial antigens differ from virus antigens in two major respects. First, bacteria are weak antigens and require much larger numbers to cause satisfactory antibody production. Second, whereas viruses require living cells to propagate, bacteria will multiply in many different environments.

In the manufacture of bacterial agents, organisms are selected, usually from field outbreaks, and grown on an artificial medium. These vaccines may be produced as living antigens but usually are used in the inactivated (killed) form called bacterins.

New technology in the production of vaccines for bacterial diseases involves isolating and using only the portions of cell walls that stimulate the diseases. The resulting vaccines are called subunit vaccines. Some vaccines against Gram-negative coliform mastitis infections cross-protect for several species of the causative bacteria. This is because the bacteria share the same subunit of the cell wall. The negligible amount of endotoxins in subunit vaccines reduces their tendency to cause fever or other adverse post-vaccination effects.

Because of weak antigenicity, an adjuvant is often added to the bacterial agent. The adjuvant acts to increase favorable immunity by slowing down absorption. Small amounts of the large mass of antigens are absorbed over a long period of time so that the animal's body will continue to produce antibodies which ward off the disease.

At various stages of production, each batch lot of vaccine is tested and must meet rigid standards which have been established by the manufacturing firm as well as by the U.S. Department of Agriculture. These standards assure that the final product is pure and safe and has sufficient potency to provide the livestock industry with sound immunizing agents.

Anti-serum

Anti-serum is produced by continually exposing an animal to a disease-causing organism (hypering). This causes the animal to develop a high level of immunity (hyperimmunity). Cells and fibrin are separated from the blood of hyperimmune animals to make anti-serum. Injecting this anti-serum into susceptible animals provides temporary protection against the disease caused by the organism used in hypering.

Anti-serums, a few of which are hemorrhagic septicemia serum, anti-bacterial serum No. 3, anti-erysipelas serum, and anti-hog cholera serum, have filled a need by serving as the sole source or as a supplemental source of antibodies to provide immediate, temporary protection to otherwise unprotected animals. A specific anti-serum prevents or treats a specific disease.

While these products have filled a need, they are not without complications and undesirable features. Frequently, anti-serum dosage has been quite large, requiring multiple injection sites, with possible accompanying irritation or shock.

Hyperimmune Globulin

Sometime ago, methods were developed to remove excess serum fluids by an evaporation process. This process reduced the size of the dose but left a concentration of serum solids which still contained a number of compounds unnecessary to the treatment or prevention of the specific disease. In addition, these compounds were often detrimental to the general health of the recipient by causing irritation and stress.

Research immunologists have known for years that the gamma fraction of blood serum globulins contained most of the protective antibodies and antitoxins.

The gamma globulin may now be taken away from the other components of anti-serum by a frac-

tionation process. Use of this complex process yields a product which is a chemically pure, bacteriologically sterile suspension of pure hyperimmune gamma globulin (antibodies) for treating or preventing the specific disease.

The following tabular matter shows the components of blood. These outlines make obvious the fact that whole serum contains many components in addition to the antibodies and antitoxins which are desired. It is also quite evident that only water is reduced in concentrated serum. Further examination reveals the unique position of hyperimmune globulin as a pure antibody, antitoxin suspension.

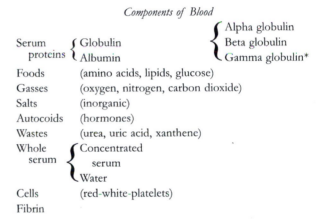

Components of Blood

Serum proteins {	Globulin {	Alpha globulin Beta globulin Gamma globulin*
	Albumin	
Foods	(amino acids, lipids, glucose)	
Gasses	(oxygen, nitrogen, carbon dioxide)	
Salts	(inorganic)	
Autocoids	(hormones)	
Wastes	(urea, uric acid, xanthene)	
Whole serum {	Concentrated serum Water	
Cells	(red-white-platelets)	
Fibrin		

*Gamma globulins are the serum antibody sources available for protection.

The specific active ingredients needed in the area of prevention are the antibodies (gamma globulin). Therefore, the other solids in serum are of no material value in the production of protection.

In the area of treatment, the truly needed active ingredient still doesn't change. The antibodies or antitoxins are needed to accomplish the same as in prevention, only in larger amounts, thus, the increased dose to provide the activating factor for agglutination (antigen-antibody reaction). This process inactivates the invading causative agent and is one method of prevention or treatment.

In most instances when serum is used for treatment, there is an extreme need for antibody activity. The involved animal is usually under the stress of the disease and many times needs additional supportive assistance. It is obvious that antibacterial drugs are indicated to assist in the battle. It is also obvious that nutritional and fluid supportive therapy is indicated. For years, whole or concentrated serum has been used to supply the nutrients and fluid, as well as the antibodies. Whole serum supplies all the solids plus water (fluid). When whole serum is used for supportive therapy for fluid (water), tremendous volumes are needed to be of any significant value. Not only is this source of water expensive, but it also can be tissue irritating, because of the number of injection sites.

The use of concentrated (evaporated) serum, other than for its antibodies, can only be of nutrient supportive use in treatment through the blood solids because the volume (water) has been reduced. The blood solids are made up of approximately six groups of ingredients. The food (protein, carbohydrates, and fats) and salt (electrolytes) portions are the only portions beneficial from a nutrient supportive view. They are beneficial, providing the sick individuals' metabolic processes are functioning and capable of utilization. Otherwise, they are detrimental and add to the already overtaxed system. The hormones present in blood solids may vary tremendously from the kind to the amount, thus having the opportunity of disrupting an already troubled system. The normal wastes present from the donors can only do one thing: Add additional burden to an already stressed sick system.

In today's animal health line, those needed supportive products are available in specific pure usable forms without the unneeded waste products of whole or concentrated sera. When therapeutic levels are needed to support the sick or debilitated animal, fluid therapy products, such as dextrose, Amino-lite, Hydro-lite, etc., can be used much more economically than the anti-serums.

PROPER CARE OF BIOLOGICALS[2]

A vaccine cannot perform at its best without proper attention from the person vaccinating. Improper care and handling of a vaccine can greatly reduce its strength and effectiveness.

---

[2]Courtesy, Myzon Laboratories, Inc., Syntex Division.

Biologicals produced under license are free from contaminants. Care should be taken to maintain that condition. Containers with multiple doses should be discarded when partially used unless they were opened in an aseptic manner and refrigerated. Empty live-virus containers should be burned. Vaccines have varying expiration dates, depending on the particular product. They should not be used beyond the stated time. This date is based on holding the vaccine under optimum conditions, such as keeping it in a cool or refrigerated place. Vaccines that are outdated may have lost part of their antigenic potential and are less effective as immunizing agents. In addition, multiple dose bottles of modified live virus vaccine should be used up within two hours of the time that they are opened, or they will lose their potency.

Most biologic products are sensitive to light and heat, and directions for their proper storage should be strictly followed to prevent deterioration. If its handling or storage has been faulty, a biological may have become inert. This is especially important in connection with highly infectious diseases. When an inert vaccine is used in such cases, both the veterinarian and the livestock owner are often unable to explain the subsequent appearance of the disease. The importance of using dependable biologicals and having them properly stored and competently administered can scarcely be overemphasized.

VACCINATING

There is nothing particularly complicated about vaccinating animals. A good hypodermic syringe, sharp needles, fresh vaccine, sterile techniques, and some way to hold the animal are the essentials. Syringes and needles should be clean and sterile. When using modified live virus vaccines, the veterinarian and the livestock owner should avoid the use of chemical disinfectants. Such chemical agents will inactivate the vaccine.

Vaccines should always be administered in accordance with the manufacturer's directions. Some are given intradermally, others subcutaneously or intramuscularly. In most instances, the material is inoculated in one place. Where the amount is more than 10 ml, as in the case of anti-serum in large animals, injecting small amounts of the material in several places is often advisable.

Only Healthy Animals Vaccinated

The amount of disease immunity an animal obtains through vaccination is strongly influenced by the health or condition of the animal at the time of vaccination. Unthrifty animals, animals incubating infectious diseases, or those heavily parasitized are poor vaccination risks. Under certain conditions, it is possible to actually produce a disease outbreak by superimposing the additional stress of vaccination.

Only persons having special training and knowledge of animal disease, and experience in the use of biological products, should handle and use biologic products and attempt the immunization of livestock. They should (1) read carefully all information on labels and direction circulars before making any injections; (2) keep the products stored under refrigeration; (3) handle them in accordance with directions; (4) use products produced under a U.S. veterinary license; (5) keep records of vaccination, including the serial number of the product; (6) follow instruction; (7) carry out exactly the recommendations of the manufacturer; and (8) vaccinate only healthy animals.

Vaccination Program for Large Animals

Recommended vaccines for *horses* are:

1. Equine encephalomyelitis (sleeping sickness): eastern and western

2. Tetanus. The toxoid will vaccinate a horse. Whereas, the tetanus antitoxin can be given in cases of a horse being wounded and it will protect the horse for about two weeks.

3. Equine influenza, twice annually

4. Equine rhinopneumonitis: the vaccine has different virus types for respiratory and reproductive protection, check with your veterinarian

Optional vaccinations for *horses* are:

1. *Streptococcus equi*: intranasally

2. *Ehrlichia risticii* (Potomac horse fever): in problem areas

3. Rabies

Recommended vaccines for *cattle* are:

1. *Leptospirosis:* five-way vaccine strains, twice annually preferred

2. Infectious bovine rhinotracheitis (IBR), bovine virus diarrhea (BVD), parainfluenza 3 (PI3), bovine respiratory syncytial virus (BRSV) and *Haemophilus somnus* for dairy cattle and beef herds

3. Clostridial diseases, especially blackleg (*Cl. chauvoei*)

4. Campylobacteriosis: in natural service herds

5. Trichomoniasis: in natural service herds

Optional vaccines for *cattle* are:

1. Brucellosis, heifers only

2. Pasteurellosis: in stocker and feedlot cattle

3. Anaplasmosis

4. Rabies

Vaccinations for *breeding swine* are:

1. Parvovirus: twice annually and all gilts must have two doses before breeding

2. Leptospirosis: twice annually

3. Erysipelas

Optional vaccines depending on disease prevalence:

1. Porcine respiratory and respiratory syndrome virus (PRRS)

2. Swine influenza

3. Neonatal pig scours: *E. coli, Clostridium perfringens,* rotavirus, and transmissible gastroenteritis (TGE)

4. Pseudorabies

Optional vaccinations for nursery to finishing *swine* are:

1. PRRS

2. Mycoplasma *hyopneumoniae*

3. *Actinobacillus pleuropneumoniae*

4. Swine influenza

Optional vaccinations for *sheep* and *goats* are for:

1. Hemorrhagic enterotoxemia (*Clostridium perfringens*, type C) for problem areas.

2. Contagious ecthyma (sore mouth) for problem areas only. Never vaccinate on clean premises.

3. Leptospirosis. Vaccinate only if a problem exists.

4. Epididymitis. Vaccinate rams only.

5. In known bluetongue areas, vaccinate sheep prior to the insect season, but do not vaccinate (a) ewes during the first 2 to 3 months of pregnancy, or (b) lambs until they are 4 months of age.

CHAPTER 6

Quarantine in Disease Control

The word *quarantine* is derived from the Italian word *quaranta*, meaning "forty," because ships and their passengers coming from a port where smallpox or other infectious diseases prevailed were detained for a period of 40 days. Today, the word, in general, means "the isolation of a person or an animal sick with a contagious disease." It also refers to a place where the sick are isolated from other animals until the danger of a contagious disese spreading has disappeared.

LEGAL QUARANTINE

In its wider application, a quarantine may be enforced against an individual animal, against all the animals or all animals of the same species, in a township, county, or state, and against those in a foreign country. A quarantine is usually enforced by the sheriff, who is occasionally aided by the militia. For state quarantines, original authority has been placed in the hands of the state veterinarian; for interstate or foreign quarantines, the authority has been delegated to the U.S. Secretary of Agriculture, who acts upon the advice of veterinarians and the chief of the federal veterinary forces. On their own initiative, livestock owners very frequently place an individual animal or groups of animals in quarantine, for the purpose of controlling the spread of a contagious disease, or for the purpose of adding new animals to a disease-free herd or flock. These new animals are kept in isolation for as long as 90 days, in extreme instances, or until the owner, after consulting with a veterinarian, is satisfied that the danger of con-

taminating the original animals with a contagious disease no longer exists.

PRINCIPLES OF QUARANTINE

In general, the principles governing an effective quarantine are similar for both individuals and groups of animals. There must be no direct or indirect contact between the animals in quarantine and those not so restrained. However, there are degrees of quarantine, depending upon the contagiousness of the disease or upon the volatility of its causative factor. For example, many parasitic diseases of the skin, including mange, are not nearly so easily spread as is shipping fever. A quarantine does mean that animals infected with any one of the very large diseases must be kept segregated in a separate barn or yard away from other animals.

Theoretically, the quarantined animals should have different attendants, their own drinking vessels and other utensils, as well as the usual barn equipment. If another attendant is not available, then the one attendant should take care of the sick group last. When chores and treatment are completed, footwear must be cleaned by immersing them in a footbath loaded with suitable disinfectant. Soap in warm water is an automatic remover of infection and in a limited manner is an antiseptic. For a highly volatile disease, such as anthrax, fumigating clothing or protecting the usual clothing by wearing rubber outer garments may be necessary.

Isolating Sick Animals

Whenever certain animals of a group contract a contagious disease, though other animals of the same group are still apparently free from it, the healthy should be moved to clean surroundings, while the diseased should remain in the old, contaminated yards and buildings. The usual practice of removing the sick and leaving the healthy in disease-contaminated surroundings is an unsound one. If following the foregoing procedure is not practical, that is, if the healthy must be retained in the yards or buildings in which the disease originated, then such places must be disinfected in a most thorough manner. Furthermore, yards and buildings housing diseased animals must have drainage away from places occupied by the healthy ones.

It is by means of the quarantine, aided by other steps, that the spread of communicable animal diseases has been controlled. It is much more effective for a condition such as leptospirosis than for infectious bovine rhinotracheitis, because the latter is due to a highly volatile contagion, while the former is almost stationary in its nature. The quarantine has not effected the high hopes held out for it at one time, nevertheless, it is a disease control measure of some value.

Possibly one of the reasons for the lack of effectiveness of quarantine in disease control is that all animals infected in a herd or flock do not show symptoms of the infection, with the result that these subchronic individuals perpetuate the disease in the healthy herd.

Another factor is that, to prevent the spread of an infection, the person delegated to apply the principles of isolation must have enough reasonably good knowledge about the nature of these invisible disease-producing elements. The veterinarian in charge of the attempt to control the spread of the disease should specifically instruct the attendant of the animals about the intelligent application of the principles involved in the quarantine.

Responsibility of Regulatory Personnel

Officers, usually state or federal veterinarians, must recognize that in establishing a quarantine they are acting for the public good. This imposes great restrictions, and usually heavy financial losses, upon the owners of the livestock. Possibly, with the single exception of restricting the freedom of an innocent person convicted of a crime, there is no other procedure where an entirely innocent individual has such a severe "penalty" imposed by legal authority. In establishing the restrictions of a more or less general quarantine, the state livestock sanitary officer is acting as prosecuting attorney, jury, and judge. The United States has been fortunate in having officers of sound judgment and high professional standards and training.

In most states, the carcasses of animals that have died of infectious and contagious diseases may not be placed in or near flowing streams, for doing so would endanger down-stream livestock. Such carcasses may not be transported over public highways—the only exception in this is, in those states having such a regulation, bonded rendering establishments, with approved safeguards, may move or transport carcasses of animals that have died of communicable diseases.

CARE OF QUARANTINE FACILITIES

Buildings, equipment, and feeding utensils used in the quarantine area must be thoroughly cleaned and disinfected, or new arrivals will become infected with the contagion. The first step is to remove all litter and refuse and then disinfect the premises.

Disinfection implies the destruction of all disease-producing germs; in a broader sense, it may also include the destruction of parasitic factors. It must be emphasized, however, that many animal diseases are contracted because of the close contact between the ailing and the well animals; or, stated another way, the premises may be no more than of secondary importance in a matter of this nature.

Disinfection of the premises will make animal quarters reasonably sanitary, but the value of such a step is limited strictly to an attempt at preventing the spread of disease on the premises, and does not in any way control the disease due to the presence of the subchronically infected animals or "carriers" in a group.

On the other hand, infected animals do contaminate the premises with disease-producing agents that are in their urine, manure, and occasionally, in their secretions and exhalations. Since the number and vigor of germs are the greatest in the immediate area and vicinity occupied by the ailing animals, disinfection should be concentrated in such places.

Disinfecting the Premises

Since organic material, such as manure and the usual animal habitation refuse, protects germs and limits the action of disinfectants, this material must be removed. First, the premises should be moistened with water or a disinfectant so as to control dust and airborne infections. This moistening should include all woodwork, followed by litter removal from night barns, night lots, feed sheds, small pastures and yards, and fences, especially those of wood construction, and from around hay and feed racks, even though they have not been in use for some time. The walls and ceilings and other woodwork in barns and sheds must be thoroughly brushed or broom swept to remove cobwebs, old scaled whitewash, and other objectionable material. All burnable material removed by these steps should be destroyed, and the remainder spread out on arable land at some distance away, so that it will be fully exposed to the sun's rays. If buildings have dirt floors, at least 3 inches of this should be removed, then a disinfectant applied to the exposed surface, and, finally, the floor should be covered with clean material, such as fresh dirt, sand, gravel, or cinders.

If there are pools of water on the premises, these are to be either drained or fenced off so that they will not be accessible to animals.

All feed in mangers and feed boxes in the rooms that are to be disinfected should be removed and burned, or it may be fed to livestock not susceptible to the disease against which the disinfectant is being applied.

If water under pressure is available, everything that has previously had litter and other material removed from it should be hosed off. This should be followed by a thorough scrubbing with scalding hot water, to which some lye has been added, to dissolve a good deal of the dirt. The hot water will kill worm eggs, and, the lye will destroy many types of germs. One pound of lye to 20 gallons of water should be used.

Water troughs and feeding boxes and racks should be scrubbed with the lye solution in readiness for effective treatment with a disinfectant solution. Since there is a valid objection to the use of odorous disinfectants in the treatment of many utensils, those constructed of metal may be satisfactorily cleaned with hot water and soap and then immersed in boiling water. Special attention should be paid to the surface and ground about and underneath water troughs. The muddy or moist soil under water troughs has been known to harbor some disease-producing germs for over a year after surface disinfection has been applied. If on low ground, water troughs should be moved to a part of the yard or corral where good drainage may be established.

The application of the disinfectant solution is best done by means of a spray pump so as to force it into all cracks and crevices in the woodwork. Applying it with a broom is a good method, but this does not force it into the wood. The addition of some whitewash to the disinfectant solution is of value, so that a person can better see where the material has been applied. Blankets, robes, and other loose paraphernalia may be disinfected by immersing them for at least several hours in the disinfectant.

The choice of a disinfectant varies somewhat, depending upon the nature of the infection. Different types of germs do not all react similarly to the same disinfectants (see Chapter 4). In general, disinfectants are more effective when used in hot solutions, and those disinfectants having a phenol coefficient of at least 6 should be used. Usually, the better disinfectants have a printed statement on the label indicating their phenol coefficient. A phenol coefficient of 6 means that the disinfectant in question is six times as effective in its action upon typhoid germs as is phenol or carbolic acid.

Disinfecting Woodwork

For the disinfection of all woodwork, saponated cresol solution, made in accordance with the U.S.

pharmacopoeia, and diluted with hot water so as to make a 3 percent solution, is effective, but its odor is objectionable in dairy barns. For outside woodwork, a cheaper and very effective disinfective is a 5 percent water solution of chlorinated lime, but it must be used in the open, as it is very irritating to people inhaling its fumes inside buildings. If a formaldehyde solution is used for inside disinfection, it should be in 4 percent strength, though for use in previously cleaned water troughs it should be painted on full strength over the inside and outside of the trough, and permitted to dry, after which drinking water may again safely be placed in the trough. A full strength solution of the gas form-aldehyde in water is known as formalin. There is 37 percent of the gas in the water. See Chapter 4, section on "Fumigating Buildings" for safety practices to be followed when using formaldehyde inside buildings.

Disinfecting Dirt Floors

Dirt floors or corrals are best, though not per-fectly disinfected, by saturating them with oil and burning it, and then turning under the top soil by plowing. Whenever an infection is very severe, it is an excellent plan to repeat the disinfection at the end of 24 hours. As stated before, the direct action of the sun's rays is a most efficient destroyer of germs, pro-vided the germs are not hidden under rubbish or in darkened places. If disinfection is not done thor-oughly, little more is accomplished, as stated by an em-inent authority, than to create a bad smell.

Disinfecting Needles

In this connection, when the disinfection of buildings is under consideration, it is appropriate to mention the importance of disinfecting hypodermic needles. The use of these needles has become so very general in obtaining blood samples from animals for laboratory diagnostic purposes, as well as in the injec-tion of vaccines, that, without disinfection or steriliza-tion, there is grave danger of spreading from infected and "carrier" animals the infections of anaplasmosis and others. One needle should be used for one animal in diseases such as anaplasmosis or bovine leukemia. Hypodermic needles are relatively cheap, at $0.15 each, so the preferred method would be to use indi-vidual needles per animal. In mass vaccination of herds, the needles should always be changed every ten animals, or sooner if a needle becomes dull, or bent. When using modified live vaccines, do not swab the needle across a cotton ball or sponge soaked in disin-fectant, as the vaccine may be rendered useless. The needle should be disinfected after each use—it should be placed in a reliable disinfectant and rinsed in clean water before it is inserted into the next animal. Storing the needles in alcohol during intervals when they are not in use is an excellent expedient. Sterilizing needles over a flame softens the metal. Boiling water will ster-ilize needles in a few minutes, but if a stilet is not kept in them, rusting is a problem.

Attention must again be called to the close rela-tionship existing between cleanliness in the care of animals and disinfection. When premises and animals are kept clean, this means frequent removal of animal excretions and debris, as well as frequent grooming. These steps mechanically remove many germs and their hiding places. Germs not mechanically removed by cleaning processes nevertheless are weakened and made more susceptible to natural destructive agencies, which, in most cases, renders them incapable of caus-ing disease in their weakened condition.

CHAPTER 7

Animal Housing and Health

Animals must frequently be housed in order to comply with the demands of domestication, for the safety of their keepers, during illness, and as protection against the elements and predatory foes. Newborn animals must be protected from being chilled. Even larger animals at times succumb to the terrific assaults of nature by cold winds, rain, snow, and ice. Sheep with heavy fleece exposed for days at a time to soaking rains frequently become affected with so-called rain rot in which patches of wool loosen from lacerated skin. Insufficiently drained and filthy feedlots are always a most prolific cause of foot rot in cattle.

Animal housing usually requires a high capital investment. Improper housing is economically unsound, and is contributory to animal ill health. It is, therefore, important to give serious consideration to the following factors: (1) location of barns, sheds, and concentration lots; (2) minimum housing requirements of animals for animal comfort; and (3) construction features in buildings that will contribute to economical heating, ventilation, and cleaning (Fig. 7-1).

Fig. 7-1. Modern feedlot facilities that provide comfort, sanitation and easy access for feed trucks.

LOCATION OF BUILDINGS

The correct location of a site for farm buildings is a matter of serious importance not only for preventing disease but also for avoiding legal confrontation due to the encroachment of urbanization. It is a mistake to assume that ownership of land conveys the right to locate animal buildings without consideration of those people living in close proximity.

A site that is remote from residences and commercial development should be selected. New buildings should be constructed far enough away from main highways and thoroughfares so that their appearance and odor will not be objectionable to passersby. The livestock unit should be located where the prevailing winds will not carry odors to the owners's residence or to close neighbors downwind.

Particular attention should be given to location when existing buildings are to be remodeled, for, if a building is poorly located, it should not be remodeled, even for use as a small livestock unit. Once power, water, feed storage, and other necessary facilities are provided for an enterprise, it becomes the nucleus for all further expansion. What once was planned as a rather small enterprise may soon grow to appreciable size, and, if it is poorly located, it may cause major problems.

Animal buildings should be placed low enough and far enough away from the farm home to avoid the flow of barnyard effluvia in the general direction of the home. Each of the two groups of buildings should preferably be on high, well-drained ground, with the advantage in favor of the human dwellings. The important point to bear in mind in erecting all buildings is that they should occupy sites higher than their immediate surroundings. If the natural lay of the land does not provide for this sort of location, then it should be man-made.

Good drainage away from buildings is so generally recognized as an essential that it is seldom overlooked. Buildings should not be near bodies of stagnant water or swamps. Usually, such locations are breeding or living places for infectious organisms and insect life that can make the preservation of health of nearby animals an extremely difficult matter.

A well-founded general belief is that buildings with the long axis in a northern-southern direction, providing an easterly exposure, are the best placed. A northern exposure is usually avoided, since it is cold and frequently damp, because of a lack of direct sunlight; however, it does offer the advantage of coolness during the hot months of the year. A southern exposure is hot and more likely to subject animals housed in such buildings to the annoyance of insect pests. An eastern exposure obviates many of the disadvantages of either a northern or a southern exposure.

DESIRABLE BUILDING FEATURES

One of the desirable features in buildings intended to house animals includes a solid, reasonably high foundation. Double-walled buildings or walls of porous material, afford better protection against cold than those otherwise constructed. Floors of broom-finished concrete, though difficult to keep clean, give the animals a good footing. The coldness of concrete floors may lead to rheumatic conditions which may be largely prevented by a thick layer of sawdust or by mechanically dried cattle manure for dairy cattle. Floors should be provided with drains to carry away and sanitarily dispose of liquid waste. This does not necessarily involve the construction of an extensive system, since it may be quite satisfactorily attained by shallow gutters directed toward a common conduit.

The ceiling of the quarters occupied by the animals, which in many instances is the floor of the loft, must be so closely constructed as to prevent any material, including dust, from sifting down on the occupants. Dust from hay lofts is rich in disease-producing factors and should not be permitted to infiltrate the atmosphere that animals are compelled to inhale.

Windows are of importance from the standpoint of both ventilation and sanitation. The sunlight they admit exposes dirt so that it may be removed. However, sunlight that passes through ordinary window glass is not destructive to germs. Furthermore, well-lighted places are more conducive to essential physiological changes in the animal. In general, having windows at the rear of animals is best, but, if they must be placed in front, they should be high enough to throw the light above the animals rather than directly into their eyes. The glare may also be controlled by the use of frosted glass and awnings. Animal buildings should have approximately one-half as much light as human habitations.

INDIVIDUAL-STALL HOUSING FOR DAIRY CATTLE

Individual stalls are becoming more and more important to dairy producers throughout the United States as a method of housing dairy cattle during the winter months and, in some cases, for the entire year. Each year, more producers are converting their loose-housing buildings to individuals stalls (Fig. 7-2).

Fig. 7-2. Individual-stall housing for dairy cattle. (Courtesy, University of California-Davis)

Advantages

The individual stall has many advantages, even if the dairy is not in an area of severe winter climate. These advantages include:

1. Cows are kept cleaner.
2. The amount of bedding used is much less than is used in a loose-housing system or a shaving pile. Savings have been reported from 52 to 75 percent.
3. Damage to udders and teats is reduced.
4. Milking time may be reduced because time is saved in washing cows during rainy, wet weather.
5. Less water is used in washing the cows. This may be a factor if there is a short water supply.
6. The area required per cow is less in the individual-stall housing than in conventional loafing sheds.

Disadvantages

There are some disadvantages to crowding animals into a smaller area. In individual stalls:

1. Older animals may tend to lie in the alleys rather than in the stalls.
2. The alleys must be cleaned daily.
3. A good method of handling wet manure is needed.
4. If bedding is used, additional hand labor may be needed to bed the stalls.
5. Some hand cleaning of stall may be needed.

Planning and Layout

A number of methods and materials can be used to make the free stalls. Stalls may be made from wood or pipe, or a combination of both, or prefabricated stalls may be purchased through equipment distributors. Regardless of the type of stall used, or whether remodeling or building new construction, the owner must consider some basic construction details (Fig. 7-3).

The roof should cover the stalls adequately and be high enough to allow a person on a tractor to clean the alleys easily. If a self-unloading wagon is to be used to add shavings to the stalls, there must be

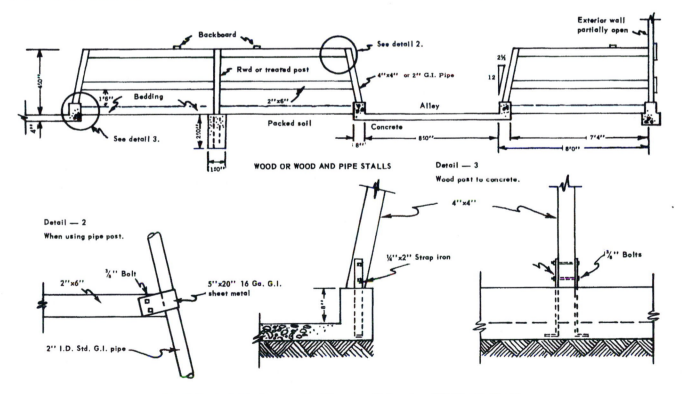

Fig. 7-3.   Individual cow stalls. (Courtesy, University of California-Davis)

enough room so that this equipment can be maneuvered. A few extra stalls should be provided so that the last cow will not have to hunt for the last stall and so that an increase in herd size is provided for.

Ventilation

Ventilation is an essential part of any cattle housing. Cattle often use the stalls nearest the best ventilation first. Louvers, openings under the eaves, or similar openings, will provide adequate ventilation if they are large enough.

The sides away from the prevailing winds and rain should be left partially open to insure adequate ventilation. Adequate ventilation also is necessary to prevent moisture from condensing on the insides of the building. Completely open stalls with only a roof are used successfully in some parts of the country.

Alleys and Curbs

The alley should be at least 8 feet wide to accommodate the cattle and cleaning equipment. Some producers prefer 10 to 12 feet. The curbs on both sides of the alley should be at least 8 inches high and preferably 8 inches wide for strength. Recommended alley slope is 3 percent for scrape-out and 4 percent for flush-down. In an excessively long alley, it may be sloped toward both ends if proper drainage is provided. For extra long barns, openings should be left between every 20 stalls or so, to enable cows to find the empty stalls more easily.

Stall Floor

Stall floors may be made of any fill material that will tend to pack and still maintain some drainage characteristics so that moisture does not stay on top. The floor should slope from front to rear. The fill material should come to about 2 or 3 inches from the top of the curb, so that the bedding used on top will not spill into the alley.

Stall Construction

Stalls may be of wood or metal, or a combination of these material. The partitions should be designed

to allow for maximum ventilation. The lower board or bar should be placed on the stall 18 to 21 inches from the floor. If the board is lower, a cow may make her hips sore from rubbing on the board when lying down.

Stall posts should slope inward, approximately 2½ inches per foot of height, to prevent cattle from bumping them as they move in the alley.

Stall Sizes	Inside Width	Length
Large breeds	48 inches	8 feet
Medium breeds	46 inches	7½ feet
Small breeds	42 inches	7 feet

To reduce the amount of manure dropped in the stall some dairy producers place a board or cable along the top of the stall, far enough from the front so that when the cow stands up she will have to back to the rear. The average distance from the withers to the rear of the cow should be used to estimate the location of the backboard. This will be approximately 23 to 30 inches from the front of the stall.

Bedding

Bedding may be of many different types of material. Some bedding materials presently used are: shavings, sawdust, straw, shredded paper, and cracked walnut shells (Figs. 7-4 and 7-5). However, some dairies use little or no bedding.

Sawdust is more satisfactory than shavings because it is not as fluffy and the cows do not paw it out of the stalls as much. Most dairies add bedding on a weekly basis, to reduce labor costs.

Fig. 7-4. Cows in free stalls bedded with sawdust.

Fig. 7-5. A mattress covered with shredded paper keeps this cow comfortable.

Free Stalls Under Shade

Although free stalls usually are considered as housing in cold, wet areas, stalls under shade are gaining interest in some states that have extremely hot weather (Fig. 7-6).

Free stalls under the shade reduce the problem of crowding and stepped-on teats and udders. They also provide shelter for the animals during bad weather in those areas. The roof must be designed so that there is a maximum amount of shade in all stalls during the heat of the day.

Free Stalls for Calves

Calves of all ages will use free stalls; therefore, they should be considered in planning calf and heifer

Fig. 7-6. Free stalls for dairy cows, open-air with shades. (Courtesy, Babson Bros. Co., Oak Brook, Ill.)

housing. Stall housing for calves and heifers will be the same as that for cows. Lighter material can be used in construction, the stalls will be smaller, and there will be less manure. The stalls can be made in portable units and moved if necessary.

The size of these stalls will vary, depending on the size of the animals. The following may be used as a guide:

100 to 300 pounds	24 inches by 4 feet
300 to 500 pounds	30 inches by 5 feet
500 to 700 pounds	36 inches by 6 feet

INDIVIDUAL CALF PENS

In most cases, the newborn calf should be placed in an individual pen soon after its birth, and it should be kept there until after it has been weaned. Using individual pens offers several advantages: (1) each calf can be observed individually, and any abnormality or sickness that may develop can be determined (for example, lack of appetite); (2) parasites can be controlled while the animal is young; and (3) exact amounts of feed can be given (Fig. 7-7).

Individual calf pens need not be elaborate. Any building that is closed on the north, west, and east sides and does not leak may be used in areas with a moderate climate. Individual pens should be about 2½ feet wide and 4 feet long. Partitions between pens should be solid so that the calves will not be able to suck each other. Floors should be of the insulate type, because other types are cold and wet. Slatted wood or welded wire with 1-inch mesh on a substantial frame

Fig. 7-7. Individual tie-pens for calves. (Courtesy, American Jersey Cattle Club, Reynoldsburg, Ohio)

elevated 18 to 24 inches above the permanent floor makes a very satisfactory device for a calf pen, for it is readily cleaned and disinfected.

BULL PENS

Bull pens may be 10 by 12 feet to 12 by 12 feet for the Jerseys and Guernseys, and 12 by 12 feet to 12 by 14 feet for the larger breeds. Partition heights are recommended to be 5 feet, 3 inches. If cows are kept in pens, the size of the pens may vary from 10 by 10 feet to 12 by 12 feet, with partitions 4 feet, 6 inches high.

LOOSE-HOUSING BARNS

Some dairy producers prefer a loose-housing barn over a stanchion barn. In the loose-housing barn, animals can move about at will (Fig. 7-8). If there is ample bedding above a firm layer of straw and manure, heat is generated and the bed is warm. Also, the animals can move about and avoid drafts. There should be concrete flooring along the mangers and the outside entrance.

These hard surfaced areas must be cleaned daily, though in other areas of the barn manure may accumulate in a firm, thick layer which can be removed to arable portions of the farm. High-producing cows may be housed in the coldest weather without losing production, though feed consumption is somewhat higher. High quality milk can be produced if the cows are kept clean by frequent grooming, and if the bed-

Fig. 7-8.   Heifers in loose-housing type of shelter.

ding is two or three times as liberal as in an individual stall barn.

Loose-housing barns are cheaper to construct and require less labor than those with individual stalls. These advantages, in the opinion of some producers, balance the unsanitary features, as compared with the individual stall setup. The secret lies in providing ample dry bedding and in keeping the cows clean. If the litter becomes wet and soggy, 10 pounds of powdered limestone for each 100 square feet of floor space should be mixed in it, and the process repeated every four weeks.

Utah State University found that for each Holstein cow at least 75 square feet of space was necessary in the loose-housing type of shelter. When only 40 square feet was available per cow, it required 3 percent more bedding, and the cows were not as clean as in a larger space allotment.

SHELTER FOR HORSES

Elaborate shelter is not necessary for a horse, but some protection from the weather is desirable. Simple three-sided buildings with shed-type roofs are adequate if the open side is away from the prevailing wind and rain. Horses may be kept in box or tie stalls. Box stalls should be at least 10 by 10 feet, with solid walls of smooth planks at least 7 feet high. A well-drained earth floor is preferable to wood or concrete. If concrete is used for flooring, it should be covered. The ceilings should be at least 8 feet high. Four-foot doors, divided horizontally at about the halfway point so the top half may be open for ventilation and light, are desirable. Metal or wood feed boxes and hay racks should be placed at a convenient height for the horses (Fig. 7-9).

When horses are kept in box or tie stalls, stalls should be cleaned daily to prevent thrush. Fly problems can be reduced by promptly disposing of manure.

Tack and feed storage should be included as a vital part of the shelter plan for horses. Horse feed must be kept clean and dry to prevent mold and spoilage. Tack should be easily accessible to the caretaker and properly stored on hangers to keep it off the floor.

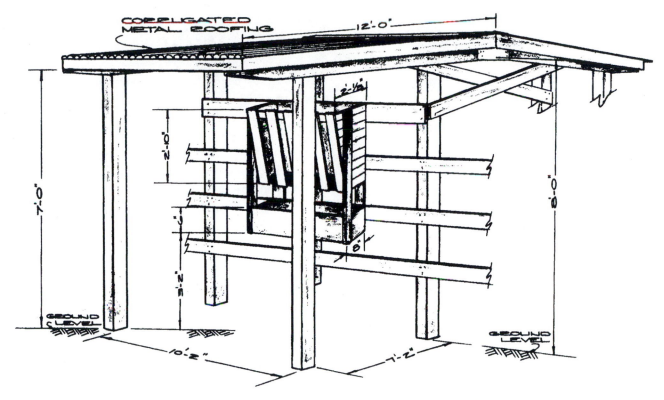

Fig. 7-9. Dimensions for a simple shelter, hay rack, and grain box for horses.

Corrals

Horses that are confined to stalls must have exercise, so the caretaker should plan a daily program of exercise for them. Corrals are desirable for handling horses. When space is available, they should be located adjacent to the stall and be large enough to provide an exercise area. Wood and pipe are preferable fencing materials. Barbed wire is undesirable because of the danger of cuts. Horses also can catch their shoes in woven wire and damage their feet. Horses often push fences out of shape by leaning on them. This can be prevented by placing a smooth strand of wire at the top of the fence and energizing it with an approved electric fence charger.

Polymer plastic or polyvinyl chloride (PVC) fences are the newest in horse fences. They (1) resemble traditional white board fences; (2) are durable, safe, and low maintenance; and (3) cannot be chewed by horses. The main disadvantage to this types of fence is the initial cost, which is about twice that of board fence.

Providing plenty of clean, fresh water should be included in plans for shelters and corrals. Tanks and automatic devices should be located where they may be drained and cleaned easily.

SPACE AND EQUIPMENT REQUIREMENTS FOR BEEF CATTLE

Feed and Water Facilities

Feeding space is determined by the size of the animal and the number of animals that must eat or drink at one time. In the following discussion, animals weighing up to 600 pounds are considered to be calves; those weighing more than 600 pounds are considered to be older cattle.

Hand Feeding

All cattle eat at the same time; calves, 18 to 22 inches; older cattle, 22 to 26 inches; and mature cows, 26 to 30 inches.

Self-feeding

Feed is always available: hay or silage, 4 to 6 inches; grain and supplement, 3 to 4 inches; and grain and silage, 6 inches.

Watering

Plenty of clean, fresh water should be provided at all times. Three to four inches of open water-tank space per head on pasture, or one automatic water bowl for every 40 cattle, should also be provided. A satisfactory winter water temperature range is 40° to 50°F; a satisfactory summer temperature is 60° to 80°F.

Feedlot Facilities

The following lot space should be provided: paved lots, 40 to 50 square feet per head, including open housing; dirt lots, 150 to 200 square feet per head, including paved aprons for bunks and waterers (more is desirable under some soil and climatic conditions).

A paved area of at least 12 feet around waterers, feed bunks, and roughage racks should also be provided, with all aprons along bunks, sheds, and waterers having a slope of 1 inch per foot. A backup bar at the base of the feed bunk 1 foot wide and 4 to 6 inches high will prevent cattle from excreting in the bunk (Fig. 7-10).

Bunk Dimensions

The bunk width should provide (1) 48 inches of eating space when cattle are fed from both sides; (2) 54 to 60 inches if the bunk is divided by a mechanical feeder; and (3) 18 inches of bottom width if cattle are fed from only one side of the bunk.

The height of the bunk floor should be 4 to 6 inches above the apron so that the apron can be kept scraped. It should be 8 to 12 inches above the apron if frozen mud, snow, etc., will accumulate.

The throat height of feed bunks should be up to 18 inches for calves, and up to 22 inches for older cattle and mature cows. It should be increased to 30 inches if hogs are run with cattle.

Roof

If a roof over the bunk is provided, it should be 5 to 6 feet wider than the bunk to protect the bunk and to provide minimum shade. It should be wide and high enough to clear cleaning equipment and to provide shelter for the bunk and summer shade for the cattle.

SPACE REQUIREMENTS FOR SHEEP AND SWINE

Ewes should be allowed from 12 to 16 square feet of barn space. Lambing pens 38½ by 46 inches are often used because they provide suitable space and make efficient use of 16-foot lumber. Fig. 7-11 shows

Fig. 7-10. This concrete slab helps to prevent cattle from defecating in the feed bunk.

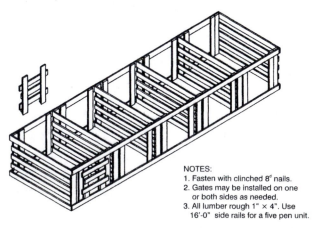

NOTES:
1. Fasten with clinched 8ᵈ nails.
2. Gates may be installed on one or both sides as needed.
3. All lumber rough 1" × 4". Use 16'-0" side rails for a five pen unit.

Fig. 7-11. Lambing pens for ewes. (Courtesy, University of California-Davis)

Fig. 7-12. Ewe and lamb in individual lambing pen. (Courtesy, Washington State University, Pullman)

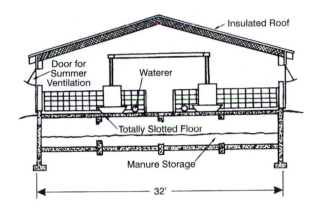

Fig. 7-14. Growing-finishing or breeding unit with a totally slotted floor. This design eliminates the need for daily cleaning of floors. (Courtesy, University of Illinois, Urbana).

how to construct a series of pens, using 1 × 4 rough lumber with 16-foot side rails. Twenty square feet of shelter space should be allowed ewes and lambs (Fig. 7-12). Feeder lambs can get by with about 6 square feet of barn space.

Sows in most commercial operations farrow in a crate, which provides around 35 to 40 square feet of floor space for both the sow and pigs (Fig. 7-13). Farrowing pens should be no smaller than 6 by 8 feet. A very popular size of farrowing pen is 5 by 16 feet.

In recent years, swine producers have adopted slotted floors over a pit of water for market hogs to reduce space requirements per animal and lower labor costs for manure handling. Six to eight square feet per pig is sufficient under such a system (Fig. 7-14).

Supplemental Heat for Pigs

Chilling is a problem that results in high death rates for newborn pigs. Pigs at birth have very low resistance to chilling because their temperature drops from 3 to 9 degrees below normal during the first hour after birth. The greatest drop is during the first 20 minutes of extra-uterine life. At the end of the first hour, the pigs' temperature starts to rise, so that by the second day it is up to an approximate normal of 102.5°F.

One of the most common methods of supplying supplemental heat is the use of 250-watt-Pyres-type heat lamps suspended 18 to 24 inches above the bedding and beyond the reach of the sow. A barn temperature of 55° to 65°F is suitable for the sow, but newborn pigs must have a temperature of 85° to 90°F in the pig nest.

Brooders are used to provide supplemental heat for pigs farrowed in pens. They contain the heat and prevent the sow from contacting dangerous electrical equipment. Heat lamps suspended to the side of a sow in a farrowing crate are commonly used to pro-

Fig. 7-13. A farrowing stall (or crate) on slotted floor.

vide supplemental heat for pigs when the sow's movement is restricted.

VENTILATION FOR HOUSED ANIMALS

Proper ventilation is an essential component of confinement housing. Ventilation of livestock buildings involves simply changing the air in the structures, but the process is complex in that many factors must be understood if the system is to perform satisfactorily.

Animals in open lots are subjected to natural ventilation and can usually move freely to another location when conditions around them are objectionable. This is not the case with animals housed in confinement: They are restricted to a much smaller area with less freedom to move when temperature varies, moisture levels increase, odors intensify, drafts occur, and prevailing conditions around them become uncomfortable. These conditions cause stress and subsequently, adversely affect performance. Therefore, controlling or partially controlling the animal's environment is necessary.

Ventilation is the key to controlled environment. Controlled ventilation is essential for:

1. *Removing moisture, odors, gasses, and airborne disease organisms.* During the winter, ventilation is primarily for moisture and odor control. It is estimated that a mature cow will exhale 18 to 20 pounds of water in a 24-hour period. A barn housing 50 cows requires a ventilation system capable of removing 115 to 125 gallons per day, or condensation will form on the walls, ceiling, floor, and windows, causing structural deterioration. Also, the floor will remain wet. In addition, the resulting damp, stale environment contributes to disease.

2. *Providing oxygen for normal respiration.* A constant supply of fresh air blends with the air in the building and thus reduces the level of contaminants entering the respiratory system.

3. *Controlling the temperature.* During the summer, the primary purposes of ventilation is to control the temperature in the building. At the same time, high velocity air moving across the animals increases the dis-

sipation of body heat. This constant flow of air is also effective in maintaining a drier floor.

4. *Providing an environment conducive to optimum performance.* The environment may be defined as all the conditions, circumstances, and influences surrounding and affecting the growth, development, and production of animals. Thus, the three aforementioned conditions are encompassed here. When these conditions are intermeshed with other sound management practices, efficient production should result.

5. *Providing good working conditions for workers.* People perform best when they are comfortable and contented. Housing design should reflect this consideration, or production will suffer. Workers spends one-third to one-half their time at their jobs. Thus, minimal distractions and discomforts allow the greatest devotion of personal efforts to their jobs.

Heat and Ventilation

Sufficient heat in a barn is necessary for good ventilation. This can be supplied by animals. A 1,000-pound cow, for example, should be available for every 550 to 650 cubic feet of barn space. Fewer cows leave the barn cold and difficult to ventilate. In general, excess heat makes the barn easier to ventilate than cold does. Drafts are harmful. A temperature of 40°F in a dairy barn will not reduce milk production if the animals are accustomed to it, but a higher temperature, which dairy workers prefer for their own comfort, is not harmful to the cows.

Air Space Required

The amount of air space to be allotted to each animal is subject to considerable variation because of conditions that cannot be controlled. When the outside air is not in a state of high motion, and when the barn is of average construction, so that doors and windows do not fit snugly, a good general rule is to set aside approximately 50 cubic feet of space for each 100 pounds of liveweight of most domesticated animals, though sheep require fifty percent more than this. The ventilating forces should be sufficient to permit a renewal of the air in the designated space at the rate of eight times per hour. Though faults in housing construction usually permit this ventilation to take place, under abnormal out-of-door conditions, the exchange of air in the building may be too violent, resulting in undesirable, and frequently harmful, drafts. A tightly constructed, adequately insulated building is easy to ventilate mechanically. Consequently, variations from this call for alternative design considerations.

Insulation

Insulation with a high resistance to the flow or passage of heat is essential to reduce heat loss and thus conserve energy and eliminate condensation on interior surfaces. The effectiveness of insulation materials is characterized by several terms, with the most common term being the "R" value, or Resistance. The higher the "R" value of the insulation, the better the insulation. As the resistance of insulation increases initially, there is a sharp decrease in the heat loss through the walls and ceiling. Greater thickness of insulation is required in a farrowing house and nursery than in a finishing building for hogs, or a housing unit for cattle. For specific insulation recommendations, the owner should consult a specialist familiar with the region.

There is an optimum temperature at which pigs perform best. In a farrowing house and nursery, supplemental heat is necessary in the winter to achieve this comfort level, but in other animal housing units, supplemental heat may not be needed, or, if needed, only for brief periods. Heat produced by the animals themselves and the supplemental heat are used to:

1. *Compensate for building heat losses.* The heat that is lost through walls, roof, ceiling, windows, floor, etc. Adequate insulation is the key to minimizing this loss and, at the same time, reducing the supplemental heating requirements. A one-time investment in the proper installation of insulation brings long-term energy savings.

2. *Warm the ventilating air.* Incoming cold air must be heated in order to increase its moisture-holding capacity. Cold air has little capacity for picking up mois-

ture. In general, moisture-holding capacity doubles as the air temperature is raised 20°F.

3. *Evaporate moisture from the floor.* Buildings with slats minimize the moisture load. Correspondingly, a reduction in heating requirements for this purpose results in additional energy savings.

Components of the Ventilating System

The ventilation system consists of the air inlet, louvers, fans and fan controls, and in some systems, an air outlet.

Air Inlet

The air inlet is an opening through which the ventilating air is uniformly distributed into the building. Uniform distribution is accomplished with a slot or box inlet (Fig. 7-15). Commercial inlets, of both types, which automatically adjust the inlet opening to correspond with the air flow at any time, are available.

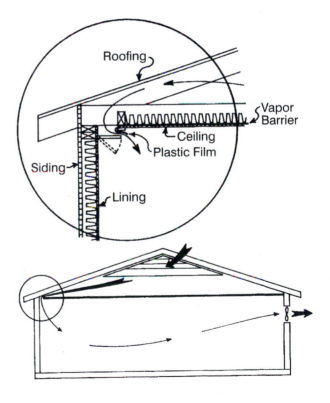

Fig. 7-15. Diagram of a slot-inlet system of ventilation. (Courtesy, University of Illinois, Urbana)

In the winter, pulling the incoming ventilation through the attic to take advantage of solar preheating is desirable. On many occasions, the incoming air on the south side of the house is 10°F warmer than the outside air. A solar collector in the attic would result in a greater temperature rise between outside air and incoming air. With fuel costs much greater now than in several years past, preheating of incoming air is certainly worthy of consideration.

Louvers

Louvers in the gable ends of buildings serve a two-fold purpose: They allow air to enter the attic in the winter for preheating, before entering the building through the air inlet, and they serve as a non-mechanical ventilator during the summer for lowering attic temperatures. The net free opening in the louvers should be 1.7 square feet for each 1,000 cubic feet per minute (cfm) of fan capacity, or 1 square foot for each 250 square feet of attic ceiling area, whichever is larger. Ventilation of the attic in the summer can further be improved with ridge ventilators and openings for air to enter the attic between the rafters.

Screens over the louvers, as well as over other openings into the attic, should be no less than ½-inch mesh and preferably ¾-inch mesh. Otherwise, they will clog with dust and become ineffective.

Fans and Controls

Ventilating fans move air against pressure caused by wind and restricted air inlets (Fig. 7-16). Consequently, this must be taken into account when fans are selected, so that the capacity can be delivered at 0.1- to 0.125-inch static pressure.

Continuous operating fans are more effective in ventilating than cycling fans on and off. The development of variable speed fans and their subsequent use provide a continuous but variable air flow. This is very effective in oscillating between the minimum and normal winter ventilation rates. Above the normal winter ventilation rate, air flow should be provided by single-speed fans controlled by thermostats. Fans, controlled by interval timers, that cycle on and off do not provide the continuous air flow. Thus, mixing of incoming air with room air ceases when the fan is off. This renders the ventilation system less effective.

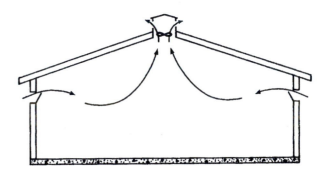

Fig. 7-16.   Diagram of an exhaust ventilation system with fan on ridge. (Courtesy, University of Illinois, Urbana)

CHAPTER 8

Hereditary Factors and Abnormalities

The study of heredity is called *genetics*. It is the study of similarities and differences that are passed from one generation of plants and animals to the next. An individual animal is usually similar in major characteristics to other members of the same species. There is, therefore, little difficulty in distinguishing one species of farm animal from another. It is within this context that the "like always begets like" concept is acceptable. There is no doubt that a mating of a bull to a cow will produce a calf rather than a pig, but indi-

viduals within a species vary a great deal from their parents and other members of their family in certain characteristics (Fig. 8-1). For example, Duroc hogs usually have very little variation in color, but feed conversion percentage of lean cuts and mature body size often differs greatly within this or any other breed of swine.

The source of similarities and differences in all organisms is due to heredity and environment. In cases other than clones of plants and identical twins

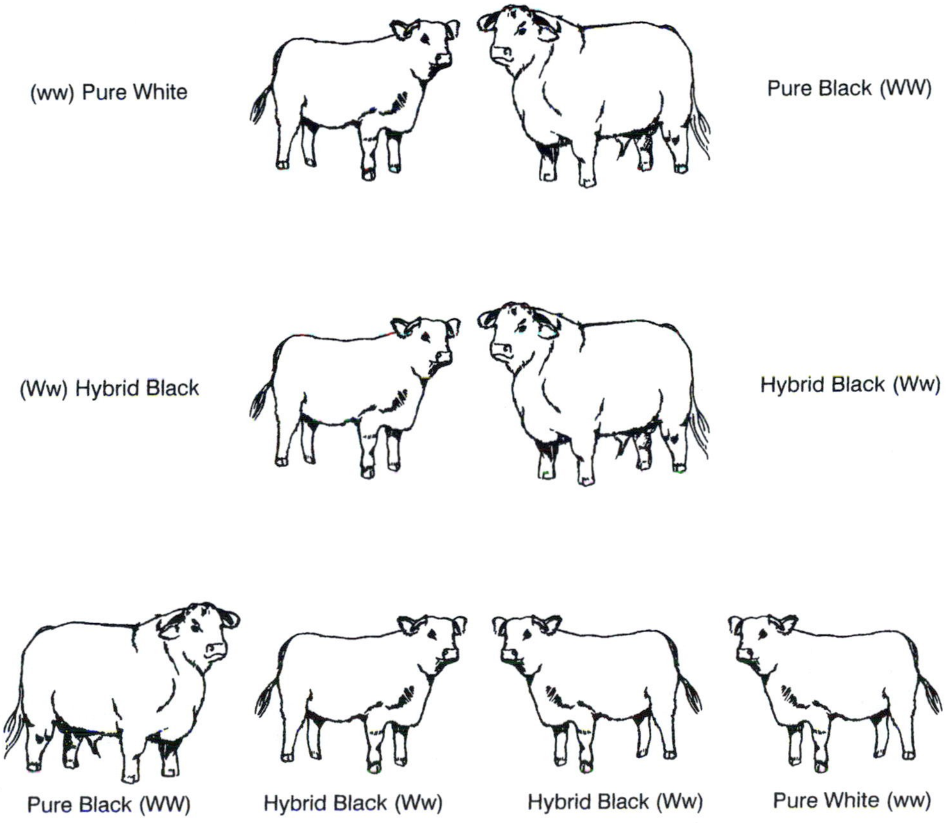

(ww) Pure White   Pure Black (WW)

(Ww) Hybrid Black   Hybrid Black (Ww)

Fig. 8-1. A cross between homozygous parents. The second cross between heterozygous parents gives a phenotypic ratio of 3:1 and a genotypic ratio of 1:2:1. (Courtesy, University of Georgia, Athens)

Pure Black (WW)   Hybrid Black (Ww)   Hybrid Black (Ww)   Pure White (ww)

93

in animals, both hereditary and environmental factors exert influences that affect the development of farm animals. The animal breeder attempts to normalize environmental conditions and/or measure their influence on specific animal characteristics in order to interpret and make sound use of breeding results.

THE CELL AND INHERITANCE

The basic unit of life is the cell. The animal body is made up of two general types of cells, reproductive and somatic cells. Reproductive or germ cells are formed in the gonads—the testes of the male and the ovaries of the female. Reproductive cells are also called gametes. When an animal reaches sexual maturity, specialized cells within the gonads become functional and begin the process of gamete formation or gametogenesis. Sexual reproduction involves the union of two gametes, the sperm and the egg. A union of the two gametes is called a zygote (Fig. 8-2).

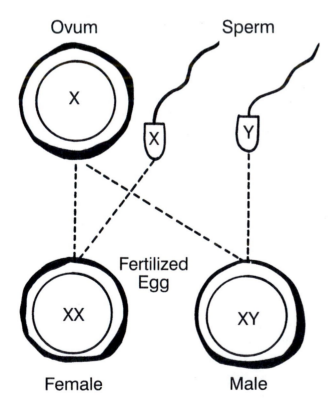

Fig. 8-2.   Sex determination at fertilization. (Courtesy, Vocational Education Publications)

Somatic cells or body cells make up the many tissues and organs of the body. They are formed by a cell division process called mitosis. The single cell formed by the fusion of the sperm and egg divides and forms two new cells. Each of these cells in turn divides, and the process of cell division continues until all the cells of the mature animal are formed.

The somatic cells of all normal animals contain a fixed number of chromosomes. The chromosomes are paired and similar in size and shape; therefore, they are referred to as homologous chromosomes. The number of chromosomes in the body cells is called the diploid number and is expressed as 2n. The number of pairs of homologous chromosomes in the body cells is different for the various species of farm animals as indicated by the following:

NUMBER OF CHROMOSOMES (1N)
SELECTED FARM ANIMALS

Animal	Chromosomes (2n)
Horse	64
Donkey	62
Mule	63
Cattle	60
Sheep	54
Swine	38

The egg and sperm of each species also contain a fixed number of chromosomes. This number, called the haploid number, is one-half that of the body cells and is expressed as 1n. Meiosis is the form of cell division which occurs in the gonads to reduce the diploid number of chromosomes (2n) to the haploid number (1n) when the gamete is formed. This enables the sire and dam to contribute equal numbers of chromosomes to their offspring and, at the same time, to maintain the constant number of homologous chromosomes characteristic of the species.

The genes are located on the chromosomes. They, too, are paired in the somatic or body cells and are duplicated with chromosomes as cells divide in mitosis (Fig. 8-3). Likewise, they follow the same pattern of reduction in number, as chromosomes are reduced from the diploid to haploid number by meiosis in gametogenesis. When fertilization occurs, genes located on the chromosomes contained in the sperm and egg are again paired and express their hereditary influence on the young animal.

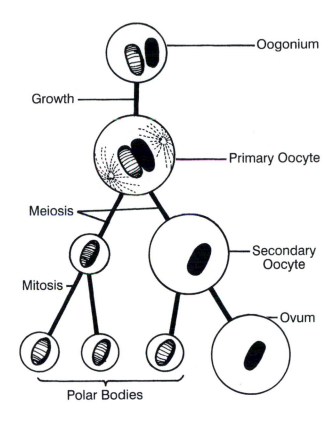

Growth

Meiosis

Mitosis

Oogonium

Primary Oocyte

Secondary Oocyte

Ovum

Polar Bodies

Fig. 8-3.   Oogenesis, using only one pair of chromosomes. (Courtesy, Vocational Education Publications)

BASIC GENETIC TERMS

Allele—An *allele* is one member of a pair of genes that have corresponding locations on like chromosomes.

Genotype—*Genotype* refers to the actual genetic make-up of an individual. For example, BB, Bb, or bb represents the genotype of three different animals for a single trait.

Phenotype—*Phenotype* refers to those physical traits that are visible. For example, color patterns, sex, horns or their absence, dwarfs, "comprest" or normal cattle, and mule-footed or the normal split-hoofed condition of swine.

Dominant—*Dominant* is a condition in which one gene covers or overshadows the expression of the other member of the allelic pair. In Angus cattle,

the gene for black hair color is dominant over the gene for red.

True or complete dominance of one allele over the other member of the pair does not occur in all circumstances. There is no dominance in the color pattern of Shorthorn cattle. Both alleles express themselves in the phenotype. A Shorthorn with a red hair coat has two genes for red (RR) and a white coat has two genes for white (WW). The roan color is a blending of the red and white (RW).

Partial or incomplete dominance of one allele to the other member of the allelic pair may· also occur. The "comprest" condition in Hereford cattle which was popular in the late 1930s and 1940s but which has now been virtually eliminated was apparently due to a semi-dominant mutant gene. The "comprest," a short, blocky, low-set animal, intermediate in size between a dwarf and a normal individual, was a result of the combination of one comprest gene and one normal gene. Two of the comprest genes resulted in a dwarf.

Recessive—The term *recessive* is used to describe a gene that is masked or over shadowed in the presence of its dominant allele.

Homozygous—*Homozygous* refers to an individual in which the members of a given pair of genes are alike, for example, BB or bb.

Heterozygous—*Heterozygous* refers to an individual in which the members of a given pair of genes are unlike, for example, Bb.

CONGENITAL ABNORMALITIES

Congenital abnormalities are caused by hereditary factors or environmental factors within the uterus that alter the normal development of the ovum or embryo and result in death or malformation of the fetus. A malformation that involves a particular organ or part of the animal's body is called an *anomaly* (Fig. 8-4). When the malformation is extensive, the specimen is referred to as a monster. Congenital defects or anomalies may be classified according to their effect

Fig. 8-4.   A congenital abnormality that is not hereditary. (Courtesy, Dr. Murray Fowler)

on the fetus or animal. Those that cause death are referred to as lethals. Anomalies that result in structural or functional impairments are called sublethals.

Anomalies of genetic origin may be caused by a single gene that affects both males and females or they may be sex limited. Some anomalies of farm animals are caused by one recessive sex-linked gene, while others result from homozygous recessive genes.

Environmental factors such as viral infections, nutritional deficiencies, hormonal imbalances, and the ingestion of toxic materials by pregnant females can cause congenital abnormalities that resemble genetic defects. Anomalies caused by environmental factors are probably more extensive in farm animals than is usually recognized.

Another category of defects in farm animals depends, to varying degrees, on both hereditary susceptibility and environmental factors. Such defects include cancer eye in cattle, scrotal hernia in swine, congenital photosensitivity in sheep, and vaginal and uterine prolapse in cattle and sheep.

LETHALS IN CATTLE

Some of the lethal characteristics observed in cattle that result in death to the calves are hydrocephalus, bull-dog calf, hereditary neuraxial edema, hairlessness, disorders of the brain, amputated limb, short spine, and mummified fetus.

Hydrocephalus may occur in calves, pigs, lambs, or foals. It is an accumulation of fluid around the brain due to a dysfunction that interferes with circulation of fluids through the ventricular system.

The *bull-dog calf* has a tremendous bulge in the forehead, a short face, and a depressed short nose. The tongue protrudes, and the palate is cleft or absent. The neck is short and thick, and the limbs are shortened. In Dexter cattle, there is a heterozygous form that can easily be recognized by the shortness of the limbs, while heterozygous animals of other breeds have a normal appearance. All the homozygous animals die, regardless of breed.

Hereditary neuraxial edema is transmitted to calves by a recessive autosomal gene. It is most common in Hereford calves and is characterized by muscle spasms and an inability of the calf to rise and remain standing.

Hairlessness is a condition in which the skin at birth is devoid of hair, though there may be some hair in the vicinity of the natural body openings and at the extremity of the tail. Sometimes there is no skin below the knees, on the ears, and on other parts of the body. This condition has been observed mostly in Holstein and Jersey cattle.

Disorders of the brain may occur in any of the species of farm animals. They include agenesis or absence of a brain and hypoplasia which means defective or incomplete development. The bovine viral diarrhea virus may cause cerebellar hypoplasia in calves.

Amputated limb calves observed in Swedish Holsteins all trace back to a famous bull. The forelegs terminate at the carpal joint and the hind ones at the hock. There are additional minor head deformities. Death takes place a few hours after birth.

Short spine refers to a backbone so shortened because of fewer vertebrae that the animal appears neckless. The tail is very short and is attached high up. Death occurs shortly after birth.

A *mummified fetus* is one that shrivels and dries up. It has been observed in Red Danish cattle produced by daughters of the bull *Oluf Godthaab* when they were mated to three descendents of the same bull. When mated to other bulls, the offspring were normal, or the bulls mated to other cows brought normal calves. The calves died about the eighth month of gestation, though carried full term. Excessive fluids in the foetal membranes (hydrops amnii) were a constant feature. Usually, the neck was short, the joints prominent, and the legs stiff.

Other lethals of cattle are (1) lameness in the hindlimbs; (2) muscle contractions of the limbs so that these members are almost wrapped around the body at birth; (3) short legs and underdeveloped claws, or more or less complete fusion of the claws; and (4) a condition in which the lower jaw is shortened and immobile, probably because of fusion of the joint.

LETHALS IN SHEEP

Lethals observed in sheep include:

1. Anencephaly or absence of a brain, which occurs in lambs whose dams ingested toxic plants early in the gestation period.

2. Myopathies that cause muscle contractions of the limbs.

3. Rigid bent fetlocks with a short, thick body, large skull, and straight wool.

4. Skeletal myopathies that result in paralysis of the hindlimbs.

5. Amputations at the fetlocks.

6. Shortened, immovable jaw.

7. Homozygous grey.

All the lethal traits in sheep quite closely resemble similarly named conditions in cattle.

LETHALS IN SWINE

Lethal hereditary traits noticed in swine are:

1. Skeletal myopathies that result in paralysis of the hindlimbs.

2. Myopathies that cause muscle contractions.

3. *Atresia ani*, in which the anus is imperforate.

4. Thick or greatly swollen forelimbs, these pigs being born in the same litter with normal pigs.

LETHALS IN HORSES

Literature has only a slight reference to lethal hereditary traits in horses. Probably the most important one is that in which the left dorsal colon near the pelvic flexure is completely closed. The principal recording of this condition relates to the offspring of a Percheron stallion that was sent from Ohio to Japan. The condition is probably a recessive trait. Veterinary anatomists, however, state that, in many horses used for dissection purposes, the lumen of the left dorsal colon is frequently very small though open and unobstructed.

NON-LETHAL ABNORMALITIES

Inherited condition in livestock that are not necessarily associated with or do not result in death of the individual are quite numerous (Figs. 8-5, 8-6, and 8-7). However, such conditions do result in economic loss to the producer.

Animals often inherit poor structural arrangement of bones in the feet and legs. When stress to these defects is applied by rapid growth, heavy weight, mounting to breed, traveling over rough range country, jumping, or fast strenuous work in the case of horses, an impairment of normal function occurs. Diseases of this nature are quite common in horses. Examples include ringbone, bone spavin, and splints. In addition, other non-lethal abnormalities include:

1. Scrotal and umbilical rupture
2. Fewer than four teats in cattle
3. Extra or rudimentary teats
4. Blindness, due to congenital cataract
5. Defects in hair and teeth
6. Extra digits or toes
7. Lack of pigment (albinos)
8. "Screwtail"
9. Syndactalism or single toes
10. Fused teats
11. Flexed pasterns
12. Entropin
13. Notched ears
14. Blind nipples
15. Cross-eye, appearing at about one year of age

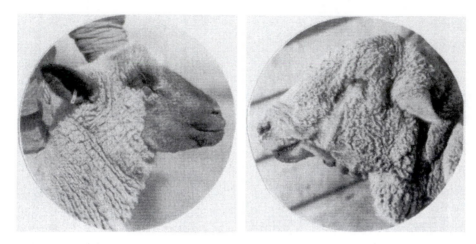

Fig. 8-5.   Undershot (left) and overshot or parrot mouth (right) jaws. When undershot, the lower jaw is longer than the upper. When overshot, the lower jaw is shorter than the upper. (Courtesy, U.S. Sheep Experiment Station, Dubois, Ida.)

Fig. 8-6.   Rambouillet yearling rams showing contrast between open- (left) and closed-faces (right). In making selections, the progressive sheep producer will discriminate against wool-blind rams and ewes. (Courtesy, U.S. Sheep Experiment Station, Dubois, Ida.)

Fig. 8-7.   An abdominal rupture in sheep.

16. Parrot mouth
17. Myoclonia congenita
18. Dwarfism
19. White-heifer disease
20. Freemartin
21. Hermaphroditism

Myoclonia Congenita

Myoclonia congenita is commonly called shakes, trembles, shivers, jumpy pig disease, and dancing pigs. It is a congenital disorder that appears in pigs at birth or shortly thereafter. The cause of myoclonia congenita is unknown. Speculation as to its cause includes a genetic defect, nutritional deficiency, and a viral infection in the dam during gestation.

Myoclonia congenita usually affects only a portion of the pigs in a litter and a percentage of the litters in a herd. It may occur, more or less simultaneously, in several herds located close to each other. The major sign of this disturbance is a tremor. The tremor may be quite mild or so severe that the pig has a jumpy or dancing appearance. The tremor often ceases entirely when the pig lies down.

The mortality rate for pigs with myoclonia congenita is quite low if they survive the first week of life.

Dwarfism

Subnormal growth and development has caused great financial losses to cattle breeders. The types of dwarfism which occurred in beef cattle within the past 40 years are comprest dwarf, crooked-legged dwarf, long-headed dwarf, "snorter" dwarf, and midget. The "snorter" dwarf has been the most costly.

The "snorter" dwarf, when first observed as a baby calf, may be mistaken for a calf with outstanding beef conformation. The blocky appearance, short, compact body, and mature look at birth are first observed, but when the calf is one or two months old, the bulging forehead, prominent eyes, protruding lower jaw and tongue, and low-hanging head will identify it as a "snorter" dwarf. Development after this age is characterized by slow growth, increasing lack of co-ordination in walk, enlargement of the middle, and heavy breathing; hence, the name "snorter." Many "snorter" dwarfs die before they reach weaning age, although some have lived for several years and a few have reproduced.

White-Heifer Disease

White-heifer disease is an anomaly of the reproductive tract of cattle. It may occur in all breeds of cattle and has no specific connection to hair coat color. White-heifer disease is believed to be caused by a single recessive sex-linked gene. It may affect any portion of the female genital tract and, therefore, has no single characteristic. Incomplete or defective development of parts or all of the reproductive tract is commonly observed in females with white-heifer disease.

Freemartin

A freemartin is a sterile heifer born twin to a bull calf. This condition commonly occurs in about 90 percent of the heifer calves whose twins are males. It is caused by a hormonal interference brought about by the circulation of blood between the two embryos. Hormones from the gonads of the twin male impair the development of the ovaries, uterus, cervix, and vagina of the female.

Diagnosis of the freemartin condition in heifer calves can be determined by a vaginal examination, using either the finger or a 6-inch test tube. Penetration of the vagina will be limited to about 3 inches or less in the freemartin heifer. A more reliable diagnosis of this anomaly can be accomplished with a blood test. The use of these simple tests can eliminate the expense of raising sterile heifers.

Hermaphroditism

Hermaphroditism is a congenital abnormality that may occur in all species of farm animals. The true hermaphrodite will have both ovarian and testicular tissue and may have external genitalia representative, to some degree, of both sexes. Pseudohermaphroditism is more common in farm animals than true hermaphroditism. It occurs most frequently in sheep, goats, and swine. The pseudohermaphrodite possesses either testicles or ovaries and external genital organs that resemble the opposite sex.

To the practical livestock producer, these inherited characteristics are of economic importance because of the reduction in productive and reproductive efficiency. The elimination of undesirable genes from breeding stock is a responsibility that should be shared by all livestock and dairy producers. Replacement stock should not be selected from families with a history of hereditary abnormalities, no matter how desirable they are from a phenotypic point of view.

Hump-Back Disease

Hump-back disease is seen in 1 to 2 percent of the Holstein steers being fed in feed yards in the Imperial Valley.

Swollen Umbilicous

Swollen umbilicous is seen in $\frac{1}{4}$ to $\frac{1}{2}$ percent of the Holstein steers being fed in feed yards in the Imperial Valley. These are not hernias, and they disappear by the time animals weigh about 700 pounds.

CHAPTER 9

Basic Husbandry Practices

Livestock and dairy producers must routinely perform basic management practices which involve minor surgery, such as castrating, dehorning, and docking. In addition, they often assist during parturition, give injections, drench, spray, and pill animals. To subject animals to experimentation by untrained caretakers is inhumane. An accepted rule of animal care is: Caretakers should perform those practices they know and can do well and secure the services of a veterinarian when in doubt. In most cases, the veterinarian will teach clients the correct way to perform basic practices that are essential to good husbandry and animal health (Fig. 9-1).

All cases of major surgery require the skill and ability of an experienced veterinarian, and no one else should attempt such operations. Nevertheless, the more knowledge and understanding an owner or a caretaker of animals has of the principles of surgical procedures and post-operative care, the better for all concerned. First, these caretakers would appreciate more fully the skill of a qualified veterinarian, and, second, they would be better prepared to render assistance to the veterinarian in nursing their animals back to health.

RESTRAINING ANIMALS

Approved husbandry practices in livestock and dairy production require that animals be restrained occasionally for medication, vaccination, examination, and alteration. Control and handling procedures should be performed with care to ensure a minimum of stress to the animals and safety for the attendants.

Cattle

Because of their size and strength, cattle need to be controlled by special methods of restraint. In the simpler procedures of controlling cattle that are thoroughly domesticated, such as the family milk cow, grasping the nostrils by means of the thumb and fingers is sufficient. In applying the same principle, the caretaker may use a metal nose clamp or bull lead. For the domesticated, though always potentially dangerous, bull a rigid staff or metal rod about 6 feet long snapped into the nose ring should be used. This forces the bull to maintain a safe distance from the leader.

Cattle that are to be branded, dehorned, bled, or castrated in the standing position should be worked in a squeeze chute (Fig. 9-2). Spraying, vaccinating, and applying "pour-on" insecticides may be accomplished

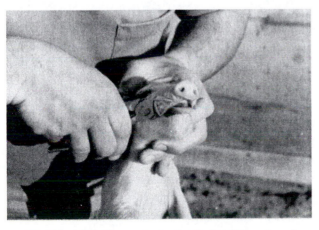

Fig. 9-1. Clipping needle teeth of pigs. Side cutter pliers are used by most swine producers to clip the teeth.

Fig. 9-2.   A squeeze chute used to restrain cattle.

Fig. 9-3.   Many management practices can be performed on cattle in a long, narrow working chute. By pulling their tails firmly to the side, the attendant can easily restrain cattle.

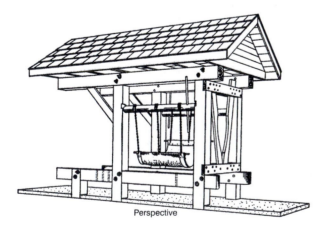

Perspective

Fig. 9-4.   Cattle stocks used to trim feet. The roof reduces the deterioration of the wood structure and belt.

by crowding cattle into long, narrow working chutes (Fig. 9-3). For control during the trimming of feet, cattle stocks or a turntable is indispensable. Both these devices should be equipped with a sling to prevent cattle from lying down. (Fig. 9-4 shows a set of stocks that are designed for a permanent location.) The roof reduces deterioration of the wood from rain and provides protection from the sun. Portable stocks that fit in the back of a pick-up truck are quite handy for agricultural education instructors and 4-H Club leaders to use at the homes of their students. They must be anchored when in use because unruly, large cattle can tip over unsecured portable stocks.

Cattle exhibitors control show animals in a blocking stand so that the groomer may clip the head and tail and clean out the hair on the brisket and underline (Fig. 9-5). These stands not only restrain the cattle but they also provide a tie station to secure the animal's head. This enables the groomer to complete the blocking assignment in a stance similar to the position the cattle will be set up for exhibition in the show ring.

Fig. 9-5.   A blocking stand used to restrain show cattle while they are being clipped.

There are times when casting an animal for proper control is desirable or absolutely necessary. On ranches, young cattle and calves are cast by means of a lariat and a rider on a trained horse. When the animal is pulled down, one attendant approaches the calf from the back side, places a knee on the neck, folds the foreleg and pulls up and back to prevent the calf from gaining traction with the leg next to the ground. A second attendant seated behind the calf, pushes the undermost hindlimb forward with his/her foot just above the hock joint, and at the same time, grasps the uppermost hindleg and pulls it firmly backward. This is an ideal position for castrating.

The casting of individual cattle of large size and strength is sometimes necessary, and, for this, a rope around the body is the standard method (Fig. 9-6). The animal is tied to a strong post or tree trunk with a short rope. Then a long rope is placed around the animal's neck and tied so that it won't slip. The rope is extended back, and a half-hitch is placed just behind the forelegs. A second half-hitch is placed so that it encircles the body in front of the hips and in the flanks. At the points where the body is encircled, the rope must not be tied. When the free end of the rope is pulled, it will readily tighten around the two places of encirclement. A strong steady pull will cause the strongest, most vigorous, and most recalcitrant animal to go down, so that its feet may be tied together. If the feet are not tied, pressure on the ropes must be maintained or the animal will get up.

The disadvantages of using the half-hitch for casting are: (1) it compresses the heart and lungs, thus affecting these organs, especially in fat cattle; (2) it compresses the organs of digestion, thus giving rise to bloat; (3) in male animals, it frequently results in damage to the penis; and (4) in dairy animals, it exerts undue pressure on some of the mammary blood vessels.

The objections against the half-hitch are overcome by the following manner of casting, which requires a strong halter, a bull lead, and a 40-foot length

Fig. 9-7.   A desirable working area for cattle. It has a small containment pen, a narrow alleyway, and a squeeze chute.

of rope. The attendant should (1) secure the animal by means of the halter or a short rope placed around the animal's neck that which will not slip and choke it; (2) place the rope over the withers and bring the ends down along the sides of the neck and backward between the forelegs to cross beneath the breastbone; (3) continue the ends along the sides so that they cross over the region of the loins and then downward and backward between the udder or scrotum and the inner surface of the hindlimbs—the two ends should emerge a short distance above the points of the hocks; and (4) pull strongly on such a hitch to bring the animal to the ground. Separate lengths of rope may be used to confine the legs.

The use of drug restraints, given by injection, has eliminated the necessity of using mechanical casting techniques for most procedures. Additionally, they make for more humane and safer control of the animal.

Sheep

Sheep are easily contained long enough for drenching or for vaccinating by crowding them into an alley or a chute (Fig. 9-8). A closer examination of the udder, feet, testicles, or sheath requires individual restraint. Sheep are small enough to be thrown and controlled quite easily. Figs. 9-9 and 9-10 show two methods of restraining sheep in a manner that leaves

Fig. 9-6.   Casting a bull with a rope. (Courtesy, American Hereford Assn.)

the attendant's hands free to examine or to treat the animal.

The most common method of individual restraint is for the attendant to turn the ewe up on her rump and cradle her between the attendant's knees. Rams and large ewes are easily restrained by carefully throwing them on their side. The attendant steps into the ram from the back side, places one foot near the hip touching the ground, and sets the other foot between the ram's hindlegs. The foot should be placed behind and just above the hock of the leg on the ground. The bend of the attendant's knee is used to contain the top rear leg of the ram.

Sheep and cattle should not be placed on their backs. Sheep occasionally inadvertently become lodged in this position. Death from congestion of the lungs and closely related conditions, such as interstitial edema or alveolar hemorrhage, may occur within minutes in sheep if they are not returned to a normal position. Lung damage is also possible when sheep and cattle are restrained for long periods of time on their sides. Good management, therefore, requires that ani-

Fig. 9-9. An attendant can restrain a ram and still have complete freedom of his hands.

Fig. 9-10. Cradling a ewe between the knees to restrain her.

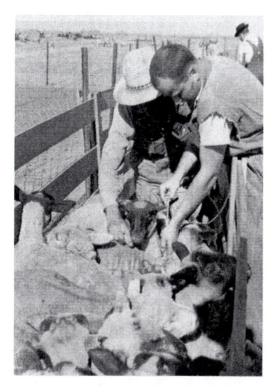

Fig. 9-8. Vaccinating feeder lambs for enterotoxemia in a narrow chute.

mals restrained in this manner be released as quickly as possible.

Horses

Proper handling of horses is unhurried, kind, and firm. It takes years of experience for an individual to develop the ability to judge an animal's disposition and to make a quick appraisal. Frequently, a horse's mood changes while being handled. The ability to spot this mood change will enable the handler to change tactics as the horse's mood changes, and thus continue to restrain the horse without much hassle. Brute force is always undesirable. If the horse's resistance is overcome by brute force, the horse will usually remember and fight much harder the next time. Whereas, if the handler talks the horse into submission, the animal will be easier to handle in succeeding sessions.

Holding the Horse

The person holding the horse should almost always be on the same side of the animal as the person who is working on the horse. Then, if the horse acts up, the handler can pull the horse away from the operator. The handler can divert the horse's attention by talking to the horse, jangling the lead, or tugging lightly on the halter. If necessary, the assistant can pull the head hard around to the side. The horse can also be restrained for short periods of time by lifting one foreleg and holding it up. Most horses are very easy to work on while they are being fed grain.

Lead Shanks

There are two types of lead shanks recommended for use on horses. They are the rope and the leather chain shank. The rope shank is stronger than the leather chain shank. It can be used to tie a horse. Chain shanks should never be used to tie a horse, but they can be used for more effective restraint. The following procedures are recommended for using a chain to restrain a horse: (1) place the chain portion over the horse's muzzle, (2) pass the chain through the horse's mouth, and (3) place the chain over the gum of the horse's incisors (Figs. 9-11a and 9-11b).

Chains should not be put under a horse's jaw because it is very easy to break the jaw when pressure is applied to control the animal.

Covering the Eyes

The handler should use one hand to cover the eye of the horse on the side the operator is working on or to blindfold the horse for easier handling. (Fig. 9-12). To administer injections, the handler can divert the horse's attention by pinching the neck, while the operator inserts the needle, or by pinching the small of the back with the thumb and forefinger, while the operator gives an injection in the rump (Figs. 9-13a and 9-13b).

Fig. 9-11a.   Using a lip chain to restrain a horse.

Fig. 9-11b.   Using a chain twitch to restrain a horse.

Fig. 9-12.   A horse with a blindfold.

Fig. 13a. Pulling the mane firmly at the same time the needle is inserted to divert the horse's attention. Note the proper position of the attendant.

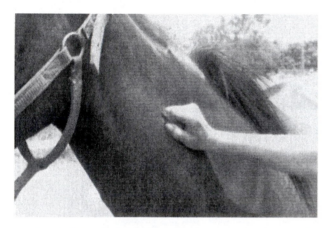

Fig. 9-13b. Pinching the skin on the neck to divert the horse's attention momentarily.

Other procedures and equipment that can be used to restrain horses include (1) grasping a handful of mane and pulling down on it, (2) lifting the horse's tail and pushing it straight up, (3) lifting one front leg and holding it up, (4) placing the horse in stocks, and using a cross tie, barnacle, scotch hobble, war bridle, side line, or twitch.

A twitch may be purchased from an animal equipment vendor or made from ¼-inch nylon rope and a short stick. About 12 inches of the rope is threaded through a hole in the end of a stick. The loop may be placed around the horse's muzzle and tightened by twisting the handle to restrain the horse.

To use a twitch on a horse's muzzle for any purpose other than to obtain temporary control over the animal is inhumane. The horse's muzzle is one of its most sensitive organs, for it is highly endowed with sensory nerves. A twitch applied unnecessarily, or for too long a time, is an abuse of power that no lover of the horse can condone.

Drugs are available which will provide complete restraint of a horse. However, these drugs should be administered by persons well trained in their use, as the type of drug and dose require careful consideration and administration.

Hogs

Boars and large sows are extremely strong and frequently vicious. A strong, solid panel wired securely at the top and bottom to a corner of a fence can serve as a squeeze for temporary restraint.

A hog snare placed in the mouth behind the tusks so as to encircle the nose may be effectively used to control hogs (Fig. 9-14). A nylon rope may also be used in the same manner. Both devices must be kept tight on the hog's nose to maintain control. Hogs confined in this manner will usually pull back to the limit of their power during the time they must submit to treatment by the attendant.

CASTRATION

Castration is an approved management practice in meat animal production. Experimental evidence reveals that intact males produce leaner carcasses, and they gain faster and more economically than castrates fed in

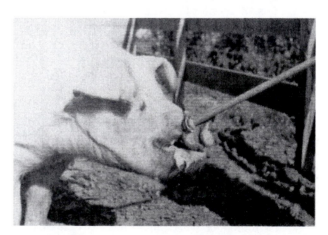

Fig. 9-14. Using a snare to restrain a small boar.

the same manner. Castrates, on the other hand, are quieter in the feedlot and easier to handle. They bring a higher price for slaughter than intact males and, in the case of cattle, grade higher. The obvious advantage of castration is to control reproduction and permit commingling of males and females.

Castration of beef, sheep, and swine is routine when basic principles are followed. Serious complications are likely to arise unless the following procedures are implemented:

1. Animals must be properly restrained to prevent undue excitement, worked promptly, and released as quickly as possible.

2. Sanitary procedures must be incorporated that will prevent infection. This includes the equipment used, the attendant, the scrotal area, and holding facilities that are free from dust and filth. Castrating equipment should be disinfected after each operation in areas where anaplasmosis is prevalent.

3. Incisions into the scrotum must be large enough and low enough to ensure proper drainage.

4. The testicle and lower portion of the spermatic cord should be removed by pulling and scraping the cord to inhibit bleeding. Emasculators should be used to crush the spermatic cords when mature animals are castrated.

5. Whenever possible, castration should be scheduled during the non-fly season of the year. When animals must be castrated during the fly season, insect repellents should be applied.

Castrating Cattle

Bull calves can be castrated from the time they are a few weeks old up to eight months of age without serious consequences. Young calves are easier to handle, usually bleed less, and heal faster than older cattle.

There are three common methods of castrating cattle. They differ only in the location of incisions made on the scrotum. In the first method, the scrotum is stretched tightly, a knife blade is inserted through both sections of the scrotum, and the entire lower half of the scrotum is opened by the incision. The calf will usually draw up the testicles when the knife is first inserted. One of the testicles is then squeezed down and pulled tightly, and the cord scraped until the testicle is released. The procedure is repeated to remove the remaining testicle. This method is best adapted for use when a calf table is used or when the calf is held on the ground by two attendants. The procedure enables the operation to be performed quickly, provides for maximum drainage, and permits normal development of the cod.

The second method involves splitting each side of the scrotum parallel to the middle line. The incision should be made on one side, and the testicle removed from that side before the incision is made on the other side. The second incision should be made over the center of the testicle, and from about the top one-third to the lower end. Extending the slit well toward the lower end of the scrotum to allow proper drainage is essential. This method is well adapted for chute castration where the bull calf is standing and the operator works from the rear.

In the third method, the operator grasps the lower end of the scrotum, stretches it tight, and cuts off the lower one-third. The ends of both testicles are exposed. Removal of the testicles involves the same techniques as described in the first method. The procedure is satisfactory and expedient for very young calves, but unsatisfactory for older cattle. There usually is a swelling and constriction above the circular incision sufficient to restrict drainage and, unless it is reopened, death may result.

The Russian method of castration is not commonly used in the United States. Experimental data show that Russian castrates perform better than steers in all phases of production, but not as well as bulls. The technique was developed by a Russian scientist to make use of testosterone for increased meat production. The part of the testicles that produces spermatozoa is removed, leaving some of the hormone producing part intact. To do this, a scalpel is inserted through the scrotum in the middle third of the large curve of the testicle and opposite the epididymis. The scalpel is rotated 180° and then withdrawn. This loosens the glandular central core of the testicles which then can be pressed out through the puncture wound (Fig. 9-15).

Two methods of bloodless castration are used on cattle. The most popular involves the use of an

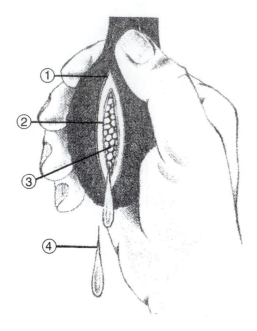

Fig. 9-15. Russian method of castration: (1) cut scrotum exposing testicle; (2) cut the membrane which covers the testicle; (3) note interstitial tissue (source of androgen for growth); (4) remove seminiferous tubules (which contain the sperm) by squeezing.

emasculatome to crush the spermatic cord through the scrotum (Fig. 9-16). The attendant should use the following procedures:

1. Move the cord to the outside of the scrotum.
2. Clamp the emasculatome about 1¼ to 2 inches above the testicle.
3. Hold the clamp in this position for a few seconds.
4. Repeat the procedures on the cord for the other testicle.

The attendant should observe the following precautions to avoid slips:

1. Make sure the emasculatome works properly.

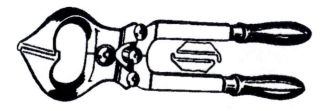

Fig. 9-16. An emasculatome.

2. Check to see if the cord has slipped out.
3. Clamp only one cord at a time.

An elastrator may be used to apply a specially designed rubber ring around the scrotum just above the testicles. This inhibits the circulation of blood to the testicles and causes them to atrophy. This method involves a possibility of tetanus and also leaves the calf with a small cod which is undesirable to many cattle feeders. The following procedures are recommended when an elastrator is used to castrate calves. The attendant should:

1. Restrain the calf on his side.
2. Use an elastrator to expand the rubber ring.
3. Press both testicles through the opened ring.
4. Release the ring and remove the elastrator.

Castrating Pigs

The material and tools needed for castrating pigs are a sharp knife, or a single-edged razor blade, alcohol or another disinfectant solution to cleanse the area where the incisions are to be made, and an antiseptic powder to sprinkle in the wound.

The holder grasps the pig by its hindlegs and places its shoulders between his/her knees. With the toes together and heels apart, the holder's knees are forced against the pig. This helps to hold the weight of the pig, as well as to hold it securely.

The incision is made at the bottom of the scrotum to allow good drainage. The entire testicle should be removed, including the epididymis.

Castrating Sheep

Lambs can be castrated with a knife, elastrator, or emasculatome. The knife is used most often to castrate lambs on range operations because many lambs must be worked in a short period of time. Ideally, lambs should be castrated from one to three weeks of age, but this is often impossible. It is advisable, however, to castrate lambs as young as possible. The following procedures are recommended for castrating lambs. The attendant should:

1. Restrain lambs by locking the hindlegs just above the hocks with the bend of the forelegs.

2. Use sanitary equipment and procedures.

3. Use a sharp knife to cut off the lower third of the scrotum.

4. Expose the testicles by gently pressing them outward with the thumb and the index finger of each hand.

5. Grasp each testicle individually and pull it out of the scrotum with a smooth, firm, deliberate motion.

6. Release lambs on their feet and provide clean holding areas or move the lambs to pasture.

The use of an emasculatome is the preferred method of castrating mature lambs, but it is a slower and less dependable method for young lambs. The instrument is used to crush the spermatic cord. The operation must be done very carefully, or failure will result. The cord should be placed within the jaws of the emasculatome before pressure is applied to close the instrument.

The procedures recommended for castrating lambs with the elastrator are the same as those reported for calves.

DOCKING LAMBS

When possible, docking should be done on a bright, warm day. The attendant should provide clean, freshly bedded quarters and keep the lamb quiet both before and after the operation.

The main reasons for docking lambs are:

1. Docked lambs are much cleaner around the tail and are less susceptible to fly strike and maggot infestation.

2. Docking improves the general appearance of lambs and sheep.

3. Docked or short-tailed ewes are easier to breed and settle.

4. The dressed carcasses of docked lambs are more attractive in appearance.

5. Packers and lamb feeders pay higher prices for docked lambs.

Whenever possible, lambs should be docked before they are 14 days old. The tail should be cut off 1 to 1½ inches from the body—near the ends of the caudal folds on the underside of the tail.

Lambs 7 to 10 days old may be docked by one person using a pocket knife. The lamb should be held in a standing position between the person's legs. The person should stretch the tail tight with one hand and sever it with a quick upward stroke. The cut should always be made from the underside of the tail to the top or woolly side, and always use a moderately sharp knife—a very sharp knife may cause excessive bleeding. Before cutting the tail, push the skin back toward the body to leave some excess skin to grow over the cut end.

The elastrator-rubberband method may be used within 24 hours after the lamb's birth. The rubberband should be stretched over the lamb's tail with the elastrator, and then slipped off the prongs. The tight band will cut off the blood circulation, and the tail will drop off in a few days.

The hot-iron method of docking is commonly used to dock lambs in the range country. The lamb's tail should be inserted through a hole in a board. The board prevents burning the lamb any place other than the tail, and its thickness regulates the length of dock. The iron should be heated to a cherry red. If properly heated, and if suitable pressure is applied, the iron will sear the wound and minimize bleeding.

The emasculatome is sometimes used for docking. The tail is clamped in the instrument, following which it is severed with a knife just outside the jaws of the emasculatome. The pressure applied with the emasculatome squeezes the main artery and prevents bleeding, but this is a slow method of docking.

Profuse bleeding may occur when the tail is cut off. To stop the bleeding, the attendant should use a rubberband or tie a string tightly around the tail stub. The lamb should be marked so the rubberband or string may be removed in about 30 to 45 minutes, and then apply a good disinfectant, such as Lysol solution, iodine, or one of the many commercial products available from livestock supply houses, to the freshly cut stump.

Fly strike may be a problem if the elastrator method is used during fly season. To reduce chances of fly problems, the tail should be cut off after the band has been on for two or three days and a disinfectant and fly repellant applied.

DEHORNING

Dehorning is practiced commonly on cattle and occasionally on sheep. Horned cattle require more space in feedlots, at feed troughs, and in transit than cattle without horns. They often inflict serious injuries upon each other. In addition, properly dehorned feeder cattle bring more per hundredweight than cattle with horns.

Age is an important factor in dehorning. It usually determines the method to be used. The horns of calves from birth to about three months of age are no more than skin appendages. Horns at this stage are loose to the touch because they have not united with the bones of the skull. During this time, calves can be dehorned quite easily with very little discomfort.

When horns have attained full growth, they are attached to the skull bones. As the animal ages, the space or cavity in the horns becomes larger, and it is soon in direct anatomical communication with a large cavity in skull known as the frontal sinus. Sinus cavities and their passageways are lined with a highly sensitive mucous membrane. The significance of this is that, as a result of dehorning, the frontal sinus is opened, and unless protected, is exposed to germ contamination. As a result, sinus infection frequently ensues in which the animal suffers excruciating pain and a serious disturbance of general health. This can be avoided by (1) being clean with the surgical procedures involved in dehorning, (2) applying an antiseptic dressing to the wound, and (3) protecting the open area from the elements, insects, and filth with a stick-on bandage.

Methods of Dehorning

The three methods of dehorning cattle that are commonly used are chemical, hot-iron, and mechanical. They all are designed to destroy the cells which surround the base of the horn. When this is success-

fully accomplished, the horns are permanently removed.

Chemical

The caustic potash method of dehorning calves is not the preferred method as the paste may run into the eyes of the animal and cause damage.

Hot-Iron

The hot-iron method of dehorning is widely used in range areas. Since the development of electrically heated branding irons, the hot-iron method is becoming more widely used in beef and dairy herds as well. Hot irons are commonly used to dehorn calves up to four or five months old (Fig. 9-17).

The hot-iron method of dehorning is reasonably rapid and practically bloodless. Irons other than the electric type are sold in sets. The size that fits the horn to be removed should be selected. The copper-capped head of the iron should be heated and applied to the base of the horn. The skin around the horn needs to be copper colored for an adequate burn. The tip of

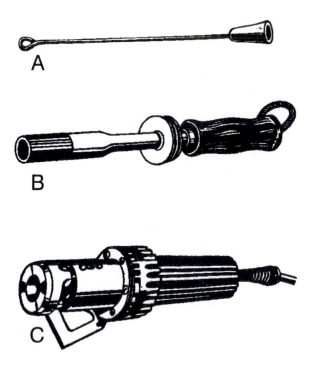

Fig. 9-17. Dehorning irons. Types B and C are electronically heated.

the horn bud does not need to be burned for the dehorning to be effective. The horn or button will slough off in four to six weeks.

Mechanical

The most common types of mechanical dehorning equipment in use are (1) dehorning tube or spoon (Fig. 9-18), (2) Barnes dehorner (Fig. 9-19), (3) keystone dehorner (Fig. 9-20), and (4) dehorning saw (Fig. 9-21). The following procedures are recommended for effective use. The worker should:

1. Restrain cattle so their heads can be held relatively still.

2. Keep all dehorning equipment clean and disinfect each dehorner after use on individual animals in areas where anaplasmosis is prevalent or suspected.

3. Select an appropriate type of dehorning equipment for the job. Dehorning tubes work best on young calves before the horns attach to the skull. Barnes dehorners are effective on calves with a horn diameter up to 3 inches. The keystone dehorner and saw are recommended for use on older cattle.

4. Place the dehorning equipment directly over the horn and remove a small ring of hair with each horn.

5. Control excessive bleeding by pulling the blood vessels located at the lower base of each horn. A heavy cord may also be used to reduce bleeding. The cord should be placed around the skull just below where the horn was removed so that pressure will be exerted on the blood vessels.

6. Sinus cavities opened by dehorning should be protected with a gauze pad.

ADMINISTERING MEDICINES TO ANIMALS

Medicines are administered to animals in many ways. The route and method of administration depend upon the results desired, condition of the animal, and the nature of the medicine to be given. The most common methods of administering medications to farm animals are drenching, pilling, injecting,

Fig. 9-18. Tube or spoon dehorner.

Fig. 9-19. The Barnes dehorner is especially adapted to dehorning calves. This type is available in two sizes. The larger one can be used on cattle up to a year old.

Fig. 9-20. The Keystone dehorner is commonly used on mature cattle and sows.

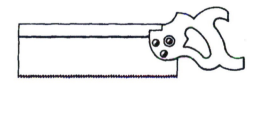

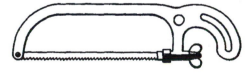

Fig. 9-21. Two types of hand saws that are used to dehorn cattle.

pour-on for topical administration, and ingestion with feed or water.

Routes of Alimentary Administration

The simplest way to give medicine to farm animals is in feed or water. This is the common route for administering anthelmentics to swine for the control of internal parasites. Cattle, horses, and sheep are occasionally dewormed in this manner also.

Liquid medicines are most effectively administered to horses by the use of an esophageal tube. A veterinarian should conduct the procedure. The tube is carefully inserted through the horse's nostril, past the pharynx, through the esophagus, and into the stomach. Note well: Great care must be taken to make sure that the tube has reached the stomach, and that it is not doubled over in the esophagus or, worse yet, located in the trachea. Severe irritation may occur if the stomach tube is not properly placed.

Drenching horses is time consuming because they may choke and aspirate the substance, causing pneumonia. Dewormers, anti-inflammatories, vitamins, and electrolytes are available in paste formulation for easy oral administration to horses. The small dosage size of about 10 ml. allows for owners to administer the medications safely. Some pills, powders, and granules can be mixed with syrup to form a paste and given orally to horses.

Cattle are more easily drenched, as a rule, than are horses because they swallow more readily (Fig. 9-22). They should be restrained to limit movement of the head. This can best be accomplished by working cattle through a squeeze chute. When a squeeze chute or head gate is not available, the operator can restrain the animal by placing his/her thumb and forefingers in its nostrils. Pressure can then be applied to raise the animal's head and open its mouth to receive the medicine. The barrel of the drenching gun should be placed on top of and near the back of the tongue when the medicine is administered. When drenching any animal, the operator should take the following precautions:

1. Never attempt to drench an animal that is down.

Fig. 9-22. Drenching cattle.

2. Do not elevate the mouth of an animal being drenched higher than the level of its eyes.

3. Direct the barrel of the drenching gun toward the center of the animal's mouth, on top of the tongue and in the back portion of the mouth.

4. Exercise caution to prevent damage to the soft palate of the mouth.

5. Do not rub an animal's throat in an attempt to help it swallow.

Sheep are probably the most susceptible of all farm animals to damage from drenching. It is not uncommon for sheep to die immediately after being drenched when improper techniques of drenching have been used. Two basic techniques are recommended for drenching sheep: the slow method and the fast method.

The slow method is safe and therefore appropriate for inexperienced caretakers of sheep. First, the operator restrains the sheep so that it can neither back away nor go forward. This can be accomplished by straddling smaller sheep and placing the operator's legs in front of the sheep's shoulders. Second, the sheep's jaw is cradled in one hand and the barrel of the drenching gun is placed in the sheep's mouth with the other hand, letting the sheep nibble and chew on the barrel as the drench is gradually administered on stop of the sheep's tongue. The operator should top momentarily when medicine starts to drip out the side of the sheep's mouth.

The fast method is used by experienced sheep handlers. Their skill is usually so well-perfected that a novice may think the technique is simple. The inexperienced should not attempt to drench sheep by this method.

Sheep are usually crowded tightly into an alleyway or chute so the operator can move rapidly from one sheep to another. The drenching gun that is commonly used has a measuring device to administer a 2-ounce dose of medicine with each pull of the lever. The gun is commonly fed from a knapsack filled with the drenching material. Hand-filled drenching guns that hold two, three, or four doses may also be used, but they are, of course, slower because they must be refilled frequently.

The operator places the barrel of the drench gun in the side of the sheep's mouth, just in front of the molars, directs the barrel to the back of the mouth, and closes the mouth and nostrils with his/her other hand as the medicine is released (Fig. 9-23).

Medicinal preparations in solid form are administered through the mouth by balling gun. This method of oral administration is used more commonly with

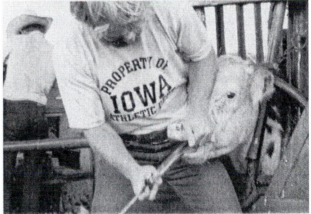

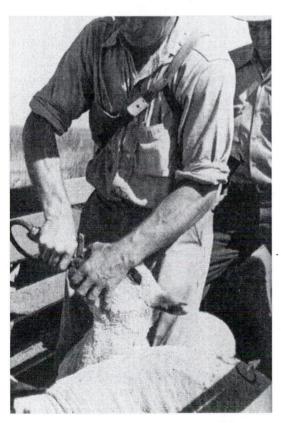

Fig. 9-23. The fast method of drenching sheep.

Fig. 9-24. Administering a bolus by using a balling gun.

cattle and sheep than with horses and hogs. It is a safer method of administering medicine than drenching and offers an added advantage in storage. The procedures are quite similar to drenching, except a pill or bolus is used in place of the liquid. The balling gun is, of course, larger in diameter and much longer than a drenching gun. Great care must be taken to insure that the balling gun is used in a gentle manner. Forceful use of a balling gun can easily perforate the back of the pharynx (Fig. 9-24).

Routes of Parenteral Administration

Meat quality begins on the farm, and medications administered parenterally can have an effect on carcass quality from scarring or abscessing. Improper injection techniques, improper routes of administration, and improper injection sites can render a medication useless, as well as damage the tissues of the animal.

Parenteral administrations exclude routes of the alimentary canal. They include intravenous (IV), intraperitoneal (IP), intrarumenal (IR), intramuscular (IM), and subcutaneous (SQ) injections.

Intravenous

Intravenous injections are made when pressure is placed on the lower region of the jugular groove; thus, the vein will swell and can be felt or seen (Fig. 9-25). A needle should be thrust into the vein—when blood begins to flow out, the needle is properly placed.

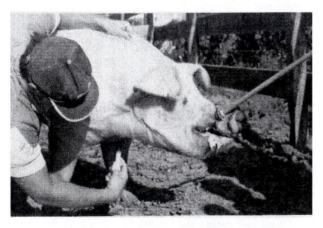

Fig. 9-25. Drawing a blood sample from a hog. The same procedure can be used for intravenous injections in swine.

Then, the syringe should be attached and pressure released so that the operator can slowly inject the solution. If the animal shows rapid breathing, staggers, etc., the attendant should withhold the solution, wait a few minutes, then proceed at a slower injection rate. A clean, sharp 2-inch × 14-gauge needle should be used for cattle.

Intravenous injections (IV) are used when delivering a medication that is irritating to the tissues, or when a drug needs to be rapidly absorbed through the body. An example is the administration of calcium solutions to treat milk fever in dairy cattle.

Intraperitoneal

Intraperitoneal injections (IP) are used in cattle and sheep, but should never be used in horses unless the strictest aseptic technique is observed.

Intraperitoneal injections are made through the skin and muscle of the body wall on the right side into the large abdominal or peritoneal cavity, below the right loin, between the hip-bone, loin, and last rib. For a 250- to 400- pound animal, a 1 to ½-inch needle should be used; for a 400- to 600-pound animal, a 2-inch needle should be used; for a 600- to 800-pound animal, a 2½-inch needle should be used; and for an 800-pound and over animal, a 3-inch needle should be used.

Intrarumenal

Intrarumenal injections are made directly into the rumen or large stomach through the skin and muscle of the body wall on the left side of cattle. The site of injection is the "hollow of the flank." In a bloated animal, this will appear as a bulge rather than a depression. Usually, the odor of escaping gas will be noted when the rumen is penetrated. A 3- or 4-inch × 14-gauge needle should be used.

One company markets a cattle dewormer that is administered intrarumenaly. This allows for a more sure dosing scheme—the animal cannot spit up any of the medicine as compared to oral administration.

Intramuscular

Intramuscular injections (IM) are made directly into the muscles of the neck. The loin, hip, ham, and leg are expensive cuts of meat and should not receive any injections, even in young animals. Improper vaccination techniques or the administration of an irritating substance IM can leave an abscess that impacts beef and pork tenderness and quality. According to a 1995 study, each improper vaccination procedure costs the industry $1.74 per head in tissue damage to the expensive cuts of beef, and $16.88 per head in hide defects. Subcutaneous injection minimizes abscessing, scarring, and other defects that can impact beef tenderness and hide quality.

IM injections in cattle, sheep, goats, and horses are made in the neck using a triangle formed from the points of the nuchal ligament, the shoulder blade, and the neck bone. In hogs, the IM spot is visualized on the neck just behind and below the ear.

A 1- or 1½-inch × 16-gauge needle is inserted deeply into the muscle. If blood begins to flow out, the needle has probably struck a vein. Thus, it should be withdrawn and inserted in a different location.

Subcutaneous

Subcutaneous injections are made directly under the skin (Fig. 9-26). The side of the neck or behind the shoulder is a good location, but care must be taken not to penetrate the underlying muscle tissue. A ½-inch × 16-gauge needle should be used. However, if a longer needle is used, it should be inserted under a fold of skin parallel with the body.

Whenever possible, according to label directions for the medicine, administer vaccines and medications subcutaneously in the neck to avoid damaging expensive cuts of meat.

Miscellaneous Routes of Administration

Two miscellaneous routes which merit attention are intranasal injections and intramammary infusions.

Intranasal

This route includes nasal inoculations commonly administered to feedlot cattle to prevent respiratory infections (Fig. 9-27). The vaccines are in a liquid form and are sprayed into each nostril, where they contact the mucosa of the nasal passages.

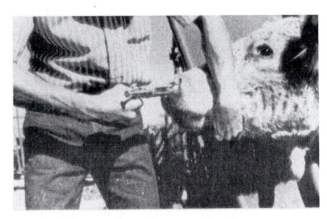

Fig. 9-27. An intranasal administration of vaccine. (Courtesy, SmithKline Animal Health Products, Philadelphia, Penn.)

Intramammary

This route of administration includes treatment for certain kinds of mastitis. Udder infusion tubes containing antibiotic ointments are inserted through the orifice of the teat into the teat canal. The antibiotic preparation should be massaged up into the udder section for penetration throughout the quarter. Separate tubes should be used for each infected quarter. The teat should be disinfected before and after each infusion.

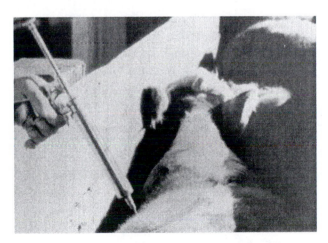

Fig. 9-26. Hypodermic injection in cattle. (Courtesy, SmithKline Animal Health Products, Philadelphia, Penn.)

TAKING BODY TEMPERATURE, PULSE, AND RESPIRATION

Temperature recordings of farm animals are taken by inserting a clinical thermometer through the anus into the rectum. An animal thermometer should be used. They are heavier and more rugged in construction than thermometers intended for human use and therefore, less likely to break when animals resist.

Two types of clinical thermometers are commonly used to take the temperature of livestock and dairy cattle. The most widely used is the mercury-loaded thermometer. It is constructed in such a manner that mercury rises in response to heat and remains stationary in the recording column of the thermometer. It is inexpensive and reliable when properly used. Many commercial producers are now using a battery-operated, transistorized, digital thermometer to record temperatures of animals. This type of thermometer provides a numerical readout of an animal's temperature in only a few seconds after insertions of the heat sensitive element into the rectum.

Normal Temperature Range

In human beings, any deviation from a temperature of 98.6°F is viewed as evidence of an abnormal condition in the body. This is not true for farm animals. The following figures indicate a normal temperature and normal range (in °F) for livestock:

	Normal Temperature	Range
Horse	100.5	99.5 to 101.5
Cattle	101.5	100.5 to 102.5
Swine	102.5	101.0 to 103.0
Sheep	103.0	102.0 to 104.0

Whenever the temperature of an individual is elevated above the normal, the animal has a fever. This should always be regarded as a possible evidence of disturbed health. However, many things other than ill health can influence the temperature. Factors such as excitement, exercise, digestion, rest, and high surrounding temperatures are involved. An animal's temperature is usually lower in the morning than later in the day. A temperature below normal in a sick animal is a very serious sign.

The Pulse Rate

The pulse is an intermittent wave in an artery. It is a result of heart action that pumps blood into the arteries and the expansion and contraction of the elastic arterial wall. The pulse rate of farm animals is subject to even wider variation than is the temperature. In general, the pulse rate varies with the size of the animal. Very small animals, such as mice, may have a pulse rate of 600 per minute, while an elephant's rate is from 30 to 45. In addition, the heart action is more rapid in young animals than in older ones. Other factors that tend to cause acceleration of heart action are excitement, exercise, digestion, and high temperature. The following figures show the normal range in pulse beats per minute for various animals:

	Range and Rate per Minute
Horses	16 to 42
Cattle	60 to 70
Swine	60 to 80
Sheep	70 to 80

Veterinarians attach a good deal of importance to the characteristics of the pulse, such as the force and fullness of the beat and the form of the pressure wave. Thus, the beat may be quick and abrupt, hard and long, slow and soft. At other times, it may be fast, thin, thready, and arhythmic. The rhythm in healthy horses and cattle is steady. If abnormal rhythm characteristics are maintained over more than a comparatively short period of time, they should be considered as evidence of a disturbed bodily function.

Sites for Taking Pulse

The sites for taking the pulse vary in domesticated animals (Fig. 9-28). In the horse, the favorite and preferred site is the external maxillary artery located at the lower border of the jaw at a point where it rests against the underlying jaw bone. If the foregoing artery is not available, which sometimes occurs when the area is diseased and swollen the artery at the inner surface and upper portion of the foreleg is quite accessible. In cattle, correspondingly located arteries may be used, or the saphenous artery on the plantar surface of the hindlimb just above the hock joint may also be used. In dairy and thin cattle, this artery is quite available. In still other cases, the coccygeal artery high up on the under surface of the tail is the preferred site. In sheep, the femoral artery high up on the inner surface of the thigh at a point where it emerges from the groin muscles is the preferred site.

Horse

Cow

Fig. 9-28. Suitable locations for taking the pulse of a cow or horse.

Respiration Rate

Respiration rate, or the breathing rate, is subject to normal variation, since it is perceptibly influenced by excitement, exertion, plethora, and size of the animal. Respiration rate is increased by fever, pain, weakness, infections, and lung ailments. The following figures indicate the range and rate per minute in farm animals:

	Range and Rate Per Minute
Horses	8 to 24
Cattle	10 to 30
Swine	8 to 18
Sheep	12 to 24

In order for one to know and appreciate the difference between health and disease, as evidenced by temperature, pulse, and respiration, constant and frequent practice is necessary. Only after repeated observations of these health functions can deviations in them as caused by disease be detected.

ASSISTING DURING PARTURITION

Parturition is the act or process of giving birth. This process occurs in three phases. The first phase is initiated by uterine contractions which are directed toward expelling the contents of the uterus. The pressure forces the fluid-filled placenta against the cervix. As the cervix dilates, the first water bag, the chorioallantois, passes through the cervical opening into the vagina and may extend through the vulva. In many cases, the bag bursts, expelling the fluid with each contraction. The second water bag, the amnion, precedes the fetus through the birth canal. It, too, may reach the external part of the vulva before it is ruptured by contraction of the abdominal muscles.

The second phase of parturition involves the actual delivery of the fetus. As the fetus moves through the cervix into the vagina, the water bags, if still intact, are ruptured. This initiates a pattern of regular straining, involving a combination of uterine and abdominal contractions to expel the fetus.

The third phase of parturition involves expulsion of the placenta. This is commonly referred to as "cleaning." The placenta or afterbirth is usually expelled shortly after the fetus is delivered. Sows may farrow pigs from one uterine horn, pass the afterbirth from that horn, rest awhile, finish farrowing, and then expel the remaining afterbirth. It is not uncommon, however, for the sow to farrow all her pigs before cleaning.

The cow and ewe usually deliver the afterbirth in three to four hours following the birth of the fetus. A range of 1½ to 7 hours is not uncommon, however.

The placenta should be considered retained if a cow or ewe has not cleaned in 24 hours. The mare is much more susceptible to uterine infection than other farm animals. A delay of two or three hours in delivering the placenta from a mare is sufficient cause for concern.

Signs of Approaching Parturition

Enlargement of the abdomen is the first sign of approaching parturition. This is usually noticeable during the second trimester of the gestation period. Cattle are referred to as "springers" at this stage, and sows are said to be "piggy." Toward the end of the gestation period, the female's udder enlarges and fills with colostrum. The pelvic ligaments relax, and the vulva swells and becomes flacid. Just prior to parturition, the abdominal muscles relax, the belly drops, and a sunken appearance may be observed just above the flanks. At this time, the female may become restless, hide out from the herd or flock, urinate frequently, and then go into labor. A sow will usually gather material to build a nest immediately before farrowing.

Normal Presentation

The normal position for the calf, lamb, and foal is forefeet forward, with the nose lying between the knees (Fig. 9-29). The back of the fetus should be up and in contact with the sacrum of its mother when

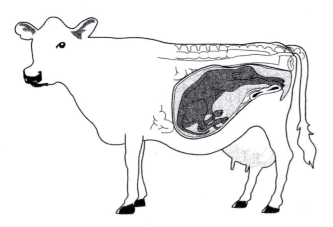

Fig. 9-29.   A calf in the uterus positioned for normal delivery.

delivery occurs. This position for delivery is complementary to the natural curvature of the birth canal and the fetus. The posterior presentation is common in the birth of pigs. When posterior presentation occurs in other farm animals, delivery is usually difficult and the assistance of the caretaker is necessary.

Difficult Parturition—Dystocia

In most instances, normal parturition in farm animals is prompt and easy. This is surprising in light of the complicated series of events that are required for normal delivery. Successful parturition is dependent upon harmonious functioning of the endocrine and nervous systems, an adequate pelvic structure in the female, and a fetus in proper position. The following are the most common causes of dystocia:

1. Improper position of the fetus (Fig. 9-30). This includes one or both legs out of position, the head back or down below the pelvic arch, breach, all four feet in the pelvis, and the fetus presented upside down.
2. Constriction of the vulva.
3. Reduced labor activity.
4. Excessive size of the fetus.

Providing Assistance

Wisdom in animal husbandry during parturition is based upon the following principles. First, the attendant must be patient enough to let nature take its course. Second, he/she must be able to recognize when assistance is needed and apply common sense to all efforts of assistance.

Generalizations about when to intervene and provide assistance are not worth much. Experience is necessary for making sound decisions in this matter. A dystocia due to a constricted vulva requires immediate assistance. The feet will protrude through the vulva during labor and recede as straining subsides. Proper assistance when this circumstance occurs calls for the attendant to take hold of the forelegs and pull while the female strains. It may be necessary to simply prevent the fetus from retreating until the next surge of labor. At that time, the fetus should be pulled.

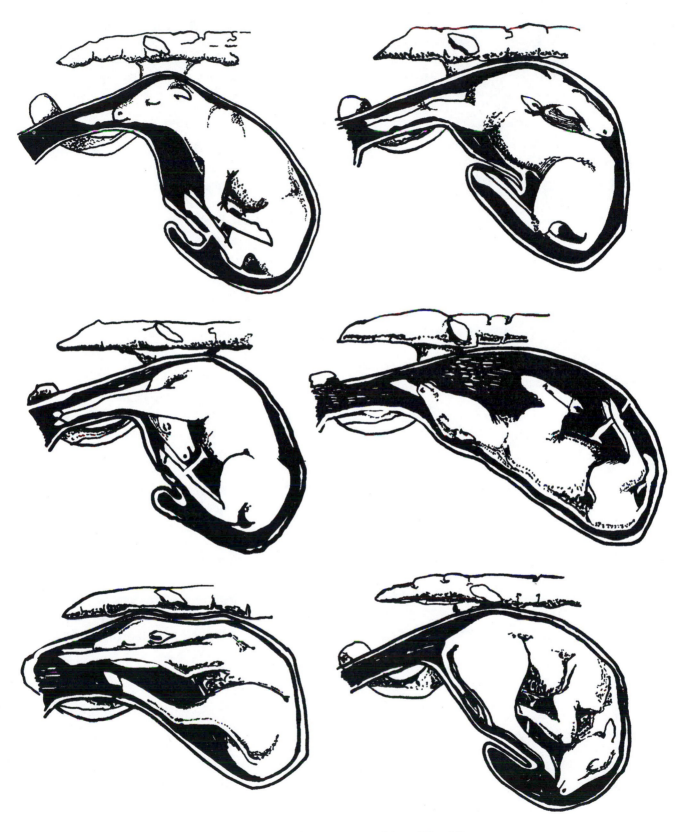

Fig. 9-30.   Abnormal presentations of the calf for delivery.

In most cases, initial assistance during parturition should be limited to regular observations of the female from a distance. Unnecessary excitement brought on by an overzealous attendant often causes reduced labor activity by the parturient female and results in a stillbirth. When a cow has been in labor four hours without progress, an examination is warranted. In like manner, a ewe or mare that has not delivered in two hours of labor will likely need assistance. The following procedures are recommended for providing assistance during parturition. The attendant should:

1. Get professional assistance if at all possible.

2. Use protective gloves. If gloves are not available, the hands should be washed, fingernails clipped and hands and arms rinsed with a mild disinfectant.

3. Clean the area around the vulva with soap and water.

4. Lubricate the arm with a water-soluble lubricant or medicated soap.

5. Gradually insert a hand through the vulva and apply pressure as the female relaxes until the fetus can be handled.

6. Determine the cause of difficulty in delivery and correct the position. This may require moving the fetus back into the uterus to position its head and forelegs properly.

7. Apply surgical chain to forelegs if necessary.

8. Pull on the fetus as the female strains. The direction of pressure should be out and downward toward the female's hocks as the fetus passes through the birth canal. Mechanical devices to assist in delivery may be required.

9. Provide attention to the newborn animal. If the newborn does not start breathing, the attendant should massage the chest, pump a foreleg, blow in the mouth, or swing it back and forth by the hindlegs.

10. Insert two chlorhexidine uterine capsules and give 100 units of oxytocin to cattle and one bolus and 40 units of oxytocin to ewes to combat uterine infection and facilitate involution of the uterus.

11. The attendant should make sure the newborn animal receives the proper amount of colostrum. Whenever dystocia occurs, the newborn animal is stressed and absorbs less of the immunity from the colostrum. Guidelines recommend providing two to four quarts of colostrum to calves in the first twelve hours of life.

FEEDING THE ORPHANED ANIMALS

The feeding of newborn orphaned domesticated animals is frequently attempted, but not entirely with success, because of the many and varied factors that are involved.

It is an established fact that, in the nutrition of young animals, milk derived from a mother of its own species is always superior to that from others. This sets up a serious obstacle in the artificial feeding of all young farm animals, excepting the calf because cow's milk is the basis of almost all artificial feeds. For example, a comparison of the composition of milk of the cow, ewe, sow, and mare is as follows:

	Percent Water	Percent Mineral Matter	Percent Protein	Percent Sugar	Percent Fat
Cow	87.2	0.7	3.52	4.9	3.7
Ewe	80.8	0.9	6.52	4.9	6.9
Sow	81.0	1.0	5.9	5.4	6.7
Mare	90.6	0.4	2.0	5.9	1.1

In addition to these ingredients, milk also contains vitamins essential for the growth of the young.

In artificial feeding, there is danger of overfeeding. The general rule to follow is to have the young animal still hungry when it has taken the allotted amount. There is far more danger in overfeeding than in underfeeding newborn animals. During the first few days of life, the feedings should be comparatively frequent—at least every two hours—and gradually reduced to three or four times per day.

For the very young animal, heating the milk to body temperature, especially during the colder seasons of the year is advisable. The attendant must exercise judgment in the matter, based largely on experience.

One of the gravest errors in the artificial feeding of young animals is a disregard for cleanliness. The feed, the utensils in which it is stored, the caretaker(s), and the handling of the feed must be as scrupulously clean as those in the artificial feeding of the human infant. Otherwise, diarrhea and digestive disturbances are certain to follow.

Young animals must receive colostrum immediately after they are born. Colostrum is lower in water content and sugar, and higher in protein, vitamin A, and fat than milk. Colostrum provides immunity against infections during the highly susceptible period of early life, because it contains a high concentration of globulin that carries protective antibodies. The blood of the newborn calf is very low in blood globulin, but the amount rises quickly after a feeding of colostrum. When skim milk instead of colostrum milk is fed, the globulin content of the blood remains low for some time.

Surplus colostrum from different species of farm animals can be frozen and stored for a period of one year or longer without losing its antibody value. Then, it may be thawed, warmed to about 100°F, and fed as needed.

The evidence is clear that young animals that do not receive colostrum, or a substitute for it, frequently die within a few days after they are born. A reliable substitute for colostrum is the oral or subcutaneous administration of from 1 to 8 ounces of blood serum. The dose depends on the size of the animal. The serum should preferably come from the orphan's mother. If this is not possible, the source of serum should come from a member of the same species.

When colostrum is not available for foals during the first 12 hours of life, 1,000 cc of horse serum incorporated in a mixture of dried whole cow's milk, dried skim milk, sugar, and water should be fed at frequent intervals.

Cow colostrum fed to lambs is a fairly good substitute for ewe's colostrum; however, lambs fed during the first 24 hours of life on sheep's serum, having incorporated in it the whole-milk mixture, will usually do better than those fed cow's colostrum substituted for ewe's colostrum.

Feeding Orphaned Foals

An adequate intake of colostrum is critical to the health of the newborn foal. The foal should receive 250 ml of colostrum every hour for the first six hours of life. Cow colostrum will suffice in cases where mare colostrum is not available. Orphan foals require care and attention best provided by a foster mare. Foals will attempt to suckle a strange mare, and the usual deterrent to fostering a foal is restraining the mare for several hours until she accepts the foal. Foster mares may be a mare who lost her foal, a mare who has a foal old enough to wean (in order to take the orphan foal), or occasionally, a mare who is producing enough milk to accept a second foal.

When hand feeding a foal, several critical differences between foals and calves must be considered. Suckling foals consume about 25 percent of their body weight daily as milk. This compares to the 10 percent of body weight that is usually fed to calves. The feeding frequency is important, as foals nurse their dams from four to seven times per hour. The directions on commercial foal milk replacers may indicate feeding four times daily, but less digestive upsets will occur if a more natural feeding schedule with smaller amounts is followed. A suggested schedule for feeding a foal milk replacer is as follows:

Days of age	1	3	7	10	12	15 on
No. of feedings	16	12	10	8	7	5
Interval between feeding (hrs.)	1.5	2	2	3	3	4-6
Volume of each feeding (L)	0.25	0.60	1.0	1.5	2.0	2.5

Feeding Orphaned Calves

Dairy calves are removed within the first twelve hours of life from their dam and fed milk in a variety of ways. Orphaned beef calves can sometimes be fostered onto another cow, or raised using milk replacer similar to dairy calves. The calf must consume colostrum within the first twenty-four hours of life to supply necessary antibodies. A colostrometer is an instrument used to determine the amount of protein and antibodies in colostrum, and only good-quality colostrum should be frozen and kept for later use in feeding calves. Commercial colostrum replacers are available from veterinarians and animal health suppliers to replace or supplement poor-quality colostrum.

Several companies manufacture milk replacers in the United States, and there can be a large variation in the quality of the replacer and hence the performance of the calves. Guidelines for the nutrient content of milk replacers include the following recommendations:

Crude protein:	15% to 20%
Crude fat:	10%
Crude fiber:	<0.5%
Ash:	<10%

Many recommendations are to feed whole milk or liquid milk replacer at 10 percent of body weight per day, i.e., a 90 lb. calf should receive 4L (4.2 qt.) per day. These guidelines may not be adequate to supply all of the caloric and protein needs of the growing calf. The calf manager needs to be attuned to the growth and health of the calves in determining the performance of milk replacers.

Energy is the most important nutrient in raising young calves. The high level of fat found in milk replacers supplies many of the calories for the growing calf. Average daily gains of 1.5 lbs. are an attainable goal in successful calf rearing programs. The caloric needs of the calf vary depending on the environmental temperature. Cold temperatures and wet surroundings dictate a higher caloric requirement for the calf. The caretaker must carefully follow the mixing directions for the calf replacer using the proper water temperature. If cold water is used, the fat may not disperse throughout the milk replacer and will gum up on the sides of the bucket or calf bottle. Calves can starve or be victims of disease because they cannot consume the energy presented to them due to improper mixing of the replacer. Milk replacers containing milk products for the protein source are preferred over plant protein products. This is due to the decreased digestibility of plant proteins in newborn calves. The protein in an excellent quality milk replacer is about 94 percent digestible, whereas, protein in a poor quality replacer may be as low as 60 percent. Poor weight gains and sickness may be observed in calves fed poor-quality replacers. Magnesium, iron, and selenium levels are often low in whole milk, and these are added to milk replacers to meet the nutrient needs of the calf. Whole milk is a good source of the B vitamins and these are added to milk replacers, as the young calf does not have a functional rumen to synthesize B vitamins.

Feeding Orphaned Lambs

The simplest method is to provide a foster mother. Usually, there is in the flock a nursing ewe with an abundant milk supply or one with a single lamb that will, with persistence and care, adopt the orphan. Smearing the or-

phan with the afterbirth or with some of the fluids of parturition results in its more ready acceptance by a foster mother. At other times, the foster ewe must be physically restrained during nursing by the orphan, or the lamb could be injured.

Feeding Orphaned Pigs

If the sow dies, or if she farrows more pigs than she can raise, the pigs may be fed by hand, or they may be distributed to other sows having litters of about the same age. The latter method is called "cross fostering" and is one of the most commonly used management practices to increase the survival rate of pigs. Due to the nursing behavior of pigs (one pig claims and nurses from one teat the entire lactation), cross fostering needs to be done in the first 48 hours of life to be successful. In herds where group farrowing is practiced, many sows will have pigs within a short period and will be candidates for creating even litter sizes through cross fostering. Most sows will accept piglets from other sows with no problems. The piglets must be allowed adequate nursing time of the mother or another recently farrowed sow to obtain colostrum. Commercial milk replacers are available to feed orphaned pigs or to supplement the sow's milk from large litters. Pigs can be hand-fed, but this is very labor intensive due to the frequent feeding patterns of pigs. Milk feeders that hold several gallons of milk replacer are good for orphaned pigs. The self-feeders allow the pigs to eat when they are hungry. The feeders must be cleaned each day to prevent scours in the pigs from bacterial overgrowth in the feeder. Another method is to supplement milk when a large litter of small weak pigs are placed onto a nurse sow.

The process of weaning pigs very early to decrease the incidence of respiratory and digestive diseases through the nursery, grower, and finisher phases has become an accepted practice. Several feed companies produce pelleted feed for pigs weaned as young as five days. These feeds contain highly digestible ingredients, such as whey, fish meal, blood meal, and plasma, and are very palatable to the young pigs. Medicated early weaning (MEW) and segregated early weaning (SEW) programs utilize these ingredients in the diets of very young pigs that are removed from the sow, and will also suffice to feed pigs orphaned due to sow death or from very large litters.

COMMON FEEDLOT HEALTH TIPS

Cattle feeding has become a high finance business enterprise, due to high feed costs and an unusually strong demand for replacement cattle. As a result, there are many caretakers responsible for the well-being of two to four million dollars' worth of cattle. It is a well-known fact that success and efficiency in any livestock enterprise are dependent upon the people working closely with the animals. Therefore, it is good management to inspect carefully all cattle daily, provide treatment under the supervision of a veterinarian, and practice good husbandry during the convalescent period.

Receiving New Arrivals

The first day in the feedyard is usually the most important for determining the ultimate health status of a load of calves. Only one or two persons should be responsible for inspecting incoming cattle while they are being unloaded (Fig. 9-31). As the cattle are unloaded, the receiving inspector should decide the following:

1. Whether any cattle are so sick that they need to be cut out and treated immediately.

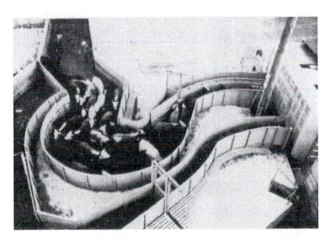

Fig. 9-31.　Excellent facilities for loading or unloading livestock. (Courtesy, Wm. Farr, Greeley, Colo.)

2. If there are enough slightly sick calves to warrant oral medication to the whole group. Oral medication should be started immediately if more than 15 percent of the calves look sick off the truck. Neo-terra crumbles tend to work best on cattle weighting less than 500 pounds, while sulfamethazine usually does a better job on heavier cattle. (*Note:* Neo-terra crumbles are topdressed on the feed; whereas, sulfamethazine is usually added to the water.)

3. Whether or not the whole load is healthy enough to process within 24 hours. Tests at the University of California's Meloland Experiment Station have shown that cattle tend to gain more and to have better feed conversions if they are processed within 24 hours of arrival. However, if cattle are weak, and/or there are many sick ones, the stress of processing will be so severe that sickness and death loss will be increased. Consequently, processing should be delayed for two to three weeks if there are many marginal calves in a load.

Feeding

Grass hay and a starting ration should be fed. Long hay should be fed to all incoming cattle to maintain their energy balance. The length of time that hay is fed with the ration should be governed by the condition of the new arrivals. If they are very healthy, the hay may be stopped within four days. However, if there are many calves with shipping fever or diarrhea, the hay should be continued until cattle are well on the road to recovery.

Watering

Water facilities should be cleaned out after each load, and they should have fresh water in them. Withholding water from the calves for one to two hours will encourage them to eat instead of tanking up on water; however, if cattle are severely dehydrated, water should not be withheld.

IDENTIFYING SICK ANIMALS

The ability of an animal caretaker to identify sick animals before they become seriously ill is one of the most important success factors in animal production. This is especially true in confinement operations.

Every livestock and dairy producer should develop a mental image of a healthy animal and systematically apply that standard to animals in order to recognize early signs of ill health. The following tips will assist the caretaker in recognizing sickness in cattle during the early stages of illness.

Hair Coat

A coat with no bloom or luster is a "dead" hair coat. If the caretaker is undecided, he/she should take a careful look around the head, eye and muzzle area. In comparison with other parts of the body, a calf's hair will have less luster in these areas. A dull, rough hair coat is a sign of nutritional stress. It takes experience and experimentation to differentiate hair coats.

Eyes

Sunken eyes are a symptom of dehydration in cattle and can be related to diarrhea, inability of the newly received animal to locate or drink water, or fever. Respiratory diseases will often cause runny or crusty discharge from the eyes.

Fill-drawn, Hollow-flanked Calves

Fill-drawn, hollow-flanked calves are not feeding or drinking. Calves that have "gone off" feed have an abnormal buoyancy to the stomach. This can be seen from behind as the calf walks away. The buoyancy is probably due to a loss of rumen contents, in general, and, more specifically, to a higher ratio of liquid to roughage. The disposition of the calf, as well as the type, is important in making a decision about fill.

Gait

The attendant should be aware of cattle that have a short, choppy stride, with their noses extended. The attendant should also watch the loosely gaited cattle which have an appearance of falling apart.

Lying Position

Calves lying with nose extended outward should be watched. Calves with pneumonia try to make themselves more comfortable in this manner.

Nose

A dry, crusted nose is a symptom of dehydration. Generally, there are other symptoms to evaluate. A mucous discharge may warrant a closer look, but by itself it is not a good indication. Panting by the calf, combined with a nasal discharge, is evidence of respiratory disease.

Joints

The attendant should be aware of calves with swollen joints and stiffness of travel.

Dull and Listless Calves

Cattle that lag at feeding are often breaking with disease or have some other symptoms, such as a fever or digestive upset. The posture of the calf, whether hunch-backed, ears droopy, or head down position, is also another indicator of ill health. Another indicator of a calf needing attention is the position of the animal in the pen. Cattle that are off by themselves are often sick and unwilling to eat the feedlot ration.

These signs should be used with other obvious signs to aid in evaluating the condition of cattle.

Grinding of Teeth

Grinding of teeth is often an indication of discomfort associated with stress. It is also characteristic of calves with acute pain, particularly gut problems and injury.

Feces on Hindquarters

Calves scouring seriously will lose their hair coat luster. This is an indication for treatment. In severe cases, there is a loss of fill, a dry nose, and sunken eyes. Dark-colored or black manure is an indication of bloody scours. This is usually coccidiosis.

Calves with no symptoms, other than a poor hair coat, often run temperatures of 105°F or higher. All other signs of ill health must be considered in combination so that an accurate evaluation of an animal's condition can be made. An exception to this might be new arrivals that refuse feed and water. However, another pen would be more beneficial than taking them to the hospital area.

Taking a calf to the hospital is expensive. Therefore, looking for indications of good health at times of indecision regarding a questionable calf is important. Licking is one of the most important indications: A sick calf does not lick itself. The healthy calf will eat and chew a cud. It will not grind its teeth and its fecal matter will be normal, according to the type of feed and time in the feedyard.

Suggestions for Riding a Pen

The rider should observe cattle before they enter the pen, because he/she can often spot signs of disturbances outside the pen. (For example, distressed cattle are often isolated to themselves.) The rider should (1) glance down regularly; (2) notice the general appearance of the manure; (3) check for blood in the manure and urine; (4) take the obviously sick cattle out of the pen; and (5) group the balance in a corner away from the gate. From the side, the rider should also observe cattle as they pass by, and then ride behind doubtful individuals, as this is the best and most deciding view. Some riders also ride around to the front of an animal, which is the poorest view.

The rider should not handle cattle roughly. If an animal breaks away, it should not be chased. The rider should go back and resume checking the remainder of

Fig. 9-32.   Sick cattle should be ear tagged or identified by some means so that accurate treatment records can be kept.

the group, for the stray will return on its own. Stubborn animals should be removed with a small bunch of cattle. Roping is usually unnecessary and adds stress to the animal's problems. From the side, the rider should check (1) the way the animal travels, (2) its fill, (3) its behavior—dullness and listlessness, (4) its hair coat, (5) its eyes, and (6) its joints. From the rear, the rider should check (1) the buoyancy of the belly and fill, (2) its way of travel, and (3) feces on its hindquarters. From the front, the rider should check (1) its nose, (2) its hair coat, (3) its eyes, and (4) its joints.

When this method of visual appraisal is coordinated with the use of a thermometer, the caretaker will improve his/her ability to identify cattle that are beginning to get sick. Cattle treated in the early stages of sickness will usually respond to treatment (Fig. 9-32).

Sick cattle should have access to an area where there is shade, feed, and water. Otherwise, they will get sicker standing in the sun waiting to be treated. The sickest cattle should be cut out first and worked separately.

TREATING SICK CATTLE

The parameters for successful treatment of bovine respiratory disease depends upon the following factors:

Fig. 9-33. The handler should provide complete data on the condition and medication given for each animal treated.

4. Pull early—the ability to clear an infection and to minimize permanent lung damage depends on early treatment of sick cattle.

5. Treat with an effective antibiotic.

The time that cattle spend in the squeeze should be minimized to reduce stress. This can be done by a well-organized treatment crew who move cattle into the squeeze quietly and quickly. They secure the animal's head, take the animal's temperature, administer the prescribed medication promptly, and record the necessary information on the health card (Figs. 9-33 and 9-34). Medicine should be readily available, and, in addition, several syringes should be filled with different antibiotics prior to the cattle's entering into the chute. Disposable syringes work well for this.

1. Laboratory data—bacterial culture and antibiotic sensitivity data so the proper antibiotic is used for the organisms present in the feedlot.

2. Cost effectiveness of treatment—the benefits need to be greater than the treatment cost.

3. Low repull rate—the first line antibiotic is effective so a low percent of animals are retreated with another antibiotic.

Common medication routes of administration and dosages of drugs that are frequently used in feedlots are:

1. Gallimycin, 2 to 4 cc per 100 pounds body weight—intramuscular.

2. Neomycin sulfate, 5 cc per 100 pounds of body weight—intramuscular or intravenous.

3. Oxytetracycline, 100 mg/cc, 5 cc per 100 pounds of body weight—intramuscular, intravenous,

LOT NO. ____		DIAGNOSIS _____									TAG NO. ____
Date	Temp.	Degree of Illness	Terra	Sulfa Sol	Sulfa Bolus	Tylan	Pen	Fura-cin	Meth-agon	Gall.	Other Treatment

Fig. 9-34. An individual health card.

or subcutaneous. Oral administration, 1 to 2 grams per head.

4. Tylan, 5 cc per 100 pounds of body weight—intramuscular or intravenous.

5. Naxcel, 1.5 cc per 100 pounds of body weight—intramuscular.

6. LA 200, 9 mg per pound of body weight—intramuscular.

7. Micotil, Nuflor, and Baytril are effective respiratory drugs that require a prescription and a veterinary-client-patient relationship. These drugs have the added advantages of single dose and subcutaneous administration to enhance meat quality.

In most cases, a much better response will be obtained if tylan or oxytetracycline is administered intravenously. (See Table 9-1 for a list of medications and their effects on various infectious organisms.)

Note well: When choosing and administering medications, always (1) read and follow the label, or (2) follow the advice of your veterinarian.

Tissue Response to Administration Routes

Intramuscular and subcutaneous administration of tetracyclines produces irritation and tissue damage. Oxytetracycline reconstituted with water or adminis-

tered in a propyleneglycol-water solvent produces the least tissue damage. Intravenous injection reduces the pain and irritation. Streptomycin commonly produces pain when injected intramuscularly. Hot and tender masses frequently develop in the areas in which the antibiotic is injected.

If any of these drugs is used for more than three consecutive days, it is likely to stress the animal, just from its toxic effects on the animal's system. Antibiotics are selective poisons that kill the infectious agents before they kill their host. Antibiotics in excessively large doses will kill any animal.

Combining Drugs in Treatment

When oxytetracycline, penicillin, and streptomycin are used in combination, the stress will not be much greater than the sum of the stresses due to each individual drug. There are other factors, however, that enter into the effectiveness of combinations of drugs when they are compared to the effectiveness of an individual drug.

There are many veterinarians who insist that only one drug should be used at a time and that certain combinations such as penicillin and tetracycline should never be used. This is correct if only one bacteria is being dealt with. The basis for their theory is

TABLE 9-1. Efficacy of Selected Anti-bacterial Materials in Certain Infectious Organisms

Infectious Organisms	Anti-bacterial Material						
	Penicillin	Strepto-mycin	Tylosin	Neomycin	Sulfa	Oxytetra-cycline	Erythro-mycin
Streptococci	x	—	x	x̲	—	x̲	x
Staphlococci	x	—	x	x	—	x	x
Salmonella	—	x	—	x	x	x	—
Escherichia coli	—	x	—	x	x	x	—
Pasteurella	x̲	x̲	x	x	x	x	x̲
Clostridium	x	—	x	x	—	x̲	x
PPLO	—	x̲	x	—	—	—	x̲
Hemophilus	x̲					x	

Inhibition: x equals good; — equals none; x̲ equals variable.

an experiment done by Lepper and Dowling in 1951. These researchers compared the effectiveness of different drugs on pneumococcal meningitis. They found that the fatality rate in patients receiving only penicillin was 30 percent, and of those given penicillin plus tetracycline, 79 percent. Mixing of a bactericidal drug, streptomycin, for example, with a bacteriostatic drug, such as tetracycline, tends to produce decreased drug activity. There are notable exceptions, however. For example, the aforementioned combination works better on brucellosis than either drug alone. In 1958, Louis Weinstein concluded that simultaneous administration of more than one anti-microbial agent was justified for four purposes: (1) treatment of mixed bacterial infections, (2) delay in the rate of emergence of bacterial resistance, (3) enhancement of therapeutic activity, and (4) therapy of severe infectious processes in which the specific cause has not been determined.

Length of Treatment

Cattle should be treated according to the label recommendations on the medication. If the animal is not responding within two days, another drug should be used. It is possible that the initial diagnosis was incorrect. An animal treated for respiratory disease may in fact have a liver abscess or another condition that requires the animal be pulled from the pen and taken to the hospital pen. Cattle that are responding appropriately usually need three days of therapy. Anti-inflammatory drugs, such as flunixin meglamine, greatly help in the therapeutic response of the animal. This drug decreases the fever and blocks endotoxin release allowing the animal to feel better and resume eating.

Care of Sick Cattle

All sick cattle should have long hay free choice. Extra bunks should be put into the sick pens and filled with hay. This will encourage the more timid cattle to eat. In addition, some milled feed should be provided. Clean, fresh water should be accessible to all cattle. The waterers should be cleaned at least three times per week.

Sick cattle can be treated more easily and faster if three sick pens are provided. Cattle that are treated once can be put into one pen, those which have been treated twice can be put into another pen, and those cattle that aren't responding can be put into the third pen. In addition, having a recovery pen for cattle to remain in for one week after being hospitalized is helpful. The recovery pen will enable cattle to get strong enough to compete with their penmates when they return to their home pen.

PART 2

Diseases Associated with Systems of the Animal Body

CHAPTER 10

The Digestive System

The digestive system consists of the organs directly involved with the ingestion and digestion of food and the egestion of unabsorbed residue (Figs. 10-1 and 10-2). These organs are divided into two groups: the alimentary tract and the accessory organs and glands.

The alimentary tract includes the mouth, pharynx, esophagus, stomach, small intestine, large intestine, and anus. The accessory organs and glands are either connected with the tract or located within its walls. They are the salivary glands, liver, and pancreas.

STRUCTURE AND FUNCTION

The arrangement of the alimentary tract within the body may be visualized as a tube within a tube.

The outer tube is made up of the skin and muscle layers that form the body wall. The inner tube is the alimentary tract. The space between the two tubes is called the body cavity. The body cavity is lined with a thin serous membrane called the *peritoneum*. The part of the peritoneum that adheres to the body wall is called the *parietal peritoneum*; the part located nearest the digestive organs is called the *visceral peritoneum*.

Mouth

The mouth is a cavity forming the first part of the alimentary tract. It is bounded by the lips in front, the cheeks at the side, the tongue below, and the hard palate above. Within it lies the teeth, the openings of the salivary glands, and the greater part of the tongue.

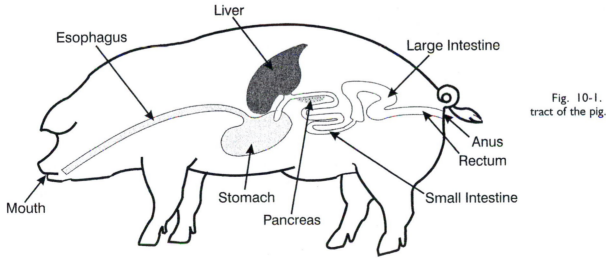

MONOGASTRIC (Porcine)

Fig. 10-1. Digestive tract of the pig.

RUMINANT (Bovine)

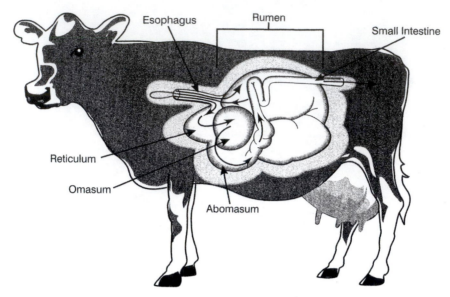

Fig. 10-2.   Digestive tract of the cow.

Major functions of the mouth and its related structures involve prehension and mastication of food, salivation to lubricate dry food for ease of passage into the stomach, and mechanisms of defense and offense.

Lips

The lips of the horse are soft and flexible so that feed may be grasped and directed between the teeth for mastication. Sheep, like horses, have very agile lips. They, unlike other domesticated animals, have a cleft upper lip, each segment being independently motile, though not prehensile. Sheep use the tongue and incisor teeth to convey feed into the mouth.

The lips of hogs and cattle are quite stiff and immobile. Hogs use their snout to root up the ground and move feed into position so that it may be scooped up with the pointed lower lip which acts much like a shovel. When hogs are not rooting, the tongue, incisor teeth, and an upward movement of the head direct feed into the mouth. Cattle use the tongue to bring feed into contact with the lower incisors and dental pad.

All domesticated animals, with the exception of the dog and cat, suck liquids with the mouth through the combined action of the lips and tongue. Dogs and cats lap liquids, using the free end of the tongue as a ladle.

Teeth

The functions of teeth are two-fold. The incisors, located in the forepart of the mouth, are used for cutting or shearing. The premolars and molars found in the cheek areas are used for grinding feed.

TEETH OF THE HORSE

The horse has 6 upper and 6 lower incisor teeth, and 12 upper and 12 lower molar or grinding teeth. All the incisor teeth are said to be deciduous, which means that the permanent ones are preceded by temporary or milk teeth.

The male horse has four additional teeth appearing one on each side of both upper and lower jaws about midway between the corner incisors and the first molars. These are the canine teeth. If canine teeth appear in the female, they are comparatively small and rudimentary.

Both the temporary and permanent incisors have regular periods of eruption and wear, so that the horse's age up to eight years may be accurately determined, and fairly well estimated thereafter by an expert (Fig. 10-3). The incisors are designated as the centrals or pincers, laterals, and corners, there being two of each in both upper and lower jaws. The milk teeth appear in the following order, and they are replaced by permanent teeth as indicated:

	Eruption	*Replacement*
Centrals or pincers	Present at birth or a few days thereafter	2½ years of age
Laterals	4 to 6 weeks of age	3½ years of age
Corners	6 to 9 months	4½ years of age

Therefore, at 4½ or 5 years of age, the horse has all its permanent incisors in full wear.

When the table surfaces of the incisors, especially the lower ones, are carefully examined, there will be found on them almost in their center, a small, dark-colored depression with a border or rim of hard, glistening white enamel. This is the dental cup, and it disappears as the teeth wear down. It must not be confused with the dental star also found on the table surface in front of the dental cup, though appearing at a somewhat later period. The dental star is permanent and of no significance in determining the animal's age.

The gradual wearing and disappearance of the cups according to a rather definite pattern in period of time enables the experienced equestrian to judge the age of the animal with a fair degree of accuracy up to 12 years. After the horse is 12 years old, accuracy in estimating age disappears, and from then on it is largely conjecture, depending upon a change in the direction at which the upper and lower incisors meet each other. In the young horse, the table surface is that of an elongated oval, with the greatest distance from

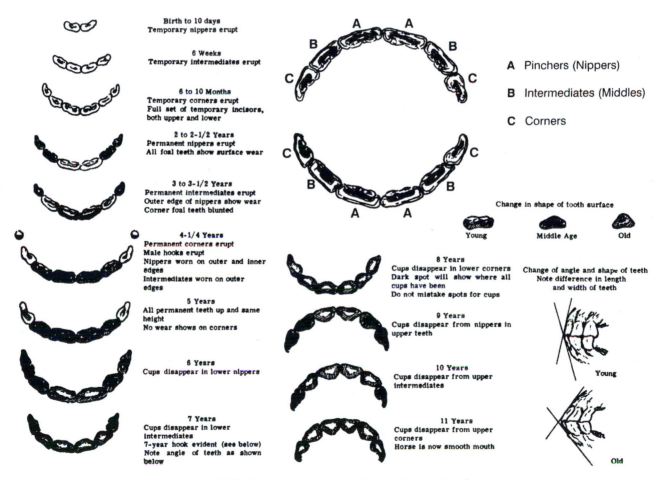

Fig. 10-3. Determining the age of horses by their teeth.

side to side. Gradually, this shape changes so as to become more nearly circular, then triangular, and finally, when the horse is about 20 years of age, the outline is again oval.

The horse has 24 molars or grinding teeth. There are six in each arcade. The three front molars are deciduous, and the three in the rear of each arcade are permanents. The three milk teeth or deciduous teeth in each arcade are gradually pushed out of their sockets during their replacement by the permanents. In other words, the temporary molars are superimposed on the permanents. The milk teeth when shed are frequently found in the feed box as "caps." The molar teeth are of negligible importance in estimating the age of the animal, though replacement and eruption take place at a definite age.

TEETH OF RUMINANTS

Cattle, sheep, and goats differ in several respects in their dentition from that of the horse. These animals have only eight incisors and all are placed in the lower jaw. The upper jaw has no incisors, though there is a very firm dental pad. Cattle differ from sheep and goats in that the incisors are loosely or movably implanted in their sockets. This must not be mistaken for a diseased condition. There are no important practical differences in the molar teeth of these animals from those in the horse.

Determining the age of cattle. The age of cattle may be determined by the eruption of permanent incisors. The first pair of permanent incisors frequently erupt at 18 to 21 months of age; the next pair at 27 to 30 months of age; the third pair at about 36 months, with the fourth pair appearing shortly thereafter. At three years of age, or possibly a little older, cattle have a full mouth of incisors. This seems to hold true, generally, for well-nourished beef and dairy breeds, though the incisor teeth of Brown-Swiss cattle appear at a somewhat older age.

Determining the age of sheep. The dentition of sheep is the same as that of cattle. Mature sheep have 32 teeth: 8 incisors and 24 molars. The central pair of incisors are termed pinchers, the adjoining teeth are

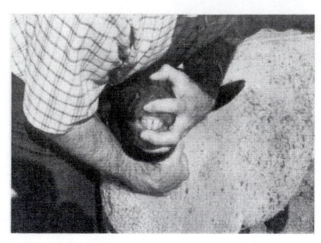

Fig. 10-4. A yearling mouth. Note the correct procedure used to part the lips and view the teeth.

called first intermediates, the next pair, second intermediates, and the outer teeth, corners (Fig. 10-4).

The age of sheep can be estimated fairly accurately until they are four years old. The lamb has small, narrow teeth commonly called milk teeth. When the sheep are 12 to 14 months of age, the pinchers are replaced by two permanent teeth which are much larger and wider than the remaining milk teeth. Each succeeding year, an additional pair of permanent incisors replace the milk teeth until the sheep has eight permanent teeth.

TEETH OF SWINE

The pig has 44 permanent teeth. There are six incisors, two canine, eight premolars, and six molars on the top and bottom jaws. In the temporary dentition, there are six incisors, two canines, and eight premolars on top and an equal number below, making a total of 32 teeth. The permanent incisors and canines replace the corresponding deciduous teeth. The deciduous premolars are replaced by permanent premolars, but no teeth precede the permanent molars.

The deciduous teeth tend to resemble the permanent teeth that replace them, thereby making it quite difficult to determine the age of hogs by examining the teeth.

The incisors located nearest the canines and the canines themselves are present in the pig's mouth at birth. The deciduous premolars and central incisors erupt during the first month. The intermediate deciduous incisors appear between two and three months. The first premolars and first permanent molars appear at about five months. The permanent corner incisors and the canines appear at about nine months.

The canine teeth, or tusks, of male hogs become quite large and project out of the side of the mouth. The upper and lower canines rub together in the process of mastication. The friction created maintains a sharp edge on the lower tusks. The extreme power in the head and neck of a boar and the protruding tusks make it extremely dangerous to work with boars unless the tusks are removed. Procedures for removing the tusks of boars are presented in Chapter 9.

Salivary Glands

The salivary glands consist of three pairs of well-defined glands. The parotid salivary gland is located back of the jaw beneath the ear, the mandibular gland in the space between the branches of the lower jaw, and the sublingual gland beneath the tongue. They discharge saliva into the mouth through excretory ducts, with openings either in the cheek opposite the third molar tooth or under the tongue.

The amount of saliva secreted varies not only in different species but also in individuals, depending on their size and on the nature of the food consumed. It has been estimated that healthy horses on dry feed may secrete as much as 10 gallons of saliva in 24 hours, while cattle may secrete 14 gallons under comparable conditions.

In general, the chief functions of saliva are (1) to lubricate the preliminary digestive tract, (2) to ease the passage of food, (3) to soften food and mechanically assist in its reduction to a less solid and more soluble form, and (4) to convert the starches of the food into sugar.

Palates

The hard and soft palates form the roof of the mouth, and the latter marks the anatomical dividing line between the mouth and the pharynx. The soft palate is in the nature of a flexible curtain that varies in length in the different domesticated animals. In horses, for example, it is so long that when food and water have been swallowed and are then regurgitated, the material exits through the nose and not by way of the mouth. In cattle, the soft palate permits food from the paunch to be regurgitated into the mouth for final mastication. This physiological process is commonly called "chewing the cud."

Tongue and Cheeks

The tongue and cheeks assist in various ways in the masticatory process. Both these organs are largely muscular, and therefore, an important function is to keep the food between the teeth and to aid in its backward movement.

The tongue consists of a fixed portion and a free portion. Its outer covering is a mucous membrane. On the upper surface of the tongue, the mucous membrane is covered with numerous projections or papillae. Certain groups of these have specialized functions to retain sapid foods on the tongue and to act both as tactile and gustatory organs.

The tongue of cattle is distinguished by its great muscular development and extreme motility, so that its tip may readily be made to enter the nostrils. The papillae on the mucous membrane are directed backward and have a thorny sheath so as to impart a very rough reaction when grasped.

The inner lining of the cheeks is also a mucous membrane. The mucous membrane in cattle and sheep contains some very prominent, backward-directed papillae, extending from the corners of the mouth to the first molar tooth, and continuing backward by a single row of papillae corresponding in direction to the border of the teeth. Black pigmented patches are common on the mucous membrane of sheep and on some breeds of cattle. These should not be mistaken for evidences of disease.

The tongue and cheeks are frequently the seat of conditions interfering with mastication. Awns of grasses become embedded here and may even fester. Sharp teeth in horses lacerate the tongue and cheeks. In cattle the tongue may be invaded by a germ, the so-called "ray fungus," which in the region of the jaws causes lumpy jaw. In the tongue, a somewhat similar organism causes wooden tongue.

Pharynx

The pharynx is the anatomical structure following the mouth which is separated from it by the soft palate. The degree of separation depends upon the development of the soft palate, from an extreme length in the horse to no soft palate at all in poultry. The pharynx is the crossroad in the animal's body. There are several distinct openings into the pharynx: (1) the opening from the mouth, (2) the opening from the nasal cavity, (3) the opening from the Eustachian tube, (4) the opening from the larynx, and (5) the opening from the esophagus. This clearly indicates that both food and air are almost continuously meeting each other and passing each other in the pharynx.

Nature has provided many safeguards to prevent misdirection of food and air as they pass through the pharynx. For example, when the opening into the esophagus is open for the reception of food, the opening into the larynx automatically closes. If perchance a particle of food succeeds in passing the laryngeal safeguards, nature has made the interior of the larynx so very highly sensitive that violent expulsive efforts connected with coughing at once expel the invading substance. The same sort of protection is not present at the opening of the esophagus; therefore, an animal can inhale a good deal of air into its stomach. Whenever the pharynx is damaged by mechanical means or by inflammation due to infection, great difficulty in breathing and swallowing usually occurs.

Esophagus

The esophagus is the tubular passageway extending from the mouth to the stomach. Measured outwardly in the horse, it extends from the region of the throat to a point corresponding to the upper third of the 16th rib. At its origin, the opening of the esophagus is just above the opening in the larynx. It continues backward in this position for a short distance and then deviates to the left side of the trachea. It continues in the latter position until it is well within the chest, and there it again assumes a position above the trachea to enter the stomach after passing through the diaphragm. The esophagus is lined by the usual mucous membrane, and, to the outside of this, for all practical purposes, there is a muscular coat.

In the horse, the upper half of the esophagus has voluntary muscle fibers, with the lower half of the esophagus consisting of involuntary muscle fibers. In cattle, sheep, swine, and other domesticated animals, the entire muscular coat of the esophagus is of a voluntary character. This means, then, that in the horse, once food has progressed beyond the first half of the esophagus, it is beyond the animal's control, though in other animals it may almost be forced back into the mouth at will.

Though the esophagus is a tubular passageway, it differs considerably anatomically in regard to its caliber. In the horse, for example, and in a measure because of its relationship to other organs, the esophagus is constricted at its origin, again at the point where it enters the chest, and finally just before its passage through the diaphragm. In cattle and swine the esophagus is dilated at its origin, that is, somewhat funnel-shaped, and again at its entrance into the stomach.

These anatomical peculiarities in the caliber of the esophagus are in a measure responsible at times for difficulties in swallowing. The horse, for example, is occasionally greedy and attempts to swallow too rapidly, so that feed is not properly moistened and the lumen of the esophagus is not lubricated. The result is that the food becomes impacted, especially at those points where the lumen is constricted. This gives rise to the so-called pharyngeal, thoracic and diaphragmatic choke. Though this choke or impaction may take place at any point throughout the course of the esophagus, it certainly occurs most frequently at the constrictions. Cattle and swine, having a funnel-shaped opening at each end of the esophagus, sometimes become too ambitious and attempt to swallow very large objects, which causes them to choke.

Stomach

The stomach is the organ into which the esophagus is emptied. So great are the anatomical differences of this organ in species of animal that two general groups are recognized: the ruminants and the non-ruminants. The former grouping includes cattle, sheep, and goats, and the latter most other animals, such as horses, swine, dogs, and cats.

The Non-ruminant Stomach

The stomach of a horse has a capacity of about 3 ½ gallons. The esophagus enters the stomach of the horse through a narrow opening, with the mucous membrane arranged around it in many folds. This arrangement offers a partial explanation for the extremely difficult vomiting in the horse, but other factors are probably involved also. The mucous membrane of the esophagus is continued apparently unchanged in the first half of the stomach, then changes to the true gastric-juice–secreting mucous membrane.

This transition from one type of mucous membrane to another is sudden, the line of demarcation being very abrupt and immediately noticeable when the stomach is opened. To the layperson, it frequently seems that a diseased condition is present, because of the marked difference in the appearance of the mucous membrane in the left and right halves of the stomach.

The darker colored mucous membrane in the right half secretes the digestive or gastric juice. This juice consists of various enzymes and hydrochloric acid and serves to break down by chemical means the complex substances consumed by the animal as feed. So abundant is the hydrochloric acid that it gives the horse a large measure of protection against the highly poisonous prussic acid formed under certain conditions of growth in many plants, especially sorghums, kafir, Sudangrass, johnson grass, and others. Food arranges itself in the stomach in layers, which do not seem to be much disturbed by water ingested at a later period. Therefore, the old precaution not to water a horse after the animal has eaten its solid feed is not well founded.

Swine have stomachs quite comparable to horses, though the esophageal opening in these animals is funnel-shaped, so that they vomit easily and almost at will. There are other important and not fully explainable factors in their vomiting, but certainly an absence of mechanical obstruction to retrograde outflow of ingesta is helpful to the process.

The Ruminant Stomach

The group of animals classed as ruminants includes cattle, sheep, and goats. These animals have four stomach compartments, known as the rumen or paunch, reticulum or honeycomb, omasum or manyplies, and the abomasum or true stomach.

At birth, the rumen, reticulum, and omasum are undeveloped. Formerly regarded in their development as esophageal dilatations, it is now known from research on the embryo that they develop from the embryonic stomach. During this developmental process, the fore stomachs lose their gastric glands and, under the stimulus of coarse feeds, they grow rapidly. When these animals are 10 to 12 weeks of age, the change is distinctly noticeable; by the time they are 1½ years old, the four compartments have reached their permanent comparative or relative sizes.

RUMEN

In cattle, the rumen is a voluminous hollow organ capable of holding 40 to 50 gallons of liquid. It is located largely in the left side of the abdomen and is in direct contact with the left upper flank. It is entered by the funnel-shaped end of the esophagus and is succeeded through a large opening by the reticulum.

The feed ingested by ruminants is mixed with a heavy flow of saliva, which is necessary for chewing and swallowing dry materials. The saliva of ruminants, unlike that of non-ruminants, does not contain enzymes to aid in the digestion of starches.

The flow of saliva is estimated to be about 120 pounds per day in a mature cow. It contains sodium bicarbonate to neutralize fatty acids produced in the rumen, thereby maintaining a suitable pH range for microorganisms residing there.

Rumen organisms. Neither the rumen nor the reticulum secretes enzymes to break down the cellulose or woody portion of the ration. This important function is performed by microorganisms that are acquired by young ruminants shortly after birth from older animals with which they come into contact. These microorganisms include bacteria, protozoa, and certain fungi. They multiply rapidly in the crevices of the reticulum and soon constitute about 10 percent of the dry matter present in the rumen.

The main functions of rumen microorganisms are to utilize cellulose, synthesize protein from nonprotein nitrogen, and synthesize certain vitamins.

Rumination. Ruminants on high roughage rations must ingest enormous quantities of comparatively non-nutritious material in order to obtain enough nourishment to sustain the body. They consume this feed hastily and with incomplete mastication. Then, when the animal is at rest, it has the physiological ability to force food from the rumen back into the mouth for more thorough chewing. It is again swallowed and finds its way into the third and fourth stomach compartments, where it is subjected to more thorough and complete digestion. This regurgitation of food with its more thorough mastication is called rumination or "chewing the cud."

When an animal becomes ill, rumination is discontinued, and the animal is said to have "lost its cud." This must not be interpreted too literally, as is sometimes done. People believing an animal has lost its cud sometimes force a greasy rag or a piece of pork into the animal's mouth in the hopes that it will resume "chewing the cud." Such practices are harmful. When the animal's general health returns, the physiological regurgitation of food will again be resumed, and "chewing the cud" will be re-established.

In a state of health, the rumen also undergoes certain peristaltic or churning motions for the purpose of mechanically aiding in the mixing of the ingesta. These movements are at the rate of two or three each minute. They may be quite readily felt by placing the hand in the left upper flank. The absence of rumen contractions can be detected in many diseases, such as grain overload, milk fever, and endotoxemia from mastitis or metritis.

RETICULUM

The reticulum or second stomach is in direct communication with the rumen by means of a large opening and is continued by the third stomach. The reticulum has a lining of mucous membrane arranged in the form of honeycomb, which accounts for its

common name. The reticulum is closely related to the heart sac in front, being separated from it by the diaphragm. Under certain conditions, the actual space separating the two organs is less than ¼ of an inch, which explains the ease with which swallowed sharp-pointed foreign bodies that have lodged in the reticulum pass to the heart, with a fatal termination.

The main function of the reticulum is its action as a screening device. Regular contractions of the reticulum cause a backward movement of the ingested feedstuffs into the rumen. A flushing action between the rumen and the reticulum allows the small food particles to become separated from the mass in the rumen—the coarse material is held in the rumen while the fine particles pass on to the omasum.

OMASUM

The omasum or third stomach is peculiar as an anatomical organ in the arrangement of its mucous membrane in prominent longitudinal leaves covered with hard elevations the size of grains of sand. This arrangement of the mucous membrane is responsible for the names "manyleaves," and "manyplies" applied to this organ. The omasum has two openings at its extremities, placing its cavity in communication with the reticulum and the abomasum.

The basic function of the omasum is to reduce the water content of the ingested material. The folds and leaves found on the inside of the omasum exert a grinding and squeezing action on the feed materials leaving them 60 to 70 percent drier than when they entered.

ABOMASUM

The abomasum is often referred to as the true stomach because it has all the characteristics of the glandular portion of the stomach in non-ruminants. The abomasum is more acid than the rumen, reticulum, or omasum. A pH of 3.5 to 4.0 is common in the abomasum which is probably due to the high acidity of the digestive juices. The feed material entering the abomasum is quite dry. Digestive juices, which contain enzymes to break down protein, add moisture so that the material passing into the small intestines is highly fluid.

The Intestines

The stomach empties into the intestines. Because of a difference in volume, the intestines are classified as small intestine and large intestine. The small intestine is arbitrarily divided into the duodenum, jejunum, and ileum. The large intestine is, in turn, subdivided into the caecum, large or double colon, and small or floating colon. The rectum, about 12 inches long in the larger animals, follows the floating colon, and it communicates with the exterior through the anus.

All the intestines have the usual mucous membrane as a lining and, to the outside of this, a coat of involuntary muscle fibers which is arranged longitudinally and circularly. In and beneath the mucous membrane of the entire intestinal tract, there are glands alone or in groups and either microscopic or macroscopic in size. The size and function of the intestines vary in different species. Those with functionally well-developed stomachs such as carnivorous animals, the dog and cat, for example, have comparatively small intestines because most of the digestive processes are completed in the stomach, and their food is in a concentrated form. Cattle have more voluminous intestines than the carnivora because their feed is very bulky, and it does not have the same degree of concentration. Horses use about the same kind of feed as cattle, but horses have a simple stomach in which the digestive processes are limited, with the result that the intestines in the horse are much more voluminous than in cattle.

After the extraction of the nutritive material from the ingested food, the residue gathers in the small colon and rectum to be expelled through the anus. The frequency of defecation varies, depending upon the species involved and the nature of the aliment. On dry feed, it is less frequent than on highly succulent material, and, in debilitating non-feverish diseases, it is increased. Foods that are very woody or fibrous have a definite retarding action on the number of defecations. The horse usually defecates from 5 to 10 times daily, cattle from 10 to 20 times, and dairy cattle as many as 24 evacuations in 24 hours.

SIGNS OF DISTURBANCES OF THE DIGESTIVE SYSTEM

Loss of Appetite

Many factors affect an animal's appetite, and refusal to eat or drink may not always be associated with disease. Overfeeding, excitement, abrupt changes in the ration, moldy or contaminated feed, and the desire to mate may cause animals to go off feed. Changes in the animal's appetite, however, should be of concern to the caretaker in order that proper managerial response can be directed to alleviate the cause or detect disease in the early stages of development.

A loss of appetite associated with other signs of disease, such as depression and fever, indicates the animal needs immediate attention.

Inability to Chew or Swallow

Diseases that affect an animal's ability to chew or swallow damage tissues of the gums, cheeks, lips, tongue, or palate. Common problems such as damaged teeth, foxtail with sharp awns, or foreign objects lodged in the mouth may be responsible for this disturbance. Many infectious diseases, however, commonly affect the mouth and surrounding areas. Vesicle-producing diseases, such as vesicular stomatitis, vesicular exanthema, and foot-and-mouth disease, produce ulcerations in the mouth, lips, and gums. Contagious ecthyma and bluetongue of sheep affect the lips, while the latter affects the dental pad and tongue as well. Tetanus produces muscle spasms of the lips and jaws, making prehension, mastication, and swallowing food quite difficult for infected animals. Cattle affected with actinobacillosis, or wooden tongue, also have difficulty of this nature.

Depraved Appetite

Animals occasionally develop a craving for foreign materials such as bones, sticks, dirt, bedding, or.

wood. This may indicate a mineral deficiency such as phosphorus. Horses and hogs often chew on wood even though they are provided balanced rations. This is probably due to boredom or restlessness.

Vomiting

Vomiting in horses, cattle, and sheep is rare. Hogs, on the other hand, vomit quite readily from a variety of causes—from the ingestion of poisonous materials, from damage to the stomach wall, or from one of several infectious diseases.

Excessive Secretion of Saliva

Herbivorous animals secrete large quantities of saliva to moisten feed and contribute to deglutition or swallowing. Excessive salivation may occur due to choke, stomatitis, abscessed teeth, or foreign objects lodged in the forepart of the mouth.

Bloating

Cattle and sheep may bloat from a variety of reasons. An accumulation of gas in the paunch causes a distention on the left side of the animal's body.

Fever

A rise in body temperature above the normal upper limit for an animal is common in most infectious diseases. A hot, dry muzzle is usually a good indication that a febrile condition exists. Animals suspected of having a fever should be checked with a clinical thermometer.

Abnormal Fecal Material

Scouring, constipation, blood in the feces, and painful elimination are signs of disturbances of this part of the digestive system. Although allowances for variation in consistency of fecal material should be made for the species involved, abnormal defecation is often an early indication of digestive disturbance.

DISTURBANCES OF THE DIGESTIVE SYSTEM

Foot-and-Mouth Disease

Foot-and-mouth disease is an acute, highly contagious, febrile disease of cattle, swine, sheep, goats, and other cloven-footed animals (Fig. 10-5). Humans are also susceptible, although there are only a few authentic cases recorded. Horses are resistant to foot-and-mouth disease.

Fig. 10-5. A Holstein cow affected by foot-and-mouth disease. (Courtesy, USDA)

There have been nine outbreaks of foot-and-mouth disease in the United States. The last outbreak of this disease occurred in California in 1929. Quarantines were quickly established, and the infection was contained within the state. Outbreaks also occurred in Mexico in 1946 and in Canada in 1952. In both cases, the United States cooperated with her neighbors to control and eradicate this disease. Although there has not been an outbreak in this country since 1929, foot-and-mouth disease is a continual threat to the U.S. livestock industry. It is widespread in Europe, Asia, Africa, and South America. This creates a perennial threat to the United States and necessitates constant vigilance if this country is to remain free from foot-and-mouth disease.

Cause

Foot-and-mouth disease is caused by a filterable virus. It is the smallest virus known to infect farm animals. There are seven distinct types and many sub-types of the virus.

Animals may be infected by one or more types or subtypes of the virus at the same time. Recovered animals may suffer repeated attacks of the disease because immunity is of short duration, and immunity to one type does not protect an animal against the others.

Transmission

The virus of foot-and-mouth disease is contained in high concentrations in the fluid and coverings of the vesicles it produces. During the febrile stage, the virus is found in the animal's blood, saliva, urine, milk, and muscle tissues. It has a remarkable capacity to remain alive in carcasses, animal by-products, and contaminated materials such as feed, bedding, equipment, and utensils. The most common means of transmission of foot-and-mouth disease is, therefore, by contact with an infected animal, but contaminated materials of any kind that bring the virus in contact with susceptible animals can spread the disease.

Clinical Signs

Vesicles, or blisters, filled with straw-colored fluids are formed in the mucous membranes covering the tongue, lips, cheeks, and palate of infected animals. Tissues between the toes and around the top of the hoof are also affected by the virus. Vesicles often appear on the teats and udders of dairy cows. The vesicles usually rupture within 24 hours, leaving raw, open sores. The pain produced by these lesions inhibits eating and, consequently, results in reduced milk flow of lactating cows, loss of weight, depression, and lameness. Lesions in the mouth cause profuse salivation in cattle, creating droolings of stringy masses of saliva. There is often a characteristic smacking of the lips due to the pain in the mouth (Figs 10-6 and 10-7).

The temperature of affected animals, especially young ones, rises rapidly during the first 48 hours and then usually falls back to near normal.

The virus may localize in various internal organs of the body and create severe consequences. Sudden death may occur from poisonous substances that af-

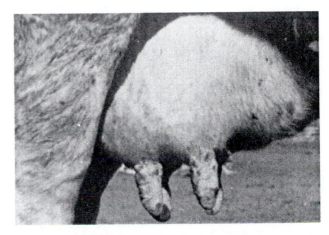

Fig. 10-6. Foot-and-mouth disease causes lesions in the mouths, on the teats, and on the feet of affected animals. (Courtesy, USDA)

Fig. 10-7. Lesions on the feet caused by foot-and-mouth disease are so painful that hogs crawl on their knees. (Courtesy,

fect the heart or lungs. Abortions of pregnant females and mastitis in lactating animals are not uncommon.

There are several vesicular diseases of farm animals that produce clinical signs so similar that diagnosis is impossible without laboratory tests and animal inoculations. Susceptibility by species of animals to some vesicular diseases is as follows:

	Cattle	*Sheep*	*Swine*	*Horses*
Foot-and-mouth disease	++	+	+	—
Vesicular stomatitis	+	—	+	++
Vesicular Exanthema	—	—	++	—

++Highly susceptible; +Susceptible; —Not generally affected.

Prevention

There are no vaccines registered for use in the prevention of foot-and-mouth disease in the United States. Prevention is based upon (1) federal restrictions on the importation of susceptible livestock and potentially contaminated by-products or materials from countries where foot-and-mouth disease exists; (2) immediate quarantine should an outbreak occur; (3) eradication of all infected or exposed animals; (4) thorough cleaning and disinfection of the holding pens, barns, equipment, etc., with a 2 percent lye solution or a 4 percent sodium carbonate solution; and (5) restocking the premises with a few susceptible animals to determine the effectiveness of the program to eradicate the virus of foot-and-mouth disease.

Treatment

There is no treatment for foot-and-mouth disease in the United States. Susceptible animals that have vesicles in the mouth or on the feet must be reported to a local, state, or federal veterinarian.

Vesicular Stomatitis

Vesicular stomatitis is an acute viral disease of horses, cattle, swine, and humans. It is commonly referred to as pseudofoot-and-mouth disease or mouth thrush. The term *vesicular stomatitis* provides a good description of the disease in the early stages. Vesicles or blisters are produced in the "stoma" or mouth of infected animals. Vesicular stomatitis is one of several vesicular diseases that affect farm animals. Others include foot-and-mouth disease and vesicular exanthema.

Cause

Vesicular stomatitis is caused by three types of virus. The general characteristics of each type of vesicular stomatitis virus are quite similar, but immunity produced in a host animal by one type does not protect against another.

Transmission

Vesicular stomatitis is not highly contagious as is its closely related companion, foot-and-mouth disease. The vesicles produced by vesicular stomatitis contain a high concentration of the virus. When they rupture, the virus is liberated in the mouth and emitted with saliva to contaminate feed, water, and equipment. Vesicular stomatitis is spread most often by direct contact. Abrasions of the mouth, feet, and teats on susceptible hosts increase the potential for inoculation when contact is made with the vesicular stomatitis virus.

Clinical Signs

The first signs of vesicular stomatitis in cattle and horses are usually excessive salivation, elevation of temperature, refusal of feed, and general discomfort. Patchy red blotches appear in the mucous membranes of the mouth, tongue, cheeks, and palate. They enlarge, fill with clear or straw-colored fluid, and rupture in a few hours. The ruptured vesicles are extremely painful. Horses may grind their teeth, and cattle may make a smacking sound with their lips in response to the pain. Vesicles may appear on the teats and feet of cattle. Milk production is reduced in lactating cows, and mastitis may occur.

In swine, the first sign of vesicular stomatitis may be lameness. This is followed by vesicles on the snout, lips, and occasionally on the tongue, between the toes, and above the hoof. The lesions are indistinguishable from those of vesicular exanthema and foot-and-mouth disease. Hogs may have a body temperature ranging from 103° to 106°F. The temperature usually drops after the vesicles rupture.

Prevention

There are no vaccines or hyperimmune serums available to United States livestock producers for the prevention of vesicular stomatitis.

Treatment

There is no treatment recommended for vesicular stomatitis. Secondary infections in the ulcerated tissues should be treated with antibiotic ointments. Diseases that produce vesicles around the feet and mouths of farm animals should be reported to a state or local veterinarian.

Vesicular Exanthema

Vesicular exanthema is an acute, highly infectious, febrile disease of swine characterized by the formation of vesicles on parts of the body. The name given to the disease provides a good description of the lesions produced by the infection. Vesicular refers to vesicles or blisters and exanthema means eruption.

Vesicular exanthema was first identified as a disease of swine in 1932. An outbreak of a disease resembling foot-and-mouth disease was reported on a hog farm near Buena Park, California, in that year. Two years later, in 1934, Jacob Traum of the University of California reported that the infective agent causing this disease was different from those causing foot-and-mouth disease and vesicular stomatitis. He named the new disease vesicular exanthema of swine.

Vesicular exanthema was contained in California until 1952. Then, an outbreak of vesicular exanthema occurred that was so rampant that swine in 42 states were infected during a 16-month period. A state of emergency was declared by the Secretary of Agriculture in August of 1952. Federal support was provided for the eradication program, including payments for indemnities and slaughter of infected swine. The number of outbreaks declined dramatically, and vesicular exanthema has been kept under control since 1956.

Cause

Vesicular exanthema is caused by at least 13 distinct types of virus. They have been identified as A_{48}, B_{51}, C_{52}, D_{53}, and E_{54}, etc. The letters designate the order of isolation, and the numbers indicate the year when they were found.

Transmission

Vesicular exanthema is spread by contact with infected animals and contaminated feed, water, equipment, and premises. The virus can be spread on the shoes and clothing of people and by vehicles moving from infected premises to areas habitated by susceptible animals. Raw garbage containing raw scraps of infected pork is the major source of transmission and the perennial threat of an outbreak of vesicular exanthema. For this reason, every state with a sizable swine population requires that garbage be properly cooked before it is fed to swine.

Clinical Signs

The fist signs of vesicular exanthema are vesicles formed on the snout, lips, tongue, and mucous membranes of the mouth. Lameness caused by lesions formed on the feet may also occur in the early stage, but severe inflammation of the feet usually occurs after the disease has progressed for several days.

Vesicular exanthema often follows a two-phase pattern. Phase one usually last for two to three days. It is characterized by the formation of primary vesicles, anorexia, listlessness, and an elevation of body temperature to 105° to 108°F. Near the end of the phase one period, vesicles erupt, and the temperature drops to 104°F or below.

Phase two occurs as the virus enters the bloodstream and secondary lesions appear on other parts of the body. This usually happens on the third to sixth day following infection. Secondary vesicles are formed on the feet, between the toes, and around the coronary band. The inflammation may be so severe that the entire hoof is sloughed off. Secondary lesions on the teats of sows nursing pigs have been reported.

When both primary and secondary vesicles rupture, raw sores are opened to bacteria, and secondary infections often complicate and delay recovery.

Vesicular exanthema, like other vesicular diseases, usually has high morbidity and low mortality. Suckling pigs are more likely to die from vesicular exanthema than older hogs. The great losses due to this disease are reduced gains of feeder pigs, loss of weight in market size hogs, abortions in bred sows, and impaired performance of the breeding herd.

Prevention

There is no vaccine approved for the immunization of swine against vesicular exanthema. Prevention is based upon strict enforcement of state laws requiring that garbage fed to hogs be properly cooked. Raw garbage should be heated to 212°F and held at that temperature for 30 minutes to destroy the virus that causes vesicular exanthema in swine. In addition, hogs or any other farm animals that have blisters in or around their mouths or noses should be inspected immediately by a veterinarian. All livestock producers in the United States have a responsibility to do their part in reporting potential threats to the animal industry.

Treatment

There is no specific treatment recommended for hogs infected with vesicular exanthema.

Contagious Ecthyma (Sore Mouth)

Contagious ecthyma is also referred to as sore mouth. It is a highly contagious disease, affecting primarily the lips of lambs and kids (Fig. 10-8). Sheep and goats over one year of age are rarely affected by contagious ecthyma, although lesions have been re-

Fig. 10-8. Contagious ecthyma in lambs with proliferative lesions on the lips. (Courtesy, College of Veterinary Clinical Medicine and Surgery, Washington State University, Pullman)

ported on the udders of ewes nursing infected lambs. Humans are also susceptible to this disease. Lesions develop on the hands and arms but are not usually serious. Contagious ecthyma is most common during the summer and fall.

Cause

Contagious ecthyma is caused by a filterable virus that is highly resistant to desiccation and other natural means of destruction.

Transmission

The virus that causes contagious ecthyma is spread quite readily to susceptible sheep by contact with abrasions on any part of the body. Transmission to people is possible; therefore, shepherds working with affected animals should take precautions to protect themselves.

Clinical Signs

Lesions usually occur first in the corner of the mouth, spread over the lips, and occasionally on the face, ears, and eyes. Scabs form rather quickly over the external lesions. In severe cases, vesicles form and ulcerate in the mucous linings of the mouth.

Occasionally, lesions are found on the feet and on the udder of ewes nursing infected lambs.

Contagious ecthyma often follows a rather benign course with very few mortalities, unless secondary infections complicate the problem. The major loss results from debilitation and reduced growth rate of affected lambs.

Uncomplicated cases of sore mouth heal in one to four weeks. The scabs dry up and drop off, and the tissue returns to normal. The virus remains pathogenic in the dried scabs over the winter and may reinfect lambs during the following season unless they have been vaccinated.

Prevention

A commercial vaccine is available for use on lambs in problem areas. The vaccine is made of finely ground scabs containing the living virus. The powdered material should be mixed with a diluent immediately before it is to be used. A small amount of the suspension is brushed over areas that have been scarified on the inside of the thigh of the lamb. Any area devoid of wool is a suitable location for vaccination. Lambs should be vaccinated at about one month of age.

A successful inoculation is apparent if pustular lesions are formed at the site of scarification and application of the vaccine. Animals that have recovered from contagious ecthyma are highly resistant to reinfection.

Since the vaccine for contagious ecthyma is a living virus and the scabs produced remain infective for several years, it should not be used in areas nor flocks where the disease is not a problem.

Treatment

Treatment should be addressed to controlling secondary infections when they occur. Antibiotic ointments are beneficial in such cases. Vaseline containing 3 percent phenol may also be used. Iodine solutions of methyl violet are recommended for treatment of the sores. The treatment must be continued until the sores are completely dried up, or relapses will occur.

The contagious ecthyma virus is transmissible to humans. Attendants should wear protective gloves when treating infected sheep.

Bluetongue

Bluetongue is an infectious, non-contagious viremia of ruminants. Sheep are more commonly infected by the bluetongue virus than cattle and other ruminants (Fig. 10-9). Bluetongue has been a serious problem in Africa for over a hundred years. It was first described as catarrhal fever and carries that common name to this date. Bluetongue was identified in California in 1952, but it was probably present in this country many years before that time. Bluetongue has been reported in many of the western states.

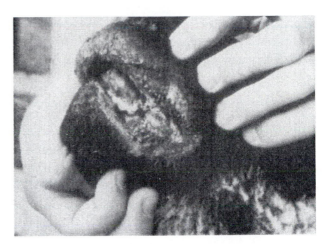

Fig. 10-9. Bluetongue in sheep. Note oral lesions. (Courtesy, College of Veterinary Medicine, Washington State University, Pullman)

Cause

Bluetongue is caused by four different strains of virus. The virus is highly resistant to drying and may live in the blood of recovered animals for three to four months.

Transmission

Bluetongue is not spread by direct contact with infected animals or contaminated materials. It can be transmitted mechanically by infectious blood. The major source of transmission of bluetongue occurs when gnats of the genus *Culicoides* carry the virus from sick to susceptible animals. The gnats are night-biting insects about the size of a pinhead. They are active during warm weather and attack sheep vulnerable to these virus-carrying insects after they are sheared.

Clinical Signs

Bluetongue is more severe in sheep than in cattle. Cattle often harbor the virus and show very little or no evidence of clinical infection.

A wide variety of clinical signs may occur from flock to flock due to the resistance of different breeds of sheep, inherited immunity of lambs from ewes that have recovered from bluetongue, and the age at which infection occurs. The fine wool breeds are considered to be a bit less susceptible than the British breeds.

The first sign of bluetongue disease may be an elevation of body temperature to 104° to 108°F for only a few days. In the acute form, this is followed by loss of appetite, depression, nasal discharge, and salivation. The mucous membranes of the mouth and nose become hyperemic or deep red in color due to the presence of excessive amounts of blood in those tissues. As the disease progresses, hyperemia of the mouth, tongue, and nose may deepen to cyanosis, which is a bluish discoloration of the tissues. The lips become swollen, and, in many cases, this extends to the muzzle, face, and ears.

After several days of fever, ulcers appear on the tongue and inside the mouth and nose. The salivation becomes thick and stringy. The nasal discharge also thickens and encrusts the nose and mouth areas. Damage to the crusts or attempts to remove them will result in bleeding, open sores.

Affected sheep often become stiff and lame. They are dull, lose weight, and take a long time to recover. Sometimes they slip their wool. There may also be a loss of a breeding season. Once in a while twisted necks result.

The bluetongue virus itself seldom kills sheep. However, it weakens their resistance to other diseases. Deaths are usually due to complicating pneumonia.

Prevention

Sheep should not be moved into known infected areas during the season in which gnats are active. Ewes, bucks, and replacement lambs over four months of age should be vaccinated at shearing time. Ewes should not be vaccinated during the first two months of gestation because a high percentage of "crazy" lambs may result.

Treatment

There is no specific drug recommended to control the bluetongue virus. Antibiotic treatment of secondary infections and the complications of pneumonia may be warranted. Good husbandry is very important in handling sick animals. Sick sheep should be isolated from healthy animals, kept in the shade, and made as comfortable as possible. They should have access to clean feed and fresh water.

Ulcerative Dermatosis

Ulcerative dermatosis is an infectious, ulcerative viral disease of sheep that occurs primarily in the western part of the United States. Ulcerative dermatosis normally manifests itself in two distinct forms. One is characterized by ulcerations around the mouth and nose or on the legs. This form is commonly referred to as "lip and leg ulceration." It may be confused with contagious ecthyma. The second form is venereal in nature. Ulcerations are formed on the prepuce and penis of rams and on the vulva of female sheep. This form resembles sheath rot or pizzle rot, but it is not the same disease.

Cause

Ulcerative dermatosis is caused by a filterable virus.

Transmission

The virus of ulcerative dermatosis is spread to the genitals by copulation. Transmission to the lips and legs is dependent upon abrasions or breaks in the skin for inoculation.

Clinical Signs

Ulcerations, whether on the lips, legs, or genitals, are covered with scabs in the early stages of the disease. Removal of the scabs reveals a raw, bleeding crater containing an odorless, creamy pus. Lesions on the face occur most often on the upper lip just below the nasal opening. The mucous membranes of the mouth are not usually affected. In severe cases, lesions may perforate the lip and damage membranes inside the mouth. Lesions on the feet occur anywhere from the coronet to the knee or hock. Animals affected with lip and leg ulcerations may show lameness and have difficulty eating.

Lesions, in the venereal form, occur first on the prepuce of the male. The ulcerations may completely encircle the entire opening of the sheath and in severe cases involve the glans portion of the penis. In some cases, swelling occurs in the preputial orifice to such a degree that the penis cannot be projected for normal breeding. The lesions on the vulva of ewes are usually less severe than those on the ram. They rarely involve the vagina.

Prevention

There is no vaccine available to prevent ulcerative dermatosis of sheep. Infected animals should be isolated, and those with genital lesions should not be used for breeding.

Treatment

Treatment is not recommended in most cases. When treatment is deemed appropriate, scabs and necrotic tissue are removed and preparations of bluestone, cresol, or formaldehyde are applied to the affected areas.

Oral Necrobacillosis

Oral necrobacillosis is an infectious disease affecting cattle. It is often referred to as calf diphtheria, necrotic stomatitis, and necrotic laryngitis.

Cause

The causative agent of oral necrobacillosis is *Spherophorus necrophorus*. This is Gram-negative, anaerobic, filamentous, non-sporulating, rod-shaped bacterium. It is commonly found in soil and manure where animals are kept or have been present. Cattle that are confined in damp, filthy areas are likely to be affected by oral necrobacillosis if they have a virus infection or have injured the oral mucosa. Calves are considerably more susceptible to the infection than older cattle.

Clinical Signs

Calf diptheria or necrotic laryngitis is the most severe form of oral infection. Characteristic symptoms are depression, salivation, anorexia, dyspnea, and coughing. The larynx becomes swollen and tender, and the temperature can rise to 106°F. Death may occur in two to seven days if treatment is not given. Spread of the infection to the lungs causes bronchopneumonia.

Necrotic stomatitis is a milder form of the disease. Typical symptoms are a temperature of 104°F, mild depression, swellings in buccal mucosa lateral to the cheek teeth, excessive salivation, and a fetid breath. When the mouth is opened, ulcers may be seen on the buccal mucosa and occasionally on the tongue. The ulcers have a raised periphery with a caseated center, and they are 0.5 to 2 cm in diameter. Deaths have occurred with this form of the infection.

The characteristic necrotic ulcers are found on the tongue, cheeks, pharynx, larynx, and trachea, and they may extend into the lungs, esophagus, or gastric compartments.

Treatment

Oral administration of sulfapyridine, sulfamerazine, or broad-spectrum antibiotics is usually effective. Intramuscular injections of penicillin are also efficacious. Early treatment will help prevent the development of chronic respiratory infections. Additional vitamin A is indicated for young calves which are affected.

Actinomycosis (Lumpy Jaw)

Actinomycosis is a chronic, infectious, non-contagious disease affecting cattle, swine, and horses. It is commonly referred to as "lumpy jaw" in cattle.

Cause

Lumpy jaw is caused by bacteria—genus *Actinomyces*, with *A. bovis*, *A. pyogenes*, and *A. suis* considered pathogenic in food animals. *Actinomyces* spp. is

Gram-positive and may occur as straight rods, coccobacillary forms, or filaments with branching.

Transmission

Actinomyces bovis is a normal habitant of the mucous membranes of the upper respiratory and digestive tracts of most farm animals. The mode of infection of *Actinomyces bovis* is not definitely known. It is probable that the infective organism enters the tissues through cuts in the mouth or openings due to the eruption of teeth. Inoculation of the teats of sows apparently occurs from bites administered by nursing pigs.

Clinical Signs

Actinomycosis in cattle may be characterized as an ostitis of the skull. It is usually an infection of the upper or lower jaw but may affect other bony parts of the head (Fig. 10-10). The bony enlargements are painful at first and may take 6 to 18 months to complete ossification. Abscesses occur around the infected bone and may erupt through the skin and discharge pus. Inflammation of the tongue, teeth, and jaws may inhibit prehension and mastication of feed.

In swine, actinomycosis is primarily centered in one or more udder sections of sows. Abscesses form in the mammary system and inhibit milk production in the affected area. Some of the abscesses rupture and discharge the pus-laden exudate.

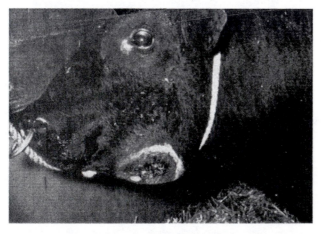

Fig. 10-10. Actinomycosis, or lumpy jaw. The abscess has ruptured.

Actinomyces bovis and brucella organisms cause two common diseases in horses known as poll evil and fistulous withers.

Prevention

No method of immunization has been devised. Clipping the needle teeth of pigs shortly after they are born is recommended to reduce inoculations in lactating sows.

Treatment

Treatment of actinomycosis in the advanced stages is usually unsatisfactory. Surgical removal of growths during the early stage followed by iodine saturated gauze packed into the wound may be beneficial. Abscesses may be opened, drained, and packed with gauze saturated with iodine.

Treatment usually includes surgical débridement and weeks to months of antimicrobial chemotherapy. Penicillin is the preferred drug in food animals, although ampicillin or amoxicillin may be used because of the broader spectrum of activity. Iodide preparations injected into the lesions, or intravenously, are not recommended in animals to be used for meat.

Actinobacillosis (Wooden Tongue)

Actinobacillosis is a chronic, infectious, noncontagious disease that affects the soft tissues and lymph nodes of the heads of cattle. Sheep are occasionally affected, and, in rare instances, people may contract this disease. The common name for actinobacillosis of cattle is "wooden tongue."

Cause

The causative organism of actinobacillosis is *Actinobacillus lignieresii*, a Gram-negative, rod-shaped bacterium.

Transmission

Actinobacillus lignieresii organisms are found abundantly in soil and manure. They commonly inhabit the mucous membranes of the digestive tract and nasal passages of healthy cattle. Inoculation is apparently dependent upon damage of mucous membranes by cuts, abrasions, or openings due to the eruption of teeth.

Clinical Signs

Actinobacillosis of the tongue is first apparent by salivation and impairment of the animal's ability to prehend and masticate feed properly. Later, the tongue becomes swollen and hard, and may protrude from the mouth (Fig. 10-11). Animals in this condition will die from starvation or dehydration unless treatment is provided.

Breathing problems and difficulty in swallowing are common signs when abscesses form in the larynx, pharynx, or soft palate of the mouth.

Growths that appear in the other soft tissues of the face and neck are usually painless and mobile to a small degree (Fig. 10-12). They enlarge, erupt, and discharge the pus-filled exudate in much the same manner as that which occurs in actinomycosis. In addition to the head and neck, actinobacillosis may infect almost any other organ and tissue of the body.

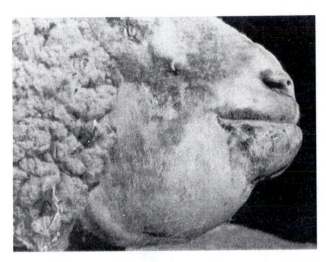

Fig. 10-12. Inflammation of the soft tissues of a sheep's mouth that is affected with actinobacillosis. (Courtesy, Dr. Murray Fowler)

The first sign of actinobacillosis in sheep is a thickening of one or both lips. The swelling usually extends up along the cheeks and jaw bone, but not the tongue. Later abscesses appear, enlarge to about the size of a walnut, rupture, and discharge their contents. The exudate mats in the wool surrounding the abscessed area, creating an obnoxious sight and a fetid odor.

Prevention

There is no vaccine available for use in preventing actinobacillosis. The goal of a prevention program is to avoid coarse feedstuffs that cause injury to the alimentary tract.

Treatment

Actinobacillus lignieresii is quite susceptible to the sodium iodide treatment. Intravenous injections of sodium iodide or the feeding of organic iodide is the recommended treatment. Surgical removal of the abscess followed by an insertion of gauze saturated with iodine is usually effective.

Treatment should be under the supervision of a veterinarian.

Fig. 10-11. Actinobacillosis, or wooden tongue. Note the extended swollen tongue. (Courtesy, Dr. Murray Fowler)

Choke

Cause

In cattle, choke is usually caused by rapid ingestion and incomplete chewing of feed. Cattle attempt to eat as much as possible in a short time so that when the rumen becomes filled, they may rest while regurgitating and rechewing the feed boluses. They frequently try to swallow firm fruits, tubers, or roots when these are available, and any of these objects may be larger than the esophagus can take. Other feeds that cause problems are apples, potatoes, cabbage, beets, turnips, and ears of corn. In addition, saliva tends to make a smooth-skinned apple or potato difficult to masticate; consequently, such objects frequently slip into the pharynx and then into the esophagus.

In horses, choke is usually caused by eating dry grain, especially oats. Greedy eaters or older horses with bad teeth and reduced vitality are the most susceptible. Shetland ponies tend to be more susceptible than other equines. The esophagus of the Shetland appears to be proportionately smaller than in the other equine breeds. Other causes of choke are dehydration, esophageal strictures, esophageal pouches, compression of the esophagus by abscesses, and ulcers of the lining of the esophagus.

The caretaker should be careful when evaluating an animal for choke, as rabies is another cause of salivation and an inability to swallow.

Clinical Signs

Bloat and salivation are the two most common signs of choke in cattle. In addition, frequent chewing movements are made in an effort to dislodge the object. The animal's attitude may range from moderate anxiety to absolute mania. The degree of bloat is correlated with both the length of time that the animal has been choking and the size of the object it has been choking on. For example, a whole apple would prevent eructation completely, and the animal would bloat very fast; whereas, a piece of apple or cabbage root would let some gas pass out, and the onset of bloat would take a great deal of time.

Cattle will extend their heads and necks and raise and lower their heads frequently, and their tongues may protrude periodically during chewing movements. An animal may cough frequently because of the excess saliva in the pharyngeal region. Frequently, the mass, may be felt in the throat. Passage of a stomach tube will confirm the diagnosis.

There is a great deal of distress in a horse that is choked. The animal may repeatedly arch the neck and draw the chin toward the sternum or alternately extend the head toward the ground. Periodically, the animal may shake its head and occasionally travel around the stall in agitation. There may be an anxious facial expression. Saliva may drool from the mouth and appear mixed with food material at the nostrils. Occasional coughing may be precipitated by the accumulations of saliva and food in the pharynx. Frequently, the mass may be felt in the throat. After 18 to 36 hours, the agitation may subside. The frequent efforts to swallow disappear, and the animal may stand quietly or get depressed. From this point on, the horse spends a great deal of time at the waterer. Passage of a stomach tube will confirm the diagnosis.

Prevention

Objects of no benefit, such as apples or cabbage roots, should not be accessible to cows. Worthwhile feedstuffs, such as beets, green corn, and occasionally potatoes, should be chopped or sliced so that the pieces are reduced in size and texture.

Horses subject to habitual choke may be benefited by a controlled diet. Soaked feeds may be helpful. Surprisingly, however, certain individuals show a greater tendency to choke on succulent grass than on dry hay. Possibly, the necessity for more complete chewing and better lubrication of the dry roughage with saliva is a factor.

Stones may be placed in the feedboxes of greedy eaters. The purpose is to slow down the intake of feed by requiring them to nibble grain from around the stones.

Pellets that are less than $\frac{3}{8}$ inch in diameter and cubes that are 2 inches square and 1 inch thick are likely to cause problems.

Treatment

Cattle need immediate relief from bloat if it is severe. Therefore, a choking animal that is severely bloated will have to have a trocar inserted into the rumen to release the gas. If the object can be felt, the operator should restrain the animal in a standing position, preferably in a stanchion with a noselead for dairy cattle or a chute for beef cattle. The operator should place his/her thumbs, fists, or extended fingers in the depression where the jugular vein is located and immediately behind the mass. Then, the operator should gradually work the foreign body back toward the mouth. He/She can also aid the animal by applying steady uniform pressure when the animal swallows or retches because the muscles of the esophagus tend to contract and relax intermittently and quite violently, thus grasping and releasing the object alternately. Extension and moderate elevation of the animal's head and neck at the same time also aid in isolating the esophagus. This manipulation should not be too extreme or continuous, however, because of the danger of causing inhalation of copious quantities of saliva.

When the foreign body has been brought near the pharynx, either the operator or an assistant should reach into the animal's mouth and remove the object from the opening of the esophagus, while the other person gives the material a final upward shove. With a mouth speculum applied and open, the tongue pulled out, and adequate pressure applied externally from both sides of the neck, the object can be rolled into the mouth without falling back into the sometimes narrow pharynx. If this method fails, a loop of No. 9 steel wire may be used. A 10- to 12-foot straight piece of wire can be bent back on itself at the middle so as to form a 2- to 3-inch loop. A wooden handle can be taped on the end opposite the loop, which can then be passed through the mouth and esophagus until it has encircled the object, then be withdrawn slowly.

Depending upon the type of obstructive object and its location, a steam hose or smooth garden hose may be used to push the object slowly into the rumen. The mass can become lodged in the thorax and make an already difficult situation next to impossible to treat. There is also danger of damaging tissues (puncturing the esophagus, mediastinum, or lungs) by injudicious use of this procedure. A larger stomach tube is more beneficial than a smaller one. The larger tube will push the obstruction, rather than slip past the obstruction and possibly perforate the esophagus.

The nature of the obstruction governs the urgency and selection of treatment of choke in horses. Grain chokes usually are the least serious. Treatment consists of tranquilizing the horse, then tying the head low enough so that water pumped into the esophagus can run back out the horse's nose. Then, a stomach tube is put down the esophagus as far as the lodged material. Water is repeatedly pumped into the tube to soften the mass. Whenever recovery is not prompt, an intravenous solution of gallons of electrolyte solution should be administered so that the animal will be able to produce enough saliva and mucous to aid in softening the mass. Endoscopes with the ability to grasp and remove foreign material may facilitate the resolution of choke in horses. Horses are apt to lodge material in the proximal esophagus.

Surgical treatment of choke is seldom necessary, and it is undesirable because it frequently causes the animal to have a constricted area in the esophagus after the surgical wound heals. This, of course, predisposes the animal to more chokes.

Bloat

Bloat is a non-contagious digestive disorder of ruminants. At times, all animals bloat to some degree, but ruminants, especially cattle, are more severely affected than monogastric animals. Bloat is an excessive accumulation of gas in the first two compartments of the ruminant stomach. It is not a result of excess production of gas but the inability of the animal to expel the accumulated gasses. Bloating animals are usually described as chronic or acute bloaters.

Cause

The specific cause or causes of bloat are highly speculative. Although much misinformation abounds about the cause of bloat, more research is needed as to its cause.

Studies indicate that acute bloat is closely associated with many factors, including animal susceptibility, type of feed, and the environment in which the animal is fed. Heredity cannot be ruled out because some animals bloat frequently, while many others which eat the same kind of feed never bloat. Also, certain cattle, especially dwarfs, are chronic bloaters. Because of the diversity of conditions under which bloat occurs, progress toward the solution of this malady has been slow.

Many theories as to the cause or causes of bloat have been advanced. Research has disproved many of the older theories and has cast doubt on others. Some of the theories that have been advanced are:

1. *Lack of coarse roughage.* This point has been labored. Finely ground feeds for ruminants may be conducive to digestive disturbances in some cases, but are not the cause of bloat. Finely ground and pelleted hay is now used successfully by many feedlots.

2. *Density of feeds.* Actually, when the rumen is functioning normally, it has two to three contractions per minute which tend to keep dense feed thoroughly mixed. Although digestive disturbances, including bloat and density of feed, may be associated, it appears that increased density within itself is not the cause of bloat.

3. *Saponins.* Extensive studies were conducted by Lindahl, et al., on alfalfa saponins. Their studies included the preparation and chemistry of saponins, as well as the pharmacological effects of this group of compounds. Their work indicated that saponins may be associated with or may contribute to bloat, but that saponins *per se* are not the cause of bloat.

4. *Excess gas production.* This appears unlikely, as the normal animals and the bloaters eat the same kind of feed.

5. *Formation of toxic substances.* This possibility has been suggested and, indeed, may cause death but does not appear to be the cause of bloat.

6. *Saliva production and/or composition.* Undoubtedly, the amount and/or composition of saliva is associated with bloat. Saliva appears to be an intermediate in the bloat prevention mechanism rather than a cause of bloat. The question, then, is why animals differ in saliva production and composition.

7. *Animal differences.* Many researchers have observed and reported differences in animal's susceptibility to bloat. Observations indicate a relationship between the bloodline of cattle and the incidence of bloat. When bloaters were mated with bloaters and to their close relatives, 50 to 65 percent of their offspring bloated.

Clinical Signs

Distention of the paunch is the classical sign of bloat in ruminants. The distention occurs over the left side and may lift the ruminal enlargement above the backline. This distortion is often less conspicuous in fat cattle than thin animals and is often difficult to detect in sheep with full fleeces.

Bloated animals stop eating, move uneasily, stand with their heads extended and their forelegs spraddled. There may be slobbering, grunting, and labored breathing as the condition progresses. If relief is not provided, the animals have difficulty standing, and later fall and die.

Prevention

The prevention of bloat is dependent upon a knowledge of its causes and upon a management scheme to reduce or eliminate those factors. This kind of information is only partially available at this time. The incidence of bloat is lessened by (1) avoiding straight legume pasture and immature legumes, (2) feeding a coarse grass hay prior to turning onto lush pasture, (3) feeding dry forage along with pasture, (4) avoiding a rapid fill from an empty start, (5) keeping animals continuously on pasture after they are once turned out, (6) keeping salt and water continuously accessible at all times, (7) avoiding frosted pasture, and (8) using proloxalene (Bloat Guard), oxytetracycline (Terramycin or Neo-Terramycin), or Laureth-23 (Enproal Bloat Blox) according to manufacturer's directions.

Treatment

Treatment is often required immediately to save an animal. Livestock producers should be able to handle this assignment in an emergency. When possible, they should use the services of a veterinarian.

When bloat is identified in the early stages, a piece of broom handle tied in the animal's mouth may facilitate eructation. Haltered animals in distress should be tied so that their front end is elevated to provide some relief to the heart and lungs.

In tight cases, a piece of garden hose should be inserted carefully through the mouth and esophagus into the rumen to relieve the bloated condition. The hose will need to be gently moved back and forth in the paunch to pick up pockets of gas. This technique may be disappointing in cases of extremely frothy bloat.

The last resort under most conditions is a rumenotomy. A trocar and cannula, if one is available, should be used. A knife will do if a trocar and cannula are not handy. The trocar and cannula should be inserted at a point equidistant from the hip, last rib, and loin edge on the animal's left side. The cannula should remain inserted in the rumen until the gas has subsided. Penicillin injections should be given for three days following trocarization to minimize infection at the puncture site. Naxcel would be the preferred drug for lactating or feedlot animals close to slaughter, due to a zero meat and milk withdrawal period for the drug.

Traumatic Reticulitis (Hardware Disease)

Traumatic reticulitis is also called traumatic gastritis, traumatic reticuloperitonitis, and hardware disease. It is an acute or a chronic mechanical injury to the reticulum. Adult cattle of all ages are likely to ingest materials that will cause hardware disease. Dairy cattle have a higher incidence of this disease than beef cattle because they consume large amounts of ground feed and hay.

Cause

Traumatic reticulitis is caused by the ingestion of sharp metal that punctures the reticulum of ruminants. Studies at the University of California at Davis indicate there has been a shift from nails to wire of the offending metal particles since the advent of balers and the use of wire-tied bales beginning about 1940. Wire now accounts for about 75 percent of the cases;

nails 20 percent; and miscellaneous objects, 5 percent. Nearly all tramp metal has been ferromagnetic. Pieces 2 to 4 inches long are most troublesome.

In the majority of cattle, metal is found only in the reticulum, although a few, particularly bulls, may also have it in the rumen. Clinical reports indicate that the problem is increasing due to the use of more chopped feeds, more feed contamination, improved diagnostic facilities, or combinations of all three.

Although there have been occasional reports of cattle deliberately eating wire, nails, or similar objects, most tramp iron is consumed with some form of feed. Hardware disease is a particular problem in cattle because of their indiscriminate eating habits, the anatomical arrangement of their stomachs, and the large amount of tramp iron available. In an experiment performed at the University of California at Davis, six 3-inch pieces of baling wire were mixed into grain fed to each of two cows. The cows cleaned up the grain and, in the process, swallowed all the pieces of wire. Forage makes up a large part of the ruminant's diet, and because of its coarse, stemmy form, it provides a good hiding place for foreign material. Hence forage is probably the source of much of the tramp iron found in cattle.

Clinical Signs

The ingestion of some metal pieces may do little or no harm. Traumatic reticulitis occurs when the sharp objects penetrate the reticular wall. The penetrating objects may pass into the pericardium since the heart is located quite near the reticulum. They may also penetrate the diaphragm. The disease may, therefore, expand quickly from traumatic reticulitis to include peritonitis, pericarditis, and pleuritis.

The first signs include anorexia, reduced milk flow, and cattle lagging behind when the herd is moved. This is often followed by an objection to move about, an arched back, and an appearance of distress. The affected animal often stands with the front legs spread wide apart, toes turned in, and the elbows rotated outward. There may be a pronounced difficulty in defecation and urination. The temperature is often elevated a couple of degrees and the respiration rate is a bit above normal. Pain is expressed by a grunting sound when pressure is applied on the breastbone and to the paunch.

Prevention

A bar magnet is recommended for controlling traumatic gastritis in cattle (Fig. 10-13). The bar magnet used is a permanent magnet, cylindrical in shape and ½ to ⅝ inch in diameter by 2¼ in length. The reticulum with its constant temperature and freedom from shock is ideally suited to maintain a magnet's permanent qualities. Magnets are most effective in preventing hardware disease. They are given to heifers or animals not previously exposed to metallic objects. It is best to take the cattle off feed for 24 hours before passing the magnet with a balling gun. The magnets are left in the animal and retrieved only after slaughter. Foreign bodies which have already passed completely through the reticulum will not be brought back. Occasionally, wires will penetrate the reticulum above its floor, and these will also escape a magnet put in after penetration has occurred.

Treatment

Conservative medical treatment may be used by veterinarians in cases in which, in their considered judgment, it is desirable. Confinement of the animal in a stall or stanchion, with elevation of the foreparts may be advised. This reduces pain and aids in walling off the penetrating object. Used with medication to control the infection, this procedure may bring about prompt recovery.

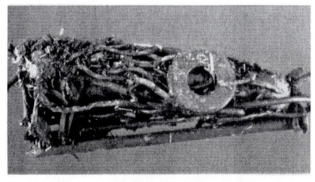

Fig. 10-13. Magnet from rumen of cow. (Courtesy, College of Veterinary Clinical Medicine, Washington State University, Pullman)

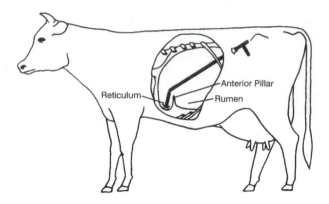

Fig. 10-14. This diagram shows the use of a metal retriever inserted through the cow's flank. Note the retriever is inside a metal tube so that sharp material will not scratch when being withdrawn.

Rumenotomy is the surgical procedure used as a method of choice by many in the treatment of traumatic reticulitis. The operation is done with the animal in a standing position and under local anesthesia. The veterinarian makes an incision in the left flank, opens the rumen, and inserts a metal retriever into the cow's reticulum to remove the foreign body (Fig. 10-14). If called early, the veterinarian is able to remove the offending object and thus return the animal to normal. Cows recover from this procedure very rapidly, and the operation has a high percentage of success.

Impaction

The ingestion of large amounts of highly fermentable carbohydrate feeds causes an acute illness due to excess production of lactic acid in the rumen. Clinically, the disease is manifested by severe toxemia, dehydration, blindness, recumbency, cessation of rumen activity, and high mortality. Rumenal impaction is called rumen overload in cattle and sheep. Impaction causes a specific type of colic in horses.

Cause

Accidental access to large quantities of whole or ground grain is the most common cause. Feeder cattle and lambs brought into feedlots and fed on excessive amounts or allowed free access of self-feeders, without a prior period of adjustment to feed, are most commonly affected. Even in adjusted cattle, the common practice of feeding to the limit of their appetite maintains many animals on the brink of overload.

The important factor in the production of the disease is the rapid fermentation of the carbohydrate in the feed by Gram-positive cocci (usually *Streptococcus bovis*), with the formation of large quantities of lactic acid which may reach a concentration of 3 percent. Rumen motility decreases as the acidity increases. In a few hours, rumen activity ceases completely. The degree of acidity that develops governs the severity of the signs produced.

Clinical Signs

The speed with which impaction occurs varies with the nature of the feed ingested. The onset is faster with ground feed than whole grain. The severity of the illness increases with the amount of feed ingested. In severe cases, clinical illness is apparent within 12 hours of engorgement. In some cases, the first sign may be abdominal pain, with kicking at the belly. There is profound depression with hanging of the head and disinclination to move. The respiration rate is usually increased, and may be accompanied by grunting. Sick animals will stop eating and will drink very little water. There may be slight bloating, but it is never severe. The nose is dry. Grinding of the teeth is a common sign, and there is usually diarrhea that is soft, light-colored, and foul smelling. In very severe cases, there may be profuse diarrhea accompanied by the passage of much mucus and some blood.

An increase in pulse rate up to 120 to 140 beats per minute is usual, and the pulse is weak. The temperature is usually below normal (99° to 101°F). Severely affected animals have a staggery, drunken gait and appear to be blind. They bump into objects and have no eye preservation reflex. The animals spend most of the time lying down as the symptoms get worse. Rapid development of severe signs usually means the animal is going to die in 24 to 72 hours if radical treatment is not used.

A history of engorgement is usually sufficient to confirm diagnosis, but if this is not available the disease can be mistaken for parturient paresis, hepatitis, or poisoning, especially with arsenic or lead. Parturient paresis (milk fever) is restricted to cows which have recently calved or ewes in late pregnancy or in early lactation.

Prevention

Seventy mg of zinc bacitracin per head per day, plus adding $\frac{1}{10}$ percent sodium bicarbonate to the ration, will tend to keep feedlot cattle from having any problems.

Treatment

In mild cases of acidosis, grain should be removed from the ration and hay with penicillin or tetracycline added should be fed.

Doses of 0.5 to 1 million units of penicillin are effective in treating sheep for acidosis, and doses of 5 to 10 million units are recommended for cattle. Alternatively, doses of 8 to 10 grams of tetracycline are recommended for cattle affected by acidosis.

In severe cases of acidosis, 2 tablespoonfuls of baking soda may be mixed with 1 pint of sterile water and administered intravenously to cattle. In addition, 1 pint of 50 percent dextrose and 40 mg of dexamethasone given intravenously will correct the staggering problem. One gallon of mineral oil should be given orally. (Baking soda, magnesium hydroxide, or magnesium carbonate should not be given orally because each causes a chemical reaction with the acid in the rumen which generates a phenomenal amount of heat. In fact, so much heat is generated that the lining of the rumen becomes blistered and separates from the underlying tissue. The result is death.)

Impaction of the Omasum

Chronic impaction of the omasum is difficult to define and is usually diagnosed at autopsy when the omasum is enlarged and excessively hard. It seems unlikely that it could cause death and is frequently observed in animals dying of disease.

Cause

Impaction of the omasum reportedly occurs when feed is tough and fibrous, particularly alfalfa stalks and loppings from fodder trees. It also occurs in sheep that are fed on the ground during drought conditions. In the latter, the impaction is due to the accumulation of soil in the omasum.

Clinical Signs

Chronic recurrent bouts of indigestion occur and are manifested by normal rumen motility, infrequent and scanty feces, refusal to eat grain, and a negative ketone test. Pain may be elicited, and the hard distended viscus may be felt under deep pressure under the ribs on the right side.

A more rapid form of omasal stasis is recorded in cattle in which the cessation of activity may be primary or secondary to other disease, particularly post calving hemoglobinuria, and calving paralysis.

The omasum is grossly distended. The leaves, and in some cases the wall of the organ, show patches of dead tissue and an associated peritonitis. Clinically, the disease is manifested by cessation of eating and defecation, an empty large intestine, moderate abdominal pain, and disinclination to move or lie down.

Treatment

Repeated dosing with mineral oil is recommended as treatment.

Impaction of the Abomasum

Cause

The abomasum becomes plugged up as part of the syndrome of vagus indigestion and in the unusual circumstance where cows are fed entirely on chopped straw. This is only likely to occur where husbandry is impoverished and where cattle are stimulated to eat large quantities of low-quality fiber feeds. Very little digestion of the straw occurs in the rumen and material that is fragmented by ruminal movements is carried over into the abomasum with rumen liquor. The mass is unable to pass the pylorus, and impaction of the abomasum occurs. Abomasal impaction can also be caused by obstruction of the pylorus with placenta, baling twine, etc., as well as functional impaction due to lymphosarcoma outside the abomasum.

Clinical Signs

The clinical syndrome is very similar to that of vagus indigestion, and, as in that disease, abomasal surgery is rarely successful. An animal with abomasal impaction appears very round and full, with the abdomen bulging ventrally on both sides, as well as high on the animal's left side. Supplementation of the diet with a high protein concentrate should encourage digestion of the straw and prevent impaction of the abomasum.

In young calves and lambs, rapid obstruction of the pylorus of the abomasum occurs after they have eaten indigestible material which includes rags and shavings. Abomasal impaction in young animals also occurs as part of a syndrome of indigestion, especially when calves are fed at infrequent intervals on overly large quantities of milk with a high content of casein. An indigestible, rubbery curd develops in the abomasum which causes persistent scouring. If the feeding mismanagement continues, the terminal stage of abomasal impaction develops. The calf becomes emaciated and has a pendulous distended lower abdomen. Fluid can be heard rattling in the abdomen when the animal is shaken. There is often a great deal of coarse straw and hair present in the abomasum, but whether this is a primary cause of the impaction or is the result of depraved appetite caused by indigestion is unknown.

Treatment

Abomasal surgery is the only satisfactory treatment, but stasis of the stomach may persist even if the impacted material is removed. The pressure should be relieved from the rumen with a large stomach tube. Surgery of the abomasum to remove foreign contents has a better prognosis in calves, as the abomasum is more accessible surgically. The value of the animal, as well as the severity and chronicity of the impaction dictate the treatment options. Salvage by slaughter is often economical in chronic cases, whereas laxatives may be effective in acute cases.

Impaction of the Large Intestine

In farm animals, the disease is common only in horses and pigs. (For a discussion of it in horses, see "Colic.") In pigs, impaction of the colon and rectum occurs sporadically, usually in adult sows which get little exercise and are fed wholly on grain. The disease also occurs in pigs which are overcrowded in sandy or gravelly outdoor yards. A special occurrence in young weaned pigs causes obstruction of the coiled colon.

Cause

Continued overloading of the colon and caecum, either, primarily, because of the nature of the feed, or, secondarily, because of poor intestinal motility, causes prolongation of the intestinal sojourn and excessive thickening of fecal material so that movement of the mass by normal intestinal motility is still further impaired. If the process is prolonged, the colon becomes insensitive to the stimuli caused by distention which normally provoke defecation. Chronic constipation results.

In pigs, the effects appear to be due largely to auto-intoxication, although the commonly occurring posterior paralysis seems more likely to be due to pressure from thickened fecal material.

Clinical Signs

The pigs become depressed and quit eating. Much of the time is spent lying down. Feces passed are scanty, very hard, and covered with mucus. Weakness to the point of inability to rise occurs in some cases. Hard balls of feces in the rectum are usually detected when a thermometer is inserted. In paralysis of the rectum, there is inability to defecate and usually some straining. The anus and rectum are ballooned, and manual removal of the feces does not result in contraction of the rectum.

Other causes of constipation, such as peritonitis and dehydration, must be considered when a diagnosis of impaction of the large intestine is made.

Treatment

One to four quarts of mineral oil.

Acidosis in Horses

In horses, acidosis usually occurs when the animals are worked very hard for long periods of time. Examples include the stress horses experience in endurance rides or during a race. In addition, horses that have severe cases of diarrhea often become acidotic.

Cause

Acidosis in horses is caused by heat exhaustion and severe diarrhea. In some cases of severe diarrhea, there is a drastic loss of bicarbonate via the intestinal tract. In addition, there is a high production of acid due to muscle contraction during strenuous exercise. Exercising horses in a hot climate can lead to a loss of 10 to 12 liters of sweat per hour, which results in massive losses of electrolytes sodium, potassium, and chloride.

Clinical Signs

The clinical signs of acidosis in horses include rapid, shallow breathing, poor appetite, weakness, lassitude, and terminal coma.

Treatment

Treatment for acidosis in horses involves oral and/or intravenous administration of sodium bicarbonate. Weakness, fatigue, and exercise intolerance in horses can be reduced by the addition of salt and lite salt to the ration twice daily. A suggested level is 2 tablespoons of salt in the feed. This will stimulate the horse to drink water and replenish the lost electrolytes.

Lactic Acidosis in Feedlot Cattle

Lactic acidosis can occur in feedlot cattle during any phase of the feeding period. The incidence of it has been emphasized in newly arrived cattle; however, it can occur in cattle that have been on feed over 90 days. It can occur in cattle on pasture, such as wheat pasture, and in cattle on high roughage feeding programs.

The emphasis and research in defining the problem have been done primarily on the refeeding and overeating syndrome which occurs more frequently when first adapting cattle to concentrates.

Cause

Acidosis is readily initiated in cattle that have been off feed and stressed. Alteration in feeding schedule or pattern, or changes in weather can initiate the problem under field conditions. The incidence of acidosis may have increased in the past few years due to greater complexity of management and larger number of cattle being fed under similar conditions. The problem, however, has existed to some degree for as long as cattle have been "pushed" on grain.

Lactic acidosis can be a problem in itself, or it can be a precursor to a series of problems correlated with digestive disturbances. Lactic acidosis can be correlated with the following problems: pneumonia, enterotoxemia, founder, bloat, diarrhea, rumenitis, liver abscesses, anorexia, anaphylaxis, and sudden death losses.

Lactic acidosis and these associated problems are all initiated by the common denominators of alterations in feeding patterns and stress. Alterations in feeding pattern can be (1) an outright removal of feed, which occurs in the marketing of feeder cattle, (2) too rapid an increase in the readily available energy level of the feed, (3) restriction of intake due to weather changes, (4) change in the palatability of feed, (5) change in the character of feed, which can be due to milling problems, (6) change in certain toxic constituents in the feed without a change in ration ingredients, such as aflatoxins, oxalic acid, and bacterial toxins.

Protozoa and Gram-negative cellulose digesters cannot tolerate drastic pH changes. They cannot live in a low pH environment. If the conditions for acidosis are produced, drastic microflora changes occur. A rapid drop in rumen pH occurs, and most of the organisms cannot survive. There is a "bloom" of one particular organism, *Streptococcus bovis*, that produces the lactic acid.

A summary of the rumen changes that occur after starvation or alteration of feeding pattern follows: (1) fermentative activity and capacity of the rumen decreases due to less nutrients being available; (2) protozoa numbers decrease, and their ability to re-establish is very slow; and (3) total bacteria numbers decrease, and the balance between bacterial species is disrupted.

The slow-growing cellulose digesters decrease, and their re-establishment is slower than the fast-growing *Streptococcus bovis* that produces lactic acid.

Clinical Signs

The first signs of lactic acidosis are usually abdominal pain with kicking at the belly, depression, hanging of the head, and disinclination to move. The respiration rate is usually increased and may be accompanied by grunting. Sick animals will stop eating and will drink very little water. The nose is dry. Grinding of the teeth is common, and there is usually diarrhea, with the passage of soft, light-colored, smelly feces. In very severe cases, there may be profuse diarrhea, accompanied by the passage of much mucus and some blood.

An increase in pulse rate up to 120 to 140 per minute is usual, and the pulse is weak. The temperature is usually below normal (99° to 101°F). Severely affected animals have a staggery, drunken gait and appear to be blind. They bump into objects and have no eye preservation reflex. The animals spend most of the time lying down as the symptoms get worse. Rapid development of severe signs usually means the animals are going to die in 24 to 72 hours if radical treatment is not used.

Prevention

The prevention of lactic acidosis in feedlot cattle is based upon sound feeding practices, reduction of stress prior to and after cattle are placed on feed, and the addition of zinc bacitracin and sodium bicarbonate to the ration.

Feedlot cattle should experience a gradual transition from high roughage to concentrate rations. They should not be denied feed for long periods of time, and drastic changes in palatability or in the preparation of feeds should be avoided.

Stress factors prior to and after cattle enter the feedlot should be reduced as much as possible. Purchasing feeder cattle that have been properly preconditioned, providing a rest period without the stress of sorting, worming, vaccinating, grub and louse control, etc., upon arrival, and providing sound feeding management are recommended to prevent lactic acidosis in feedlot cattle. In addition, 70 mg of zinc bacitracin per head per day and $\frac{1}{10}$ percent sodium bicarbonate should be added to the ration until new cattle are safely on feed.

Treatment

The following treatment for mild cases of lactic acidosis is recommended. All grain should be removed from the ration and hay with penicillin, tetracycline, or zinc bacitracin added should be fed. Doses of 0.5 to 1 million units are recommended for cattle. Alternatively, doses of 8 to 10 grams of tetracycline are recommended in cattle. Seventy mg of zinc bacitracin may be used.

In severe cases, 2 tablespoonfuls of baking soda should be mixed with 1 pint of sterile water and administered intravenously to cattle. In addition, 1 pint of 50 percent dextrose and 40 mg of dexamethasone administered intravenously will correct the problem of staggering. One gallon of mineral oil should be given orally. (Baking soda, magnesium hydroxide, or magnesium carbonate should not be given orally in severe acidosis because each causes a chemical reaction with the acid in the rumen which generates a phenomenal amount of heat. In fact, so much heat is generated that the lining of the rumen becomes blistered and separates from the underlying tissue. The result is death.)

Peritonitis

Inflammation of the peritoneum is accompanied by abdominal pain which varies in degree with the severity and extent of the peritonitis. Tenderness on palpation, rigidity of the abdominal wall, constipation, and asystemic reaction are the typical manifestations.

Cause

The most common causes of peritonitis are penetration of the abdominal wall and peritoneum or perforation of the alimentary or genital tracts.

Penetration of the abdominal wall can occur when hunters or jumpers become impaled on fences or other barriers which they failed to clear. Harness horses often become impaled on broken sulky shafts or on damaged fences in racetrack accidents. The abdomens of rodeo horses and cow ponies occasionally are perforated by the sharp horns of belligerent bovines.

Traumatic reticuloperitonitis in cattle and goats, perforation of an abomasal ulcer in cattle, and perforation of ulcers of the ileum in pigs are among the common causes of peritonitis. Surgical corrections of intestinal obstruction and of difficult foaling are two commonly performed operations in the horse which may lead to peritonitis.

Fatal peritonitis usually results if the continuity of the gastrointestinal wall is interrupted for any reason and the contents escape into the peritoneal cavity. This may occur as a result of gastric rupture, intestinal torsion, intussusception, parasite damage, breeding accidents, injudicious use of an enema tube, or rough palpation of the colon.

Clinical Signs

Animals acutely affected with peritonitis have an elevated temperature and exhibit marked depression. They assume a rigid stance and are reluctant to lie down. The abdominal muscles are contracted, and the abdominal wall is taut. Groaning heard at each shallow expiration is accentuated when affected animals are forced to move. Early constipation is commonly followed by profuse diarrhea. Dehydration is evident even though animals may consume large quantities of water. Hematologic studies early in the course of the disease usually reveal the presence of elevated white blood cell count, characterized by an increased number of neutrophils. The pulse is rapid and the mucous membranes are congested. Loss of body condition is marked with death often occurring in a few days. Mild cases of peritonitis exhibit the same signs to a lesser degree and recover without problems.

Treatment

When the cause can be determined, treatment should be aimed at correcting it. If perforation of the abdomen has occurred, corrective surgery should be performed. If a specific infectious disease is identified, it should be treated accordingly. Regardless of the cause, practically all cases requiring treatment should receive broad spectrum antibiotics systemically.

Displaced Abomasum

In this disease, the abomasum is displaced from its normal position on the abdominal floor either to the left or to the right or into an anterior position. In left displacement, a sac of the abomasum comes to lie in a position behind the omasum and to the left of the rumen. The greater curvature of the abomasum passes under the rumen, is impounded between the rumen and the left abdominal wall, and lies in the left lower flank. This form of the disease occurs at or soon after calving and is characterized by a capricious appetite, reduced fecal volume, and secondary ketosis. In anterior displacement, the clinical picture is very similar to that of left displacement, but the abomasum or the major part of it is displaced anteriorly and comes to lie between the reticulum and the diaphragm. In right displacement, the abomasum is displaced to the right and is found lying between the liver and the right abdominal wall, and, in severe cases, it may extend as far backwards as the pelvic brim. It is identical to the disease described by some authorities as dilatation of the abomasum and by others as subacute abomasal torsion.

Cause

Many factors can contribute to a displaced abomasum (DA) in a cow. The incidence of displaced abomasum is greatest in the periparturient period within a few weeks of calving. Nutritional and metabolic factors have been identified, and a large percentage of the cases of DAs can be prevented through sound nutritional and calving management.

A low-fiber, high-soluble carbohydrate ration contributes to the formation of a displaced abomasum in a recently fresh cow. The drop in rumen pH alters the motility of the digestive tract. Atony follows, which allows the abomasum to fill with gas and displace. The rumen microbes will adjust to a high grain ration. The increased incidence in DA in a fresh cow can be prevented by acclimating the cow to the higher grain levels fed during lactation, compared to the dry period. Sometimes, a low-fiber diet is relative, or the fiber level will appear adequate when the ration is formulated, but mixing or feeding errors on the farm can destroy the fiber. When the cow receives grain from a computer feeder and is expected to consume forages free choice, a relative shortage of fiber can occur if the forages are not palatable, but the cow continues to eat the large grain meal. Another instance of relative fiber shortage contributing to a displaced abomasum is through feeding finely chopped forages, such as haylage or silage. The rumen requires that a majority of the forage be long particles—2 to 3 inches long for maximum rumen health.

Ketosis, which occurs because the intake of feed energy is less than the energy drainage from milk production, is another contributor to the incidence of DAs. Ketosis occurs as the cow mobilizes body fat rapidly to meet energy demands and the elevated blood ketones further suppress the appetite. This leads to a displaced abomasum in some cases. Fat cows are more prone to DAs as they usually have smaller appetites and more fat stores to mobilize.

Milk fever, both clinical and subclinical, is linked to an increase in cases of DAs. This is due to the effects upon the contractility of smooth muscles. The flaccid stomach that is not moving is prone to displace. Milk fever further suppresses the appetite of cows, leading to poor dry-matter intake and ketosis.

A displaced abomasum can follow other illnesses, such as retained fetal membranes, mastitis, and lameness.

Clinical Signs

Intermittent loss of appetite and an abnormal appetite are common symptoms of a displaced abomasum. Weight loss is rapid and marked; however, the temperature, pulse rate, and respiratory rate are within the normal range. The patient is dull and listless and has a gaunt appearance. An obvious bulge caused by the distended abomasum may develop in the anterior part of the lower left paralumbar fossa, and this may extend up behind the costal arch almost to the top of the fossa. The swelling gives a resonant sound if a person thumps it with a finger. Feces are usually small in volume and pasty, but periods of profuse diarrhea may occur. Milk production decreases rapidly.

Usually the rumen can be felt in the left paralumbar fossa, but in abomasal displacement, the rumen cannot be felt in the left paralumbar fossa because it has been pushed medially by the displaced and distended abomasum. The following method works well for diagnosing an abomasum displaced to the left: An imaginary line should be drawn from the ventral portion of the left paralumbar fossa in a slightly downward and curved direction to the left elbow. Auscultation along this line at each intercostal space will disclose the characteristic "tinkle" (pebble-in-a-well type of sound) if the abomasum is present in this position. In some cases, the sound will be heard readily, while in others—those animals with a more active rumen, particularly when there is only a slight lower displacement, it may take from 5 to 15 minutes to detect the sound.

A history of recent or approaching parturition, together with the presence of the signs described, should enable one to diagnose most cases. Other diseases to consider are traumatic peritonitis and pyelonephritis.

Treatment

There are several treatment options for cattle affected with a displaced abomasum, although the treatment options differ for a right displaced abomasum versus a left displaced abomasum. Simply rolling the cow over to replace the stomach in cases of a left dis-placed abomasum has been used for many years. The majority of the displacements will recur if the abomasum is not surgically tacked in the proper position. Several surgical approaches are used, and the success rate of return to function and milk production is good for an uncomplicated left displaced abomasum. Right displacement of the abomasum is characterized by torsion, and the prognosis is often not as favorable due to compromise of the circulation to the abomasum.

Surgical methods of repairing displacement of the abomasum are:

1. Left standing paralumbar abomasopexy
2. Right standing paralumbar omentopexy
3. Right ventral (recumbent) abomasopexy
4. Rolling and placement of toggles into the abomasum

Bovine Viral Diarrhea

Bovine viral diarrhea is an acute contagious disease of cattle. It was first recognized in New York in 1946. At the present time, it is a problem in all areas of the United States and in most other countries of the world.

Cause

Bovine viral diarrhea spreads readily by contact. There is evidence that it can be spread by vectors, including the footwear of people moving between livestock operations. The virus is present in the feces and urine of affected animals, and susceptible animals can be infected by drenching with virus-containing material, as well as by spraying into the eyes or nose. Bovine viral diarrhea can be transmitted by injecting virus-containing material intravenously, intramuscularly, or subcutaneously. At the present time, animal reservoirs, other than cattle, have not been identified.

Clinical Signs

Bovine viral diarrhea occurs in several forms. In the severe or acute form, there is a rapid onset of fever, ranging from 103° to 108°F. There is often a dry cough and discharges from the nose and mouth. Erosions appear on the lips, dental pad, and other mucous membranes of the mouth. Lameness, symptomatic of laminitis, occurs in some herds. Animals may walk with a stiff gait, or fail to move unless forced. Hoof rings, typical of founder, often appear as an aftermath of the disease. Diarrhea appears in three to seven days; the temperature may last up to four weeks. During the fever stage, the feces are hard and dry. Mucus usually appears just before the development of diarrhea. A rapid loss of weight occurs. Milk production drops markedly and may not return to normal until the next lactation.

Abortion is another feature of viral diarrhea. The virus affects the fetus in the uterus. In experimental cases, the virus crossed the placental barrier as early as the 58th day of pregnancy, and up to, and including, the 7th month. The effects of the virus depend upon the age of the fetus. During the first four months, the fetus probably will die and be aborted; (at this time it is easy for abortion to occur and not be observed) or mummification can occur instead of abortion.

Fetuses infected during the second trimester of pregnancy usually survive but may have incomplete development of the brain and/or lungs and be hairless.

Fetuses affected during the last third of pregnancy will probably undergo a mild form of the disease, recover, and have a high level of antibody at the time of birth and before the ingestion of colostrum.

A milder form of viral diarrhea is more common than the severe or acute form just described. The onset of symptoms is sudden; the temperature will rise to 103° to 105°F; pulse and respiration are increased; there may be a nasal discharge and cough; bloody diarrhea does occur; some animals have ulcers in the mouth. This form of bovine viral diarrhea, commonly diagnosed as shipping fever, occurs fairly often in young animals being moved into feedlots.

Subclinical bovine viral diarrhea quite often occurs in young animals and is evidenced by the appearance of antibodies in the serum. There is often some form of a mild fever that is insufficient to attract the attention of the owner. These cases occur where the disease is established in the herd and are the result of spread within the herd. Perhaps a subclinical case

turns into the acute form due to the stress associated with transportation.

Chronic bovine viral diarrhea occurs in herds where the subclinical form exists and where nutrition and management are marginal. It is characterized by emaciation, poor appetite, and slow growth. The animals may not always have diarrhea. The course of the disease ranges from two to six months, and approximately 10 percent of these animals will die.

Prevention

Vaccination is possible by the intramuscular administration of modified live or inactivated vaccine. Two precautions should be observed: (1) don't use the live vaccine on pregnant cows because of possible abortions and birth defects; and (2) don't vaccinate calves under 6 months of age because it may be ineffective due to the temporary immunity from colostrum of immune dams.

Treatment

There is no specific treatment. Antibiotics and sulfonamides are commonly given to combat secondary infection. Oral administration of balanced electrolytes and fluid is indicated to rehydrate animals with diarrhea.

Transmissible Gastroenteritis

Transmissible gastroenteritis was first identified by Doyle and Hutchings in 1945 at Purdue University. It is a very serious threat to any swine producer who farrows pigs. Transmissible gastroenteritis is usually a relatively minor disease in adult animals but a very serious one in the newborn. This is because very young animals, especially pigs, are not able to mobilize sufficient body reserves of nutrients and fluids to cope with the losses caused by diarrhea.

Cause

Transmissible gastroenteritis is caused by a virus. The growth of all viruses is actually a function of the body cells they invade. Because of the intimate relationship between viruses and cells, the development of practical anti-viral drugs has been difficult. Therefore, there are no drugs that will affect transmissible gastroenteritis, or most other viruses.

Though there is no "cure" for transmissible gastroenteritis, a number of steps can be taken to reduce the chances that it will strike a herd, to stop or delay its spread if it does infect a herd, and to reduce the losses of pigs that become infected. This requires a detailed knowledge about the way the virus spreads, about its effect on pigs, and about immunity and other defense mechanisms of the pig.

Transmission

Although only a few, perhaps even one, virus particle may start a transmissible gastroenteritis infection in a pig, the stools of infected pigs may have as many as 10 million virus particles per cc. The virus-laden fecal material sticks to hands, feet equipment, feed, or anything else and spreads the disease to susceptible pigs. Some virus particles may get into the air in droplets, but this is only important in confined spaces such as in a farrowing house. A few feet of open air space will prevent particles from jumping from one individual farrowing house to another.

The virus is quite fragile. It is killed very rapidly by sunlight and warm temperatures but survives for years when frozen and kept in the dark. This is why it tends to spread from herd to herd more readily in the winter than in the summer.

Clinical Signs

An outbreak may start in any age group of pigs. If not recognized and prevented, it spreads rapidly to all the other hogs in the herd. Growing and finishing pigs usually have a yellowish diarrhea that lasts for one to three days in each pig but may take a week or so to spread through a large group. Pigs over four weeks of age may be infected but not actually have diarrhea. Death losses in these pigs are rare and only occur if some additional stress, such as a severe cold or a secondary bacterial infection, is added to the viral infection.

Very young pigs usually vomit at the first sign of disease. This is followed by severe watery diarrhea, dehydration, and death. The diarrhea on the first day of illness may be so watery that it may not be noticed on casual observation. Death may result in nearly 100 percent of the pigs infected on the first day of life, but it may be much less if they are spared from infection for even three or four days after birth.

Sows infected near or at farrowing time seem to react more severely than other mature swine. Some sows vomit, have diarrhea and fever, and cease to milk. They will usually resume milking if the pigs survive and continue suckling.

Prevention

Since viruses are not susceptible to drugs in general, immunization has been an especially important means of controlling viral diseases. There are some unusual problems in immunization against transmissible gastroenteritis, however. If sows are infected by being fed transmissible gastroenteritis virus during pregnancy, they develop antibodies in their colostrum and milk. These antibodies are able to tie up virus particles in the gut of a pig suckling the sow so they cannot infect epithelial cells. This provides protection to pigs only while they continue to suckle the immune sow. Planned infection of sows with virulent virus can produce useful immunity, but there is a danger of spreading the virus to other swine in the same or other herds. It should be done only when transmissible gastroenteritis is known to be already present in a herd and only under the direction of a veterinarian.

Treatment

There is no treatment recommended for transmissible gastroenteritis.

Enterotoxemia

Enterotoxemia is an acute, non-contagious, highly fatal auto-intoxication of sheep, primarily, but it may afflict goats, cattle, swine, and foals. It is most common among young animals, especially lambs, that are nursing high-producing dams or those being pushed for market on lush pasture and/or high concentrated rations. Enterotoxemia is commonly referred to as overeating disease, pulpy kidney disease, apoplexy, milk colic, "struck," and hemorrhagic enterotoxemia.

Cause

Enterotoxemia is caused by the absorption of exotoxins produced in the intestinal tract of susceptible animals by anaerobic, rod-shaped bacteria called *Clostridium perfringens*. There are six types of *Clostridium perfringens* that have been identified. They are referred to as types A, B, C, D, E, and F. Types B, C, and D are important in farm animal health.

Type B enterotoxemia is known as lamb dysentery. It is a problem in several countries in Europe, Asia, and Africa. It is not a common disease in this country. Lamb dysentery in the United States is caused by various strains of *E. coli* organisms. *Clostridium perfringens*, type C, most commonly affects calves and occasionally lambs and baby pigs. It was first described as the cause of an acute intoxication of mature sheep in England. Type C enterotoxemia is called "struck" in adult sheep and hemorrhagic enterotoxemia in calves, lambs, and pigs. *Clostridium perfringens*, type D, is the cause of enterotoxemia, known commonly as pulpy kidney disease and overeating disease of sheep. Calves and goats are also occasionally affected by *Clostridium perfringens*, type D.

Clinical Signs

Enterotoxemia is an acute intoxication; therefore, death often occurs before clinical signs appear. General signs often include listlessness, weakness, recumbency, prostration, coma, and death.

SHEEP

Enterotoxemia most frequently affects nursing lambs from a few days old to 10 weeks of age. Singles are more likely to be affected at this age than twins. Feeder lambs are most susceptible shortly after being placed on good pasture or heavy grain rations. Mature sheep are also susceptible to enterotoxemia when they are allowed to feed on an abundant supply of highly nutritious feed.

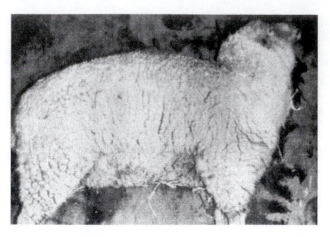

Fig. 10-15. Enterotoxemia in sheep. (Courtesy, College of Veterinary Medicine, Washington State University, Pullman)

Sudden death is common for sheep of all ages that are affected with enterotoxemia. Big, vigorous, fast-gaining lambs that were apparently healthy at night may be found dead the next morning. When observations are possible, affected sheep may jump, fall to the ground, go into convulsions, and die quickly (Fig. 10-15). For those that live longer, cerebral signs are common. These may include opisthotonos, when the head is bent backwards and the back arched; circling motions, when the head is pulled to one side; or pushing against stationary objects, when the sheep's head is depressed. Excitement, prostration, coma, and death usually follow. Although recovery does occur in some cases, the mortality rate is normally high for enterotoxemia.

CALVES

Calves under three weeks of age, nursing high-producing dams, are most likely to be affected. Harsh weather conditions at calving time seem to have an influence on the incidence and severity of enterotoxemia in calves. Calves may show distress, signs of colic, bloody diarrhea, subnormal temperature, and quick death.

PIGS

Young pigs affected by enterotoxemia may show signs quite similar to edema disease of swine. Big, healthy-appearing, fast-growing pigs may be found dead. Those living longer often have posterior paralysis and bloody diarrhea, followed by prostration, spasms, coma, and death.

Prevention

Vaccination, using the appropriate type of *Clostridium perfringens* vaccine, is usually the most desirable approach to the prevention of enterotoxemia. Vaccination of bred ewes and cows provides a passive immunity for their offspring that lasts five to six weeks. Two vaccinations are required at four-week intervals. The last injection should be given to cows about two months before parturition and to ewes two weeks to one month before lambing. Annual booster vaccinations should be administered to maintain the female's ability to pass on immunity to her offspring.

Lambs should be vaccinated with 5 cc of *Clostridium perfringens*, type D, two weeks before going on a full feed fattening program. Injections should be made subcutaneously at sites that will not damage the carcasses should abscesses develop. Choice locations include the bare skin of the inner thighs and forelegs.

Management practices that bring lambs on full feed gradually are also recommended to prevent enterotoxemia of lambs. Aureomycin in the feed reportedly reduces the incidence of enterotoxemia in fattening lambs.

Treatment

Treatment of animals affected with enterotoxemia is likely to be unsatisfactory due to the acute nature of the disease. Antitoxins may be used to provide short-term protection during an outbreak. A more common procedure, however, is to withhold concentrates, place lambs on roughage rations only, vaccinate for enterotoxemia, maintain the roughage ration for an additional 7 to 10 days, and then gradually add the concentrate portion of the ration until full feeding has been established.

Hair Balls

Cause

Hair balls are accumulations of hair, fibrous vegetable matter, and mineral material. The hair enters

Fig. 10-16.   Hair balls removed from the stomachs of cattle.

Fig. 10-17.   Tramp iron, hair balls, and mineral concretion found in the stomachs of cattle.

the animal's stomach as a result of its licking and grooming habits. The hair is moved around by contractions of the stomach, and gradually a small pellet of hair is formed. The pellet enlarges into a small ball, and the ball increases in size as additional material is plastered to it. Hair balls are usually found in the abomasum of calves (Figs. 10-16 and 10-17).

Clinical Signs

There are no definitive signs by which the presence of hair balls in the stomach can be determined.

They are usually harmless and apparently cause no inconvenience to large animals.

Treatment

There is no treatment recommended for hair balls.

Colic in Equines

Colic describes an acute indigestion, evidenced by severe abdominal pain that originates in the stomach or intestines. All animals suffer at times from colic, but the horse is more prone to this ailment than other farm animals.

Cause

Colic may be caused by windsucking, eating spoiled grain, impaction of the stomach or intestines, spasmodic cramping of the intestines, twisted intestines, and intussusception.

Impaction can be caused by any of the following: (1) eating too much grain, (2) bad teeth, (3) eating very coarse hay, (4) suddenly being changed from soft hay to coarse hay or vice versa, (5) eating hay that is not completely cured; that is, it is still quite moist and difficult to digest, (6) insufficient water intake, (7) lack of exercise, (8) internal parasites, (9) old age, and (10) eating sand.

Spasmodic cramping of the intestine is usually caused by large quantities of very cold water being given to a horse. It may also occur when small quantities of cold water are given to a horse that is tired or not in good physical condition.

A twisted intestine usually follows an attack of colic. It occurs when the horse is rolling in pain. This is usually the most painful form of colic and the pulse will be higher than with other forms of colic. It almost always will go over 100 if the intestine is twisted. Therefore, any time a horse with colic has a pulse of more than 100, immediate surgery is recommended.

Intussusception occurs most often in young horses. This is a condition when the smaller part of the intestine telescopes inside the larger portion of intes-

tine. It is the result of abnormal intestinal movement. Worms have been indicated as the cause.

Clinical Signs

Horses with colic may show alternating periods of pain and distress with periods of relief. Abdominal pain is evident by the horse's groaning, pawing the air or its belly, looking back at its sides, lying down, sweating, rolling, or sitting up on the hindquarters like a dog. There is usually an increase in pulse rate and respiration. The horse shows no inclination to eat, and there are no bowel movements. An examination of the eyes may show the conjunctiva engorged with blood.

Prevention

The following suggestions are recommended to prevent colic in horses: (1) the horse's teeth should be kept in good condition; (2) feed should not be changed suddenly; (3) the horse should be exercised regularly; (4) plenty of clean water should be provided; (5) clean, dry hay that is not extremely coarse should be fed; (6) salt should be provided free choice; and (7) the horse should not be allowed to eat off the ground; (8) a proper parasite control program should be followed.

Treatment

Treatment of colic should be under the supervision of a veterinarian. The attendant should walk the horse until the veterinarian arrives. This will keep the horse from lying down and rolling. It will also make it easier for gas to pass through the intestines.

If the horse is too tired and too sore to walk, it should be permitted to lie down but be prevented from rolling.

A veterinarian will perform a physical examination to attempt to determine the cause of the colic and outline a plan for therapy. A rectal examination may reveal twisted intestines, impactions of the large colon, or other abnormalities. The passage of a nasogastric tube will allow the veterinarian to try to aspirate fluid from the stomach, as proximal gastroenteritis has a large amount of reflux, and these cases are non-surgical.

The intensity of the therapy depends upon the intensity of the pain in the horse and the results of the physical examination. Some cases of colic always require surgery, and other causes can be managed medically through the use of laxatives and pain-relieving medication. Mineral oil administered through a nasogastric tube is a standard colic therapy to relieve simple obstructions. Non-steroidal, anti-inflammatory drugs, such as flunixin meglumine, are good as pain relievers and block the effects of endotoxin on the circulatory system. Stronger pain control may be required in some cases and the owner and veterinarian must always discuss the possible need for surgery in horses with uncontrollable pain.

Protozoal Diarrhea

Protozoal diarrhea is an infectious disease of horses of all ages which is caused by a flagellated protozoan.

Cause

The causative agent is a motile flagellated protozoan that appears on direct microscopic examination to be $\frac{1}{3}$ to $\frac{2}{3}$ the size of a strongyle egg. The organism has been identified as *Trichomonas fecalis*. The best material for microscopic examination is the fluid which has been expressed from feces collected from the rectum.

Spread of the infection from one horse to another, as indicated by clinical experience, may be erratic, presumably because of variations in resistance, stress, nutrition, pathogenicity, and other factors.

Clinical Signs

The acute condition is characterized by a temperature of 104° to 108°F, rapid pulse, red conjunctiva and oral mucosas, dryness of the mouth, varying degrees of dehydration, and abdominal pain at the onset, with lack of normal intestinal sounds. Later, there are increases sounds of intestinal liquid as the diarrhea develops. The facial expression may be anxious, and the nostrils slightly flared in severe cases. Sweating and weakness are prominent as the climax is reached in severe cases. Founder, varying from mild to extremely severe, and peripheral edema occur in a portion of the cases. The foul odor from the mouth and from the feces is characteristic.

Chronic syndromes will vary in appearance from those which are nearly asymptomatic to those with severe diarrhea. The body temperature in these cases is normal or nearly normal, and the appetite will vary from fair to good. The chronic cases with severe and protracted diarrhea are unique in that, as a rule, after the initial illness (which may go unrecognized by the caretaker), the horse is bright and active and has a fair to good appetite even though it scours profusely. If the diarrhea persists, young horses may become stunted and appear to be malnourished or severely parasitized.

Founder, varying from fairly severe to mild may be the only general sign in some chronic cases. There are asymptomatic carriers, and there are patients which show clinical signs limited to thinness, reduced ability to work, mild anemia, and general unthriftiness. This is true, at least to the extent that the condition of these horses is less than normally expected of healthy horses under the same level of nutrition and husbandry. In many cases, the feces are softer than would be expected from normal horses on dry feed. The excreta usually forms in loose balls or piles like cow manure.

Treatment

Treatment of the infection leaves much to be desired, although most acute cases may be saved and chronic cases eventually salvaged. Treatment of acute cases is, to a large degree, symptomatic and involves the use of intravenous fluids and electrolytes in quantities sufficient to combat the degree of dehydration. Erythromycin is administered intramuscularly in peracute cases if for no other reason than to combat secondary invaders. Protectives such as kaolin, pectin, and bismuth subnitrate via a stomach tube are indicated. Copious quantities of mineral oil repeated frequently are of particular value in the specific effect on the protozoa. Smooth muscle relaxants are of value in relieving pain and reducing spasms and hyperactivity of the bowel. Atropine, either injected or added to the kaolin, is great for this.

Aspergillosis

Cause

Aspergillosis (mycotic diarrhea) is a persistent diarrhea of foals and young horses. It is caused by *Aspergillus fumigatus*. This is a mold which produces blue-green spores in a culture and gives a typical color to the colony. A portion of a mold colony placed on a glass slide beneath a cover slip will reveal interlacing hyphae and numerous conidiophores. The conidia shatter easily and usually are scattered throughout the field. They are detected readily in infected tissue or in culture.

Clinical Signs

A common sign of aspergillosis is a persistent diarrhea of foals or young horses which occurs following oral administration of antibiotics or sulfonamides. Mycotic diarrhea is usually associated with a heavy parasite infection. The affected animals are in fair condition, alert, have a good appetite and temperatures within normal range. The diarrheic eliminations have no characteristic odor and do not respond to conventional treatment. Often, no specific pathogen other than *Aspergillus fumigatus* can be isolated from the feces on culture. Since this organism is commonly found in moldy hay, it is readily understandable that it can enter the intestinal tract.

Aspergillus has been incriminated in a few cases of equine abortion. The abortions occur during the 7th to 10th month of gestation, and the mares show no evidence of illness. Cultures of fetal tissues reveal pure colonies of *Aspergillus fumigatus*. In three reported cases of mycotic abortion, two fetuses had lesions in the lungs, intestines, and skin. Fungi were isolated from each area. In the third case only, pulmonary lesions were observed from which *Aspergillus* was cultured.

Treatment

Symptomatic treatment with astringents and intestinal emulcents will not control the cause of diarrhea, and failure of such treatment to produce at least a temporary remission should prompt suspicion of a mycotic infection.

The treatment of intestinal mycosis involves the administration of 20,000 units of nystatin twice a day for a period of 10 days and the inclusion of acidophilus milk via drench or stomach tube. Attempts should be made to prevent reinfection. The feed should be changed and the animals' quarters cleaned. The bedding should be removed, burned, and replaced by fresh material. Although these attempts, due to the ubiquitous nature of *Aspergillus*, probably will be unsuccessful, the reduction of a good percentage of infective spores should aid in preventing reinfection.

Potomac Horse Fever

Cause

The causative agent of Potomac horse fever is *Ehrlichia risticii*, a rickettsial organism.

Clinical Signs

Signs are fever (102° to 108°F), depression, loss of appetite, colic, edema of the underline, and stocking of the limbs. These symptoms are usually followed within 48 hours by the onset of diarrhea which, in severe cases, is watery and explosive.

A high percentage of horses with Potomac fever develop founder (laminitis).

Pregnant mares may have transplacental transmission. A mare may get sick in August and then abort a foal in November, or she may have a weak foal that dies or have a foal with respiratory disease.

Some horses show only an extreme toxemia and die within hours if they are not treated.

Treatment

The diagnostic test makes rapid treatment possible. Correct diagnosis is important since the symptoms of Potomac horse fever often mimic those of salmonellosis, yet the drug tetracycline, which is helpful against Potomac horse fever, can be deadly to horses with other diarrheal disorders.

Treatment consists of large volumes of intravenous fluids, tetracycline, and supportive treatment to control fever and reduce laminitis.

Prevention

A successful vaccine is available, which is administered intramuscularly in two doses, repeated 21 days apart, followed by an annual booster shot. So, if Potomac horse fever is present in a state or area, all horses therein should be vaccinated against the disease.

Retained Meconium in Foals

Cause

Meconium is the fecal material that accumulates in the lower bowel of the fetus during gestation. Occasionally, this material is not passed, and impaction or constipation occurs in the large intestine. Retained meconium is a greater problem in foals than it is in calves, pigs, or lambs.

Clinical Signs

Retained meconium is a type of impaction colic in foals. The young foal will act colicky, strain in an effort to defecate, and twitch the tail.

Prevention

There are no recommended preventative measures for retained meconium.

Treatment

An enema with 1 to 2 quarts of warm, soapy water is recommended.

Edema Disease of Swine

Edema disease of swine is an acute disease of pigs, usually from 4 to 14 weeks of age. It is widespread in areas where swine populations are high and may occur at any time during the year. Edema disease may be confused with hog cholera, acute swine erysipelas, pseudorabies, or several toxemias of swine, including warfarin poisoning. This disease of swine is also referred to as gut edema, gastric edema, enterotoxemia, and edema of the bone.

Cause

The disease is dependent upon colonization of the small intestine by *E. coli* that possess the ability to produce a toxin, the edema disease principle (EDP).

Injecting healthy pigs with the intestinal contents of pigs known to have this disease will reproduce edema disease. Stress associated with weaning, changes in feed, castration, and vaccination may promote the rapid growth of *Escherichia coli* organisms in the intestines, and thereby increase the amount of toxic materials that can be absorbed.

Clinical Signs

The first indication of edema disease of swine is often the sudden death of apparently healthy, thrifty pigs between the ages of 4 to 14 weeks. Older pigs may be affected, however. It is not uncommon for edema disease to occur shortly after pigs have been weaned, vaccinated, castrated, or changed to a new feed ration.

The first observable signs of edema disease include a mild listlessness, a wobbly gait, and a slight reduction in appetite. Most affected pigs maintain a normal temperature, but a few may have temperatures up to 105°F. Pigs that are mildly affected usually show signs of recovery in 36 to 48 hours. Those more severely affected usually die in 6 to 24 hours but occasionally survive for as long as 5 to 7 days.

The more dramatic signs of edema disease in the advanced stages consist of a lack of coordination, aimless wandering or a circular pattern of walking, and running into objects as if the pigs are unable to see. The incoordination may be followed by muscular tremors, convulsions, prostration, and death. Prostrate pigs often flail their legs in a running motion.

Edema of the eyelids is a common sign. Tissues in the ears, face, and jowl may also be distended by accumulations of clear or blood-tinged material. Post-mortem examinations often show edema of the stomach walls and may show edematous lesions in the mesentary folds of the large intestine. Edema of the brain is probably responsible for the impairment of motility and for the circling that often occurs in the course of this disease. Edema is not uncommon in the lungs. When the swelling involves the upper respiratory area, voice changes occur, and the pig's squeal sounds like a hoarse bark.

Hemorrhagic lesions are quite common in a pig affected by edema disease. Skin lesions that may appear to be flea bites may be observed on the belly and portions of the legs that are sparsely covered with hair. Post-mortem examinations often show hemorrhagic areas in both the large and small intestines and in the stomach. Hemorrhages may also be found in the kidneys, spleen, lungs, and heart of an infected pig.

Prevention

There is no vaccine recommended for the prevention of edema disease of swine. Management strategies to reduce stress during the time pigs are most susceptible to edema disease should be employed. This could involve castration prior to weaning, vaccination, when required, following weaning, gradual changes in rations for pigs from 4 to 14 weeks of age, and antibiotics to inhibit *E. coli* multiplication.

Treatment

Treatment of this disease has not been very successful. When an outbreak occurs following a change in feed, pigs should be placed back on the original feed if possible. Some producers feel that they lessen the incidence of the disease when there is an outbreak by administering an antibiotic.

Lamb Dysentery

Scours or diarrhea in any farm animal is an abnormal softening or thinning of the feces. Under normal conditions, moisture is extracted from material passing through the large intestines. The fecal material becomes more concentrated and takes on the characteristic shape for each species in the posterior portion of the colon. The feces of normal lambs are pellet-shaped. When a disturbance in the digestive tract occurs due to infection, parasitism, stress, or faulty nutrition, including irregular feeding, the body reacts by diluting and eliminating the material as rapidly as possible.

Lamb dysentery is an acute, highly infectious enteric disease of newborn lambs. It is most common in shed-lambing operations with poor sanitation. Lamb dysentery is not a common disease problem when ewes are lambed on the range.

Cause

The disease is caused by the bacterium *Clostridium perfringes* Type B.

Clinical Signs

Lamb dysentery is a disease that occurs when lambs are from one to two days old. It rarely occurs in older lambs. The first sign is usually a light-yellow–colored diarrhea. The color of the diarrhea may darken as the disease progresses. Affected lambs become weak, disinterested in nursing, and depressed. They stand with their backs arched and ears drooped and may show indications of colic. Untreated lambs or those that do not respond to treatment usually die within a couple of days after symptoms appear.

Prevention

Vaccinate all ewes twice, six and three weeks before lambing. Must be type "B" vaccine or "BCD" combination.

Treatment

There is no highly effective treatment, once lambs are affected. Various drugs such as sulfonamides, antibiotics, antidiarrheals, and anthelmintics may save a few.

Winter Dysentery of Cattle

Winter dysentery is an acute infectious disease of cattle. Both beef and dairy cattle are affected. Winter dysentery occurs most frequently in mature cattle that are housed during the winter months. Winter dysentery is commonly referred to as black scours or winter scours.

Cause

The cause of winter dysentery is unknown. Recent research indicates the disease may be caused by a virus, or by both the bacterial and viral infections occurring simultaneously.

Clinical Signs

The first indication of winter dysentery is a sudden diarrhea in one animal or, at most, a small percentage of the herd. The feces are foul smelling, watery, and brown to black in color. In three to five days, the disease seems to explode through the herd.

Affected animals often have normal temperature, respiration, and pulse rates. They may continue to eat normally, but milk production will drop dramatically. The diarrhea usually becomes darker as the disease progresses. It often contains blood and mucus.

Sick animals often show signs of colic. They may kick at their bellies, switch their tails, lie down, and get back to their feet frequently. They become dehydrated and lose weight rapidly. In severe cases, the animals may become so weakened that they are unable to rise.

The course of winter dysentery usually lasts only three to four days for each individual. It may require a weeks or 10 days to move through a herd. The mortality rate is low unless secondary complications occur.

Prevention

Isolate replacement animals. Separate from the herd any animal suffering from an acute attack of dysentery.

No specific vaccine is available.

Treatment

Most cattle with winter dysentery do not require treatment since they recover within a few days. Severely affected animals can be given oral astringents along with fluid and electrolyte therapy.

Swine Dysentery

Swine dysentery is known by a number of other names, including bloody diarrhea, vibrionic dysentery, hemorrhagic enteritis, bloody scours, black scours, and bloody flux. It affects swine of all ages, sometimes causing many deaths, sometimes exceeding 50 percent, although the average is about 25 percent. The higher death losses come when nothing is done to stop the disease.

Cause

Swine dysentery appears to be caused by the anaerobic bacterium *Treponema hyodysenteriae*, though it probably acts synergistically with other anaerobic bacteria normally present in the digestive tract.

Clinical Signs

A day or two after the onset of diarrhea, mucus, and/or blood may appear. When blood is present in the feces, the term "hemorrhagic dysentery" is commonly used. In younger pigs, the blood is easily recognized; but in older swine, the blood may have a dark color—hence, the term "black scours." However, in many outbreaks of swine dysentery, the bloody diarrhea is not obvious. The appetite may remain good in the early stages. Later, pigs may become depressed and lose their appetite; some may even refuse to eat, although they will usually drink. Dehydration occurs, with emaciation showing in the form of sunken flanks. Death comes suddenly, sometimes before the diarrhea is noticed. Some animals may recover completely from the disease; in others the scouring may return. After the initial outbreak, losses may occur in varying degrees for several months—pigs or breeding stock may develop severe symptoms, with sudden deaths. There may be a tendency toward runty pigs, with some of these occasionally succumbing.

Prevention

Avoid public stockyards and auction rings, isolate newly acquired animals, and practice rigid sanitation.

Treatment

Therapeutic use of antibacterial drugs is effective if treatment is started early. Water medication is preferred at first. Because drug resistant strains may be present, it is essential to choose a drug to which the organism is sensitive. Bacitracin, carbadox, lincomycin, nitroimidazoles, tiamulin, and virginiamycin are commonly used therapeutic agents.

Good management and nursing will help. Milk seems to aid recovery. There is a tendency for relapses following treatment.

Scours in Foals

Diarrhea or scours is a common digestive disease in foals. In most cases, diarrhea is not as severe to the foal as it is to other farm animals. One type of scours called "foal heat scours" occurs when the foal is 7 to 10 days of age or when the mare comes into foal heat.

Cause

Various factors may cause diarrhea, including the mare's first heat after foaling, dietary changes, parasites, and infectious agents such as bacteria or viruses. Most cases are mild and self-limiting, but the infectious diarrheas can be life-threatening and cause significant economic loss.

Recently, equine rotavirus has emerged as a significant cause of foal diarrhea. Rotavirus-induced diarrhea is generally seen in foals under three months of age and is of particular concern in the very young foal where the ensuing dehydration can be fatal.

Clinical Signs

In most cases, scours in foals is mild. It is usually characterized by foul smelling, watery diarrhea and a disinterest in nursing for 24 to 36 hours. In more severe cases, diarrhea may be profuse, the foal may refuse to nurse and may show signs of colic. Foals may become dehydrated and weakened to the point they fail to stand. A complicating disease such as pneumonia is often involved in terminal cases.

Prevention

Sanitation constitutes the best prevention. Foaling stalls should be thoroughly cleaned prior to each foaling, and the mare's udder should be washed. Staff hygiene should be strictly enforced. Foals should receive adequate colostrum. To date, no equine vaccine has been developed for the prevention of rotavirus.

Treatment

Diarrhea can most effectively be treated if discovered early. Treatment should be determined by the cause. If severe diarrhea persists for more than a day, fluids and electrolytes should be administered before the foal becomes too dehydrated. The veterinarian may also administer an antibiotic and/or a gut-soother such as Kaopectate.

Scours in Pigs

This disease is also known as colibacillosis, baby pig diarrhea, or white scours. It can be highly fatal in baby pigs. Infection occurs in the first few days of a pig's life; it seldom appears later.

Cause

The organism usually found in the fatal blood-poisoning cases is *Escherichia coli*, a normal inhabitant of the intestinal tract. When pigs are chilled, particularly at farrowing, lowered resistance permits abnormal multiplication and spread of *Escherichia coli*. Improperly fed sows and poor farrowing conditions contribute to the spread of infection. One of the routes of entrance may be through the navel cord at birth, and the disease may spread among pigs in the litter through ingestion of droppings.

Clinical Signs

Watery, yellowish-white diarrhea appear within a few days after the pigs' birth; the pigs become listless and weak, and they lose weight rapidly. Secondary infections, such as blood poisoning, pneumonia, and infection of the abdominal lining are common. Mortality may reach 100 percent. In those pigs which survive, the feces which coat the tail often dry to a hard cake, with a portion of the tail frequently sloughing.

Prevention

The following practices should be used to control colibacillosis in pigs: An effective sanitation and disinfection program should be initiated and maintained to break up bacterial build-up in the farrowing quarters. In herds of sufficient size, sows that will farrow should be grouped together to build up a resistance to organisms common to the group. This immunity will then be transmitted to their pigs through the colostrum.

Broad spectrum antibiotics, sulfa drugs, or a combination of antibiotics and sulfas may be administered in the feed or drinking water. In herds with a history of colibacillosis, sows may be vaccinated with a commercial bacterin or an autogenous bacterin prepared for a specific herd. Sulfa drugs, aureomycin, terramycin, neomycin, and nitrofurazone are often recommended for treatment of coliform scours in pigs.

Treatment

Treatment, however, should be under the direction and supervision of a veterinarian. He/She will determine the cause and administer the most appropriate treatment.

Scours in Calves

Calf scours is a disease of newborn calves that is caused by one or a combination of the following factors: (1) faulty nutrition, (2) stress, and (3) infectious organisms. Scours is one of the most serious health hazards of young calves. In addition to disrupting the digestive process, scours dehydrates and weakens the calf, making it susceptible to pneumonia and other infectious diseases.

Cause

Enterotoxigenic K 99 + *Escherichia coli*, rotavirus, and corona-like virus are the common causative agents. The *E. coli* cause diarrhea in calves from 1 to 3 days of age, the rotavirus and cornavirus cause diarrhea in calves from 5 to 15 days of age.

Clinical Signs

The clinical signs of calf scours vary widely. Calves with infectious scours that have not received colostrum or those from dams deficient in vitamin A may die shortly after birth without showing any of the classical signs. Other calves that follow this peracute course and live longer may appear normal at birth but go into shock a few hours later. They quickly become too weak to stand and therefore ingest insufficient colostrum to be of any benefit to them. They usually have a very cold nose and very cold extremities, a rapid, weak heartbeat, and enlarged painful umbilicus. White, watery scours with a fetid odor is common when the calves live 48 to 72 hours.

In less acute cases of infectious scours, clinical signs include anorexia, dehydration, a gaunt appearance, a tucked up abdomen, sunken eyes, and a rough hair coat. The temperature rises to from 103° to 106°F, and there is profuse white to yellowish-brown watery diarrhea with a fetid odor. Respiratory involvement, including rapid breathing and a nasal discharge, may be noted.

Other cases often follow a first-stage elevation of temperature to 103° to 104°F that lasts for one to two days. A second stage, one to three days later, has a more severe rise in temperature (104° to 106°F) that is followed by scouring, anorexia, and dullness. When calf scours follows a chronic pattern, the calf loses condition rapidly, becomes pot bellied, and has a rough hair coat and dirty britches due to the accumulation of fecal material on the rear end.

Subclinical cases may involve only a slight increase in body temperature, some scouring, and recovery without detection.

Prevention

The most effective preventive measures of calf scours involves the following three practices: (1) reduce the degree of exposure of newborn animals to the infectious agent (Fig. 10-18), (2) provide resistance with adequate colostrum and optimal husbandry, and (3) increase the resistance of the newborn by vaccination of the dam two to six weeks before parturition to stimulate antibodies which are then passed on to the newborn through the colostrum.

Fig. 10-18. Calf scours can be minimized by individual pens that are kept clean. (Courtesy, Maddox Dairy, Riverdale, Calif.)

Treatment

Treatment of severely affected diarrheic calves should include: discontinuing feeding milk for 24 to 48 hours; giving fluids orally and by injection to combat dehydration; administering gastrointestinal protectants; and giving antibiotics orally and by injection. The choice of the antibiotic should be made by the veterinarian, for in many areas the bacteria associated with calf diarrhea are resistant to many of the available drugs.

Necrotic Enteritis

The so-called necro form of swine enteritis is caused by a combination of infective organisms, poor sanitation, and possibly other stresses. Necrotic enteritis is characterized by patches of dead tissue in the intestine. It usually affects swine up to 100 pounds in weight.

Cause

In most cases of necro, the bacterium *Salmonella choleraesuis* has been implicated. This bacterium is usually widespread and thrives on premises with poor sanitation. When *Salmonella choleraesuis* plus stress is present, an outbreak of necro is likely. Rations deficient in niacin, pantothenic acid, and riboflavin can be contributing factors.

Salmonella choleraesuis is not the only organism involved in necro. When poor sanitation, faulty nutrition, and other stresses coincide with the intestinal damage caused by *Salmonella choleraesuis*, conditions in the gut become ideal for an invasion by other organisms.

A number of experiments have shown that pigs heavily parasitized with large roundworms in addition to *Salmonella choleraesuis* developed more severe intestinal damage than pigs that were worm free.

Clinical Signs

The condition generally begins with a higher than normal temperature, reduced appetite, and grayish diarrhea. Although the temperature may return to normal in a few days and appetite may improve, the pigs become unthrifty, continue to lose flesh and become weak. Prostration and death usually follow.

Prevention

Internal parasites, especially swine roundworms, must be controlled. A properly balanced ration should be provided, with arsenilic acid added to the feed or water. An effective sanitation program must also be implemented.

Treatment

Treatment for necrotic enteritis should be under the direction and supervision of a veterinarian. The principal causative organism of necro is also responsible for causing salmonellosis, which has symptoms very much like hog cholera.

Broad spectrum antibiotics, sulfa drugs, bacitracin, and nitrofurazone are often used to treat necrotic enteritis.

Coccidiosis

Coccidiosis is a parasitic disease of cattle, sheep, and swine. It is usually a greater problem in dairy calves than beef calves because the former are often crowded into small, easily contaminated lots, while the latter are raised on pasture. When feeder lambs, calves, or cattle are crowded into contaminated feedlots, where ingestion of large numbers of coccidia is possible, an outbreak of coccidiosis is likely to occur.

Most mature cattle, sheep, and swine harbor coccidia in their intestinal tracts and shed oocysts in their feces. The oocysts are ingested by young animals, which usually have a low-grade, unapparent infection of coccidiosis throughout their lives. This low-grade infection renders them resistant to coccidiosis until their resistance is lowered by severe predisposing factors.

Cause

Coccidiosis is a parasitic disease caused by protozoan organisms known as coccidia. Each class of animals harbors its own species of coccidia; thus, there is no cross infection between animals.

The development of coccidiosis is greatly influenced by unsanitary conditions, stress due to weaning, shipping, overcrowding, abrupt changes in feed, other diseases, and inclement weather. Birds are carriers of coccidiosis. Therefore, an increase in the numbers of birds around farm animals increases contamination of feedstuffs and the potential for an outbreak of coccidiosis.

Clinical Signs

Coccidiosis is most commonly a problem of young animals. It often occurs two to three weeks after sheep or cattle arrive in the feedlot. Diarrhea is usually the first sign, although some species of coccidia do not produce a great deal of diarrhea. In mild cases, animals may show only a slight loss of appetite and thin, watery feces. Severe cases of coccidiosis are characterized by profuse diarrhea, often stained with blood and mucus. The animals stand with their backs arched and strain repeatedly in an effort to pass fecal material. The animals become weak, listless, and dehydrated. Pneumonia may set in to complicate the situation. Severely affected animals usually die within four to six days after they become sick. Those that survive 10 days to 2 weeks usually recover unless they have complications from secondary infections.

In swine, coccidiosis is characterized by diarrhea that may be followed by constipation, emaciation, dehydration, and anorexia. Blood in the feces is not a common sign of coccidiosis in swine.

Prevention

Avoid feed and water contaminated with the protozoa that causes the disease. Segregate affected animals. Remove and properly dispose of manure and contaminated bedding daily. Drain low, wet areas. Keep animals in a sunny, dry place.

Several ionophore drugs are marketed to prevent coccidiosis, and they can be included in feed, mineral preparations, and milk replacer products. Adding ionophores to the feed/mineral ration of developing cattle has been proven to control coccidiosis and increase the rate of gain in heifers and feedlot cattle. These feed additives are not approved for use in lactating cows in the United States.

Treatment

Amprolium (Amprol) and a decoquinate are effective. They are approved for use in beef and dairy calves, chickens, and turkeys.

Salmonellosis

Infections with *Salmonella* organisms occur in two forms. One is an infection of the genital tract that causes abortions in mares and ewes. The other is enteric. Paratyphoid dysentery affects all farm animals. It is more common in young animals, but it also affects older animals that have been weakened by viral infections, starvation, or stress due to inclement weather, shipping, etc.

Salmonella organisms, when ingested in contaminated food, are highly infectious to human beings. The infective process whereby *Salmonella* bacteria pass from animal feed to animals then to human beings is explained by the following excerpt from the Bureau of Veterinary Medicine DHEW Publication 74-6013:

The food-producing animal may become infected when it eats salmonella-contaminated feed. The infected animal then goes to slaughter with healthy animals and in the processing operation the meat from both animals may become contaminated. Cross-contamination in the preparation of food then takes over to threaten humans with infection.

But the salmonella cycle does not stop with human infection. The animal side of the story continues at the packing plant, on the farm, and even at the restaurant or supermarket. These are the sources of raw material for the animal by-product rendering industry. The renderer takes dead animals from farms, non-food portions of slaughtered animals from packing and slaughtering plants, along with meat scraps from supermarkets and restaurants, and converts them into a dry protein product that is used as an animal feed ingredient.

The source of the salmonella problem in the rendering industry is the raw material. Most raw material is contaminated with salmonella when it enters the rendering plant. The rendering process, however, involves, cooking the raw material at temperatures that will destroy salmonella. This leaves the finished dry protein product essentially free of disease-causing salmonella. Unfortunately, this dry product is subject to accidental recontamination through exposure to the raw material in the plant. If the dry protein is recontaminated with salmonella organisms it will contaminate animal feeds when it is added to them as a protein supplement. Animals which eat this contaminated feed may become infected and set the salmonella cycle in motion again.

Cause

There are over a thousand serotypes of *Salmonella* that commonly inhabit the environment of both human beings and farm animals, and most of these organisms are capable of causing digestive disturbances. The most common types of *Salmonella* for the following species of farm animals are: swine—*Salmonella choleraesuis*; sheep—*Salmonella typhimurium*; cattle—*Salmonella typhimurium, S. dublin, S. entertidis, S. choleraesuis, S. newport*; horses—*Salmonella typhimurium*.

Clinical Signs

Affected animals show marked depression, a loss of appetite, a high fever, and watery, fetid diarrhea. The diarrhea often contains streaks of blood. Pregnant females may abort. As the disease progresses, the animals become progressively weaker, and in the final stages, they are unable to rise. The death loss is heaviest during the first week of the disease.

Prevention

Prevention of *Salmonella* infection is virtually impossible; rather, control is more realistic. The major sources of *Salmonella* introduction into a herd are (1) the introduction of new animals into a herd, and (2) contaminated feed. Quarantining new animals allows time for fecal shedding of *Salmonella* to stop. Wet feed and inadequately cooked feed support the growth of *Salmonella*.

Treatment

Broad spectrum antibiotics are used parenterally to treat the septicemia. A mixture of trimethoprim and sulfadiazine is effective for the treatment of salmonellosis in calves. Ampicillin also may be used for the treatment of septicemic salmonellosis in all species.

Colitis X

Colitis X is a peracute diarrhea in horses.

Cause

The cause is unknown but it may be associated with endotoxic shock. Many, but not all, affected horses have a history of stress or upper respiratory infection one to three weeks prior to onset of the disease, which has led to the postulation that the disease is associated with adrenal corticoid exhaustion.

Clinical Signs

The first clinical sign is a sudden onset of watery, fetid diarrhea accompanied by a fever. The diarrhea is rarely bloody. Within a short period of time, the infected horse becomes severely dehydrated and goes into shock. Then, the pulse becomes thready and the mucous membranes become pale. Animals affected by colitis X frequently die.

Prevention

Horses that are in poor physical condition should not be submitted to unusual stress conditions. Sound management, including proper nutrition, adequate exercise, parasite control, and other factors related to health, undergirds a preventative program with colitis X.

Treatment

Due to the peracute nature of colitis X, treatment should be handled by a veterinarian.

Prolapse of the Rectum

Rectal prolapse is an eversion of the posterior portion of the rectum through the anus. Young animals are most commonly affected, but mature animals with severe diarrhea, tenesmus (ineffective painful straining), and atony of the rectal muscles and the anal sphincter are likely to prolapse.

Cause

This condition is usually due to intestinal irritation which causes prolonged straining or pressure on the rectum, forcing it through the anus. The following factors may contribute to the problem of rectal prolapse: internal parasites, strenuous coughing, diarrhea, inherited weakness of the rectal wall and anal sphincter, and malnutrition.

Clinical Signs

When the rectum is first everted, the mass will resemble a small pinkish-red ball that is soft, shiny, and moist. If the situation is not corrected, the mass hanging from the animal's anus will enlarge, becoming hard and discolored.

Treatment

When the prolapse is detected early, treatment usually involves cleaning, lubricating, and returning the everted mass. Stitches will be required to prevent the prolapse from reoccurring. One method of retaining the swollen rectal material is to place two stitches across the anus. One stitch should be located in the upper one-third and the other in the lower one-third of the anus.

A second method of suturing an animal with a rectal prolapse is to insert a heavy cord around the anus in several places so that the two ends of the cord are left together in a drawstring fashion. The ends of the cord can be pulled and tied to restrict the area of anus. Care should be exercised to insure that ample room is left for excretion of fecal material.

The cause of straining should be determined and treated. Animals usually respond faster if they are separated from the herd or flock and placed in warm, clean, dry quarters for a few days.

In neglected cases in which the prolapsed material has been damaged, frozen, or hardened due to exposure, amputation of the everted mass will be necessary. For pigs and lambs a small-gauge hose may be notched, cleaned, oiled and inserted through the eversion into the anus. Elastrator bands should then be placed over the mass of rectal tissue to inhibit bleeding. The rubber bands should be placed near the body and in the notch on the tube that was inserted in the anus. The prolapsed material should be cut off, leaving the tube and rubber bands in place until the excised tissue is healed.

Atresia Anus

Cause

Atresia anus is a congenital abnormality in which no anal opening was formed. In some cases, there is an absence of the rectum also. When female animals are affected by atresia anus, the rectum often empties into the vagina and defecation occurs via this route.

Clinical Signs

The first sign of atresia anus in baby pigs is usually an enlarged abdomen. Examination reveals there is no anal opening. Other young animals unable to defecate may show signs of colic, uneventful straining, and distress.

Treatment

An artificial anus must be made. The end of the bowel must be located and sutured to the skin of the newly formed anal opening.

CHAPTER II

The Genitourinary System

The urinary and reproductive systems are, in fact, very different systems which share a common terminal passageway for semen and urine in the male and for urine in the female.

THE URINARY SYSTEM

Structure and Function

The primary function of the urinary system is to eliminate waste materials from the blood by collecting, storing, and excreting urine. This is accomplished by a system made up of two kidneys, two ureters, the bladder, and the urethra.

Kidneys

The principal excretory organs of an animal's body are the two kidneys. Their functioning is preceded by removal of waste products from the body cells which are deposited in the blood system and transported to the kidneys for separation and excretion. Although the kidneys' function is primarily that of excretion, kidneys also have an important role in regulating the concentration of water and inorganic salts in the blood.

The kidneys are located under the lumbar vertebrae. They are bean-shaped in pigs and sheep and lobulated in cattle. The horse has one kidney shaped somewhat like a valentine heart; the other is bean-shaped.

Ureters, Bladder, and Urethra

The ureters are two thick-walled muscular tubes that connect the kidneys to the bladder. Urine is continually in production in the kidneys and passes down the ureters into the bladder.

At the juncture of the ureters and bladder, there is a valve that prevents a back flow of urine and under normal conditions inhibits the passage of any microorganisms that get into the bladder from moving up into the kidneys.

The urinary bladder is a muscular organ composed of highly elastic tissue that stores urine until it is eliminated. After a certain amount of urine has been accumulated, muscles in the bladder wall dilate and stimulate nerve endings to set in motion a process that causes the bladder to contract, the sphincter valve to relax, and the urine to pass from the bladder into the urethra and out of the body.

The urethra is a canal extending from the bladder to the external opening of the urinary tract. It serves in both sexes as a passageway for the elimination of stored urine from the bladder. In males, it also functions as the terminal route for secretions of the reproductive organs.

THE REPRODUCTIVE SYSTEM

Structure and Function of the Male Reproductive System

The function of the male in the reproductive process is two-fold. First, the male must produce a sufficient quantity of viable sperm cells and second, deposit them in the female reproductive tract during the time she is in estrus. The inability of the male to perform the first function, that of producing viable spermatozoa, results in infertility. The semen of fertile males that cannot or will not mount can be collected, in some cases, and used for breeding purposes through artificial insemination.

The male reproductive system is comprised of the penis, two testes, which are housed in the scrotum, epididymis, vas deferens, and the accessory glands (Fig. 11-1). The reproductive organs of male farm animals are similar in function, but they vary a great deal among the species in size and shape.

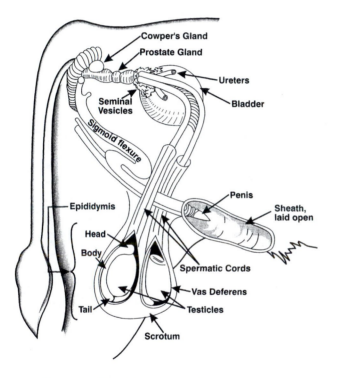

Fig. 11-1.   Reproductive tract of the bovine male.

Penis

The penis is the male organ of copulation and urination. It may be divided into three parts: the glans penis or free end, the body, and the roots which are attached to the pelvis. The urethra extends from the bladder along the lower portion of the body of the penis to the glans. It serves as a passageway for both spermatozoa and urine. In the horse and the ram the urethra extends beyond the glans penis. This free portion is called the urethral process.

The penis consists of cylindrical masses of erectile or cavernous tissue held together by fibrous or connective tissue, an outer covering of skin, and a network of blood vessels and nerves. When the male becomes sexually stimulated, the cavernous tissue becomes engorged with blood, resulting in erection.

The penis of the horse has a higher proportion of cavernous tissue than fibrous tissue. Therefore, upon erection, the penis of the horse expands a great deal in diameter and length. The opposite is true of boars, rams, and bulls. Their penises are constructed with a greater amount of connective tissue than erectile tissue. The diameter of their penises does not expand upon erection like those of horses.

The boar, ram, and bull have an S-shaped sigmoid flexure of the penis. At the time of mating, their penis also fills with blood and becomes turgid, the retractor muscle relaxes and the penis straightens and extends beyond the sheath for penetration of the female reproductive tract.

Testicles

The testes or testicles are paired oval structures that are suspended within the scrotum by the cremaster muscles. Each testicle contains thousands of microscopic tubules. These seminiferous tubules are lined with cells called spermatogonia which produce the sperm. Other cells within the testicle produce the male sex hormone testosterone.

The testicles originate in the abdominal cavity of the body, near the kidneys. They descend just before or shortly after birth through a passage in the abdominal wall known as the inguinal canal into the scrotum. In most cases, the inguinal canal closes, following descent of the testicles. In cases where it fails to close or where it reopens under stress, portions of the intestines protrude through the canal into the scrotum. This condition is known as an inguinal hernia or, in lay terms, a scrotal rupture. Occasionally, one or both testicles fail to descend and remain in the body cavity. Such a male is termed a *cryptorchid*.

Scrotum

The scrotum is a sac or pouch in which the testes are suspended outside the body cavity. The scrotum has two functions. It provides protection, and it regulates the temperature of the testicles. The scrotum maintains the temperature of the testes between 1° and 10°F below the body temperature. It does this by relaxation or contraction of the scrotal muscles. For instance, in cold weather, these muscles contract and pull the testes closer to the body for added warmth. The thermal regulator function of the boar is not as effective as that of other farm animals due to the location of his scrotum.

Epididymis

The epididymis is a coiled tube that is attached very close to the testicle. It is composed of three sections: the head, body, and tail. Sperm produced in the seminiferous tubules of the testicles are transported by way of the vasa efferentia to the head of the epididymis. In the epididymis, fluids are absorbed from the sperm suspension, and the sperm go through further stages of development or maturation. They are stored in the tail of the epididymis until ejaculation.

Vas Deferens

The vas deferens is a muscular tube that attaches to the tail of the epididymis and extends to the urethra. At the time of ejaculation, spermatozoa are propelled from the epididymis through the vas deferens into the urethra. In the case of bulls, rams, and stallions, the vas deferens widens and becomes larger near the attachment with the urethra. These glandular enlargements are known as the ampullae. They contribute fluid to the semen and may serve as temporary storage for spermatozoa.

Accessory Glands

The accessory glands of the male include the ampullae, seminal vesicles, the prostate gland, and the bulbo-urethral or Cowper's glands. These glands produce secretions that make up the greater part of the ejaculate or semen. Semen serves as a medium for spermatozoa.

The seminal vesicles are paired glands located near the base of the bladder. They secrete fluids which contain substances necessary to keep sperm alive. These glands produce about one-half the volume of semen in bulls and rams and about one-fourth the volume of semen in boars and stallions.

The prostate is a single gland which surrounds the urethra between the seminal vesicles and the Cowper's glands. It produces a thick fluid which imparts the characteristic odor to semen.

The bulbo-urethral or Cowper's glands are paired bodies which secrete a clear lubricating material to cleanse the urethra of urine prior to ejaculation.

Semen

Semen is the composite of sperm and the fluids of the accessory glands. It helps activate sperm motility, provides a transport medium for sperm in the female reproductive tract, and coagulates and forms a mechanical barrier to prevent loss of sperm from the female reproductive tract.

The volume and concentration of sperm vary widely from one class of farm animals to another. For example, the stallion and boar produce sperm in low concentration and seminal fluids in large volumes. The stallion ejaculates from 75 to 150 cc of semen per service, while the boar ejaculates 125 to 500 cc, with an average of about 200 cc. Because of the large volumes ejaculated, a boar's mature sperm, those capable of fertilizing the ova, can be depleted due to overwork. Recovery is fast, usually 24 to 48 hours. On the other hand, the bull and the ram ejaculate small volumes, with a higher concentration of sperm. The bull ejaculates from 1 to 14 cc per service, while the ram ejaculates ½ to 2 cc per service.

Structure and Function of the Female Reproductive System

The role of the female in reproduction is much more complex and extensive than that of the male. She not only produces the reproductive cell but also provides nourishment and a suitable environment within her body for the growth and development of her offspring from conception to parturition.

The female reproductive system is composed of two ovaries, two Fallopian tubes, the uterus, vagina, and vulva (Fig. 11-2).

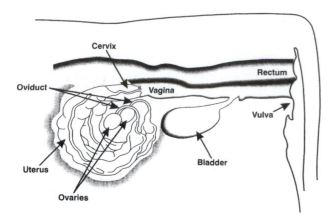

Fig. 11-2. Reproductive tract of the bovine female.

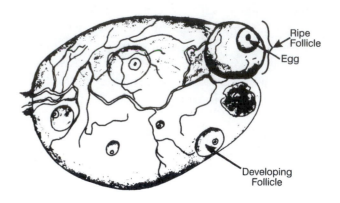

Fig. 11-3. Cow ovary during heat.

Ovaries

The two ovaries are located within the body cavity immediately behind the kidneys. The primary function of the ovaries is to produce the reproductive cell or cells and manufacture the female sex hormone.

The size and shape of the ovaries varies a great deal by species of farm animals. The mare has bean-shaped ovaries, the sow has ovaries shaped like very large blackberries, cattle and sheep have oval- to almond-shaped ovaries (Fig. 11-3).

Ovulation is the process by which the mature follicle, a blister-like protrusion on the ovary, ruptures and releases the ova (Fig. 11-4). In polyparous animals, like the sow, several follicles may rupture and release ova from one or both ovaries over a period of several hours. For this reason, two matings, spaced 12 to 24

hours apart, are often recommended to increase litter size in swine.

After ovulation has occurred, the walls of the follicle collapse and a blood clot forms in the antrum. At this time, certain of the cells which surround the follicles start to form the corpus luteum or "yellow body." The corpus luteum secretes progesterone, a hormone important in maintaining pregnancy. If conception does not occur, the corpus luteum degenerates, and a new estrous cycle begins. However, if pregnancy occurs, the corpus luteum remains functional until parturition is completed.

Fallopian Tubes

The Fallopian tubes or oviducts are paired tubes that transport ova from the ovaries to the horns of

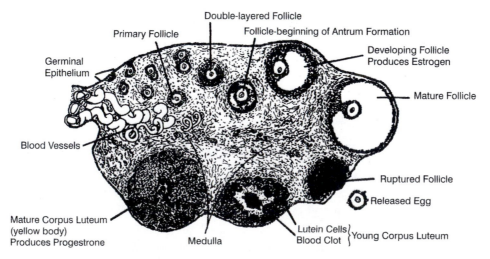

Fig. 11-4. Ovarian stages during the estrus cycle. (Courtesy, Purdue University, West Lafayette, Ind.)

the uterus. The upper portion of each tube expands to form a funnel-shaped structure called the infundibulum or fimbria. The infundibulum receives the ovum and guides it into the Fallopian tube.

The ovum is not equipped with self-propulsion-like spermatozoa; therefore, it must be propelled through the tubes to the uterine horns. This is accomplished by muscular contractions and hair-like processes in the walls of the Fallopian tubes. Fertilization of the ova by spermatozoa usually occurs within the Fallopian tubes.

Uterus

The uterus of a farm animal consists of a body, two horns, and a cervix. The size and shape of each part of the uterus varies a great deal by species. The body of the uterus is quite small, only about 2 inches long in the sow, cow, and ewe. It is much longer in the mare. Fetal development occurs in most species within the uterine horn or horns. During the latter stages of gestation, the horn housing the single fetus and horns that contain several fetuses become much greater in diameter and length than before fertilization.

The attachment of the fertilized egg to the wall of the uterus is called placentation. Ruminant animals have a cotyledenous type of placentation. In this situation, there are from 70 to 120 special areas of attachment. The sow and the mare have a diffuse type of placentation. In this case, there is a loose attachment at no particular area. The sow placentation occurs on or about the 10th to 15th day of pregnancy.

In all types of uteri, the blood of the mother and of the fetus circulate in entirely separate channels. There are no direct neural connections between the mother and the fetus. All nutrients and metabolic products must pass through both the fetal and maternal membranes of the placenta. This is commonly referred to as the placental barrier.

Cervix

The cervix or neck of the uterus is basically a sphincter muscle that controls the passageway from the vagina into the uterus. It is the opening through which sperm must pass to fertilize the egg. It is also the opening through which the young must pass from the uterus during parturition.

The cervix is about 1½ inches long in the ewe, 2 to 3 inches long in the mare, about 4 inches in length in the cow, and almost 6 inches in the sow.

The cervix of the cow and the cervix of the ewe possess circular folds which make it difficult to pass a pipette during artificial insemination. The cervix of the mare has longitudinal folds, while that of the sow has interlocking knobs. The construction of the cervix of sows and mares permits ejaculation into the uterus. The boar screws his penis into the cervix of the sow, and the stallion ejaculates with such force that semen is deposited within the uterus.

The cervix forms a closed door to the pregnant uterus. The cervix has glands which secrete a viscous, stringy mucus which forms a bactericidal plug over the cervix opening. If this plug is ruptured, abortion usually follows immediately. The cervix furnishes a lubricant for the vagina and vulva of the open tract.

Vagina

The vagina is that portion of the reproductive tract that lies between the vulva and cervix. It serves as a sheath for the penis during copulation and a receptacle for semen in the case of bulls and rams. The vagina also functions as a part of the birth canal. The vagina accommodates the urinary system in the elimination of urine. The terminal portion of the urethra opens into the vagina.

Vulva

The vulva is the external part of the female reproductive tract. It functions as the terminal part of the birth canal and the initial organ for copulation. The clitoris is located in the lower part of the vulva. It is composed of erectile tissue and has the same embryonic origin as the penis of the male.

SIGNS OF DISTURBANCES

The major signs of disturbance of the urinary system are anuria, dysuria, and hematuria. Anuria is the cessation of urination caused by an obstruction of the urethra. Dysuria is painful or difficult urination. Hematuria is a condition where blood is voided with the urine.

Signs of disturbances in the female reproductive system are:

Anestrus	Metritis
Abortion	Pyometra
Retained placenta	Vaginitis
Sterility	Dystocia
Nymphomania	Mastitis
Imperforate hymen	Agalactia

Signs of disturbance in the male reproductive system are:

Sterility	Cryptorchidism
Lack of libido	Ulceration of the prepuce
Hypoplasia of the testicle	Occlusion of the prepuce
Orchitis	Scrotal hernia
Epididymitis	Hematoma of the penis

PROBLEMS OF BREEDING ANIMALS

The value of a breeding herd is largely nullified if the semi-annual or annual crop of young animals is not up to a high average. Each female cannot be expected to reproduce up to a maximum efficiency of 100 percent, although this is the breeder's ideal; and the more this is departed from, the less profitable is the production adventure.

Sterility

Sterility in either the male or female is defined as the inability to reproduce its kind. It is also designated as infertility and barrenness. Some animals are said to be barren, though actually they conceive, but expel the young when it is still so small that the act passes unobserved. One authority has introduced the phrase "pregnancy wastage" as one of the conditions that may be included under the general designation of sterility. The livestock producer views sterility as a condition in which the normal mating of apparently healthy animals is not followed by conception. If a certain male fails to settle the females bred to him, he is usually the one at fault. On the other hand, if a female does not settle to the service of a certain male, though other females do conceive from him, the female is the one at fault.

Sterility may be the result of any one or more of a large number of conditions, including age, environment, ration, season of the year, hybrids, disease, and various other factors. Some of these factors are such as to result in permanent sterility, while others are of a more temporary character. Reproduction cannot take place if a diseased condition of the genital organs exists. To continue, for example, to subject a female with an abnormal discharge from the genitals to repeated services in the hope that she will settle is basing expectations on false premises. Healthy young animals cannot be incubated in, or produced from, sexually diseased parents.

Nutrition

The nature, amount, and kind of feed needed by a pregnant animal is a topic frequently discussed. The feed of a pregnant animal need not differ in any respect from the well-balanced adequate ration of the non-pregnant, with the possible exception that, as pregnancy advances, more feed must be offered to meet the needs of the growing young, as well as of the mother. (Some of the essentials of good feeding and the constituents that are most frequently absent, such as vitamins and minerals, are presented in Chapter 2.) In addition, if the growing and developing young animal has improper nutrition and management, it cannot develop into a healthy, vigorous breeder.

Good management under reasonably sanitary conditions, sufficient exercise, and exposure to direct sunshine must not be overlooked. Excessive fatness, especially in breeding animals, will so lower the breeding efficiency of an animal that its value for this purpose is inhibited or lost. High "show condition" and high breeding efficiency simply do not mix.

Body Condition Scoring

The practice of body condition scoring cattle and hogs has become an effective tool in managing the nutritional needs of the animals in different stages of gestation and lactation. Dairy cattle and hogs are scored on a system of 1 to 5, with a score of 1 being very thin and a score of 5 being obese and excessively fat (Fig. 11-5). Beef cattle are routinely scored on a nine-point scale. Optimum body condition scores are associated with early returns to estrus post-calving and high conception rates. The loss of excessive body condition between calving and rebreeding should be prevented by sound nutrition programs. Breeding and milking efficiency are maximized with the optimum

Fig. 11-5. Body condition score 2 (top), 3 (middle), and 4 (bottom).

body condition score of 2.5 to 3.5 for dairy cattle and swine and 5.0 to 6.0 for beef cattle.

Best Breeding Season

The livestock producer is frequently perplexed by attempting to solve the problem of the best breeding season. When carefully analyzed, there is a conflict between nature's time and the producer's desire to regulate the process for financial benefit. In nature's plan, cattle are usually bred rather late in the spring months when the female has had an opportunity to replenish her general condition by feeding on lush vitamin-rich vegetation. In this plan, parturition occurs early during the next spring months. Ewes are bred in the late fall to enable them to lamb when the weather is mild and grass is plentiful in the spring.

For economical reasons in the domesticated animals, cows are frequently bred so as to calve late in the fall and thus be in maximum milk production during the winter months. Ewes, if they can be made to come in heat, are bred to lamb in the late summer or fall when winter pasture is available. Spring lambing is desirable in some areas to take advantage of climate and in some areas to take advantage of climate and feed. Parturition at unseasonable periods necessitates special care and attention to safeguard the well-being of both the dam and her offspring. The dam usually has a shorter than normal time of sexual rest which predisposes her to breeding problems, and the young animal is born during a season of the year when inclement weather predisposes respiratory diseases that are incidental to the unnatural conditions of housing.

Age

The age at which animals may best be used for breeding is a question often raised. For cattle, current recommendations are to breed heifers when they have reached 75 percent of their mature size. The age at which this size is obtained depends greatly upon the nutritional program designed for growing heifers. One problem of very accelerated growth in heifers, where the average daily gain approaches 2 lbs., is fat deposition in the udder. The fat deposited in the prepubertal heifer will have a negative impact on the future milking ability of the heifer.

The onset of puberty, estrus cycle, duration of estrus, and time of ovulation for farm animals are presented in Table 11-1.

Normal Gestation

The normal duration of pregnancy in farm animals varies. There are many factors that apparently exert an influence upon the length of time that the young shall remain in the uterus. Diseased conditions frequently cause a premature termination of pregnancy. Sexual ill health may also prolong the

TABLE 11-1[1]

Animal	Onset of Puberty	Estrous Number of Days After Parturition	Length of/or Frequency of Cycle	Duration of Estrus	Time of Ovulation
Mare	1–1½ years (10–24 months)	9 days (7–11 days)	20–22 days	3–5 days variable	1–2 days before or near end of estrus
Cow	6–18 months	15–60 days	18–21 days	16–19 hours	After estrus 12–19 hours
Ewe.	8–15 months	6–7 months (next breeding season). Nutrition and temperature have important influence	15–18 days	12–30 hours	End of estrus
Sow	4–7 months	3–6 days anestrus[2] till pigs are weaned	18–21 days	1–2 days	Toward end of estrus

[1]From *Animal Sanitation and Disease Control*, 6th edition, by R. R. Dykstra, The Interstate Printers & Publishers, Inc., Danville, Illinois, 1961.

[2]Anestrus—no noticeable heat.

intrauterine existence. When a disease occurs in the uterus, the fetus is improperly nourished and remains unborn for some time beyond the usual duration of pregnancy. In addition, the offspring of some sires remain in the uterus longer than those of others. Furthermore, there appears to be greater irregularity in females that give birth to their first young. The state of nutrition of the mother, the sex of the unborn (usually longer for males than females), and even the severity of the weather may have an influence upon the length of pregnancy.

The gestation period for farm animals is as follows: (1) the mare's ranges from 340 to 350 days, with a maximum of 400 days; (2) the cow has an average gestation of over nine months, with a minimum of 260 days and a maximum of 300 days; (3) sheep and goats carry their young for an average of 149 days, about five months, with a minimum of 145 days and a maximum of 157; and (4) the sow has an average gestation of 114 days, or 3 months, 3 weeks, and 3 days, with 108 days as a minimum and 120 as a maximum.

SOME REPRODUCTIVE PROBLEMS OF THE FEMALE

Persistent Hymen

The hymen is a fold or a thin, circular membrane in the genital tract located a short distance forward from the outer opening of the vagina. It is occasionally present in very young females and may be persistent up to the time of the first attempted breeding. The so-called persistent hymen may be perforate or imperforate, either of which may result in preventing coitus. Furthermore, imperforate genital secretions may accumulate in the vagina so as to distend its cavity. The persistent hymen is readily overcome by surgical means.

Freemartin

This is a barren, imperfect female born twin to a male. The condition is most frequent in the case of a heifer calf born twin to a bull. It may also occur in sheep, goats, and swine. The freemartin heifer, to the casual observer, usually appears normal and may even come into heat, but seldom does she settle to service. The early transfer of blood elements between the bull and heifer fetuses allows hormones produced by the male fetus to suppress the development of the uterus and vagina in the female fetus. In an examination of 74 twins (one a male and the other a female), 66 gave evidence of a common blood supply, and, of these, all the heifers were barren. Eight did not have a common intrauterine blood supply, and all the heifers later settled to service. Freemartins are incurable and should be slaughtered at an early age. Freemartins can be detected by inserting a ½-inch lubricated tube into the vulva of the newborn calf. A heifer with a normal vagina will allow insertion of 5 to 7 inches of the tube, whereas a freemartin will have an undeveloped vagina and only 1 to 2 inches will insert.

Non-specific Infections

General or non-specific infections of the female genital tract are prolific causes of barrenness. Usually, pus-producing organisms are the cause. Nature wisely provides that neither the male nor female element can thrive in diseased surroundings. This important fact is very frequently ignored, as there are repeated attempts to impregnate an animal having genital organs so badly diseased that there is a persistent discharge of pus or other abnormal matter. Some animals do not present clear-cut evidence of these infections, though their failure to conceive after several services by a potent healthy male should arouse suspicion. Such animals should be subjected to a careful examination by a veterinarian versed in the pathology of the genital tract. Other than a collection of discharge of abnormal matter, infection or disease is evidenced by localized or general redness, swelling of the involved area, and occasionally by partial denudation.

At the time of the birth of their young, in all species, a certain degree of non-specific infection of the uterus takes place. Occasionally, the infection is more intense so that it does not clear up naturally, though there is not always outward evidence of it.

In the case of cattle, the first noticeable sign is that after as many as three matings there still is no conception, though other cows served by the same bull conceive promptly. Many of these cases conceive if a day or two following service the female is given an intrauterine injection of a solution of an antibiotic. Veterinarians are fully familiar with the technique. If, after such a treatment, conception does not take place, it indicates that the infection is too severe for the antibiotic to overcome. This method of handling is recommended only for apparently normal animals that fail to settle to at least three services. If it does not produce results, the animal should, in most cases, be sent to market for beef purposes.

Normally, when a cow in heat is served, the sperm and ovum will meet and join in less than an hour after service. This union takes place in the tube extending from each of the two ovaries to the horns of the uterus. Two or three days is required for the fertilized ovum to reach the horn of the uterus for permanent attachment and development. Therefore, if the antibiotic treatment is used within a day or two following service, there will be no interference with the fertilized ovum.

In the mare, antibiotics and sulfas have been helpful in treating infections of the genitalia, but they do not control all the many infections in the mare's organs of generation.

Cystic Ovarian Disease

Cystic ovaries present a serious economic problem in the U. S. cattle industry as cows have prolonged calving intervals. Cystic ovarian disease does not occur in the mare, but it can be one of the leading causes of infertility in the sow. Cystic ovaries in cattle are closely associated with periparturient stresses, such as twin births, retained fetal membranes, milk fever, and large losses of body condition in the early postpartum period. Cystic ovaries are characterized as follicular cysts, luteal cysts, and cystic corpora lutea. Follicular cysts and luteal cysts result from the failure of ovulation of the egg from the follicle. Therefore, the normal progression of hormonal changes does not occur to allow the cow to cycle again in 21 days. Cystic corpora lutea occur after a normal ovulation, yet contain a fluid-filled cavity 10 mm or larger. Cows with cystic corpora lutea usually can conceive and maintain their pregnancies to term.

Clinical Signs

Cows with cystic ovaries may show nymphomania, or be anestrus. Nymphomania cows show frequent, irregular, prolonged estrus and often seek other cows in heat. Cystic ovaries can be detected by palpation per rectum of structures on the ovary over 2.5 cm. Rectal palpation may reveal an enlarged fluid or mucus-filled uterus.

Treatment

Some cysts spontaneously resolve, although managers of high-producing herds will treat cysts when they are diagnosed. GnRH (gonadotropin releasing hormone) injections have been used in cattle with success rates up to 80 percent. Manual removal by digital expression during rectal palpation is not the preferred method, as blood clots can form on the ovaries, which can hamper the normal release of eggs and movement to the oviduct for fertilization.

Ovarian Dysfunction

Ovarian dysfunction, which is a normal condition in those animals approaching old age, is occasionally observed in animals of breeding age. Its exact nature is not understood, though it doubtless is the result of failure of some portion of the ductless or endocrine gland system to function. Normal sexuality is not an expression from the genital organs alone, but the manifestation of an intimate relationship of several glandular and other structures. Failure of one of these to function is followed by a disturbance in the physiological reactions of the others. In such an event, the ovaries become small, hard, and apparently functionless, so that heat periods are in abeyance and barrenness follows.

Anestrus, or the absence of heat, may also be caused by the presence of infections in the uterus. An additional important contributor to anestrus in cows is the large, body-weight loss that occurs during the early part of lactation, which inhibits the normal reproductive cycle.

Sows will not cycle and produce normal ovarian structures when lactating. The removal of pigs stimulates a hormone surge in that the ovaries form follicles and the sow will normally be in heat five days after weaning. Mares are seasonally anestrus, and do not normally cycle during the winter with the short day length. Mares can be maintained under lights to stimulate normal ovarian activity and be available for breeding in January and February.

Retained Fetal Membranes

The fetal membranes are frequently unduly retained or fail to become detached from the walls of the uterus. This accounts for the retained placenta or afterbirth. In the mare, it is considered normal if the afterbirth is discharged within 30 minutes following the birth of the foal. If the cow retains her placenta beyond 12 hours after calving, it is considered to be abnormal. Animals normally giving birth to more than one young expel the membranes after each uterine horn is emptied.

Retained fetal membranes (RFM) can be very costly economically due to the association with lowered milk production, decreased appetite, fever, and possibly abomasal displacement in dairy cows. RMF causes increased days open by increasing the time to first service, reducing the first service conception rate, and increasing the services per conception.

Causes

RFM is caused by a disturbance in the normal loosening of the placenta. Increased incidence of RFM has been associated with:

Dystocia	Premature parturition
Twin births	Induced births
Abortion	Vitamin E and selenium deficiency
Vitamin A deficiency	Milk fever
Placentitis	Prolonged gestation

Treatment

Manual removal was the earliest and most widely used method of treatment. Research has consistently demonstrated that manual removal traumatizes the uterine lining, decreases the natural uterine defense mechanisms, and reduces subsequent fertility. Oxytocin, prostaglandins, injecting the umbilical arteries with saline, and intrauterine antibiotics have been used. No treatment is the current recommendation for treatment of retained fetal membranes in cattle. The cow should be observed for signs of systemic illness and the temperature taken daily until the membranes are expelled. If the cow develops a fever, consult with the veterinarian about the appropriate drugs to reduce the fever and control the uterine infection. RFM in the sow is treated with oxytocin. Mares can develop life-threatening infections from RFM, so a veterinarian should be consulted immediately when a mare fails to expel the placenta within an hour of the birth of a foal.

Prevention

All cases of RFM cannot be prevented, but a good nutritional program balanced for energy, protein, calcium, phosphorus, and trace minerals will help prevent many of the nutritional related cases. The use of body condition scoring in cattle, swine, and mares will help to monitor the success of the nutrition program.

Only a few of the more common conditions responsible for temporary or permanent infertility have been mentioned; however, enough information has been presented to indicate the complexity of the problem. If there ever was a situation requiring the attention of an experienced, skilled person, then this is that situation.

SOME REPRODUCTIVE PROBLEMS OF THE MALE

Failure to reproduce vigorous offspring may be due to a diseased condition of the male. This problem is not nearly so frequently recognized in the male as in the female. It is a comparatively easy matter to diagnose barrenness in a female simply on the basis that she does not reproduce, either by failing to settle to service or by giving birth to immature and non-viable young. The male normally serves many females. If all these fail to conceive as a result of these services, the male is undoubtedly at fault. The male may settle some of the females, though frequently the offspring is born prematurely, or retained for a prolonged period, or so weak that vigorous life outside the dam is not possible. Some of the conditions responsible for partial or complete sterility in the male are mechanical causes, low vitality, sexual dysfunction, orchitis, and infectious diseases.

Mechanical Causes

Mechanical causes are sometimes responsible. There may be adhesions between the penis and the adjacent tissues of the sheath that prevent protrusion of this organ. Lacerations of the orifice of the sheath due to rank grass or brush may result in an occlusion that will prevent ejection of the penis. In other instances, because of a previous injury, the penis during erection is so bent or deflected that successful penetration of the vagina is impossible. Tumors on the penis may also be included in this grouping. However, some of these conditions respond quite well to surgical correction.

Low Vitality

Low vitality in the male due to an inadequate diet can cause reproductive problems. Rations low in vitamins and minerals, underfeeding, overfeeding, lack of exercise, and comparable conditions are conducive to low sexual vigor. In many instances, a restoration of vitality may be brought about by appropriate sane handling and feeding of the affected male.

Sexual Dysfunction

It is well recognized that sexual vigor in the male and female is not dependent on the functioning of any single organ. It is a combination of stimuli and reflexes originating in several endocrine glands located in different parts of the body. If the function of any one of these structures does not coordinate with the others, there is a disturbance of the entire group, and dysfunction results. The result is sterility in varying degrees. Mating problems can occur from improper exposure and training of the adolescent male, such as improperly grouping a young boar with a large, dominant sow. The environment for breeding may also contribute to mating problems, such as slippery flooring or extreme heat and humidity. Young bulls should be pasture bred to 25 cows; mature bulls can cover 35 cows.

Orchitis

Inflammation and degeneration of the testicles due to mechanical injury, or as a result of some diseases, or accumulations of pus in some portion of the genitals may result in the failure of secretion of vigorous living sperms and in infertility of the affected male (Fig. 11-6).

Fig. 11-6.   A ram with orchitis.

Infectious Diseases

Numerous specific diseases may lodge in the testicles or other portions of the genital system and destroy the procreative ability of the male. Diseases such as tuberculosis of the testicles, actinomycosis of the testicles, trichomoniasis, brucellosis, and others less common are, in the light of present knowledge, considered to be incurable, though in regard to brucellosis and trichomoniasis, there appears to be some hope for a cure.

In many of the diseases just mentioned, spermatozoa under microscopic examination are found to be slow or motionless. In normal semen, spermatozoa move in the seminal fluid in swirls of motion. Slow semen is entirely without potency. Some of these semen conditions improve as the male's general condition gets better, or under the influence of sunlight, exercise, and good feed. Sometimes, a semen examination should be given, supplemented by gonadal biopsy, when the male is suspected of being at fault in the breeding program. Veterinarians do work of this nature as an almost routine procedure in suspected male impotency.

Since the question is frequently raised, it should be emphasized that, on the basis of present knowledge, there is no indication that the nutritive requirements of breeding males are any different from good feeding practices required for any other animal. The reproductive requirements of the male are not specific. Any well-balanced ration will automatically influence the reproductive organs in the same way that other organs of the body are affected.

DISTURBANCES OF THE GENITOURINARY SYSTEM

Brucellosis

Brucellosis is a disease that used to be a severe problem in human beings and cattle. The disease is known as undulant fever when it affects people. In addition, brucellosis affects goats, swine, dogs, and horses. The incidence of brucellosis is less than 0.1 percent, according to 1996 USDA statistics.

Cause

There are three brucella species: (1) *Brucella abortus* (formerly called *Bang bacillus* because it was isolated by Dr. Bang in 1897), which infects cattle and humans; (2) *Brucella melitensis*, which infects goats; and (3) *Brucella suis*, which infects swine. The bacteria are small, non-motile, non-sporing, Gram-negative coccobacilli.

The brucella organisms cause infection by being ingested, through skin abrasions, by contact with the conjunctiva of the eye, and by contamination of the uterus when females are bred. The most common source is by ingestion of contaminated fluids from an aborted fetus. Brucellosis is usually transmitted to a non-infected herd by the introduction of infected animals.

Bovine Brucellosis

Cause

The primary infectious agent in bovine brucellosis is *Brucella abortus*, but cattle are occasionally infected with *Brucella melitensis* or *Brucella suis* when they are exposed to infected goats or swine, respectively.

Clinical Signs

The most common sign of brucellosis in cattle is abortion after the fifth month of pregnancy. Cows usually abort only once. Retained placenta following abortions and a reduction in milk production are also frequently observed. Orchitis is the only sign of brucellosis that may be observed in bulls. The seminal vesicles, ampullae, epididymis, and testicles may be infected. In such cases, brucella organisms are commonly passed in the semen. Inflammation and abscesses in the testicles may occur. In dairy cows, the milk ring test is used for diagnosis, and, in other cattle, blood serum is tested for the presence of brucella agglutinins.

Prevention

A cattle owner can prevent brucellosis from entering his/her brucellosis-free herds by raising all replacement heifers or purchasing all replacement cattle from brucellosis-free herds. If this is not feasible, then heifer calves should be vaccinated with strain 19 or RB 51 *Brucella abortus* vaccine.

Vaccinating with strain 19 will not prevent brucellosis, but it will increase the number of cattle in a herd that are immune to brucellosis. This procedure will minimize the spread of brucellosis within a herd and significantly decrease the number of infected cattle. In addition, all replacement cattle should be tested for brucellosis before they are added to the herd. Dairy and beef heifers should be vaccinated at 4 to 10 months of age with the RB 51 *Brucella abortus* vaccine by an accredited veterinarian.

The current method of brucellosis control in the United States consists of testing the milk of dairy cattle and testing the blood of all breeding animals that are sold. If reactors are found, they are traced back to the herd of origin, and the entire herd is tested. Reactors found in herds suspected of having brucellosis are branded, tagged, and removed for slaughter. This process is repeated until no reactors are found on two successive seroagglutination tests.

Semen that is being used for artificial insemination must have an antibiotic added to it.

Treatment

Numerous drugs have been recommended for the control of bovine brucellosis, but none have been successful.

Porcine Brucellosis

Cause

Porcine brucellosis is caused by *Brucella suis.*

Clinical Signs

The clinical signs of porcine brucellosis are abortion in sows that have been bred for more than 80 days and enlarged lymph nodes. Diagnosis is made by blood testing.

Prevention

There is no vaccine for porcine brucellosis; therefore, the disease must be eradicated.

Treatment

There is no treatment for porcine brucellosis.

Ovine Brucellosis

Cause

The cause of ovine brucellosis is *Brucella melitensis*, which is not much of a problem in sheep, and *Brucella ovis*, which causes epididymitis in rams. Rams become infected in the conjunctiva of the eye. The infection becomes systemic and the localizes in the epididymis. The infection inhibits the flow of sperm, and sperm granulomas form as a result.

Prevention

When *B. meletensis* is diagnosed in a herd, slaughter of the herd should be considered. Rev 1 vaccine for female sheep and goats has been recommended. Young rams should be maintained separate from mature rams.

Treatment

There is no treatment for ovine brucellosis.

Equine Brucellosis

Cause

Equine brucellosis is caused by *Brucella abortus*.

Clinical Signs

The clinical signs of equine brucellosis are fistulous withers and poll evil. Fistulous withers is an abscess over the withers. Poll evil is also an abscess that is located on the top of the head of a horse. Both types of abscesses ripen and rupture, spreading infectious material and providing an open sore for additional infection.

Prevention

Horses should not be stabled in areas that are known to be contaminated with *Brucella abortus* bacteria.

Treatment

The treatment for equine brucellosis is surgical removal of the draining abscess.

Leptospirosis

Leptospirosis is a disease that is important in humans, dogs, cattle, swine, and horses. It is of little significance in sheep and goats. The organism that causes leptospirosis is a tightly wound spiral with a hook on each end. The technical term for this bacterial form is *spirochete*.

There is a wide host range for leptospirosis that includes cattle, sheep, swine, horses, goats, and humans. The disease is widespread throughout the United States and occurs during all seasons in both sexes and all breeds. All ages are affected, but the highest mortality is in the young. Reservoirs include swine, rodents, and wild animals. Swine and cattle can be urinary carriers for months. The disease spreads more rapidly when there are concentrations of livestock. Infection can spread from one premise to another by animal movement, or it can spread in the water of moving streams. Also, the infection may be spread by natural breeding, but antibiotics mixed with semen prevent infection during artificial insemination.

Cause

Leptospira pomona is the main cause of leptospiral infections in the United States. However, *L. canicola*, *L. icterohemorrhagicae*, *L. hardjo*, *L. bratislava*, and *L. grippotyphosa* have been isolated also. Transmission occurs through direct contact with infected urine, placental fluids, or milk. Venereal transmission is common in *L. bratislava* in swine and *L. hardjo* in cattle. The leptospires penetrate exposed mucous membranes and water-softened skin.

Bovine Leptospirosis

In the United States, bovine leptospirosis is due primarily to *Leptospira pomona* and is characterized by hemolytic anemia, abortion, or failure to secrete milk, but with a high percentage of the infected animals showing no symptoms.

Clinical Signs

The incubation period for bovine leptospirosis is 7 to 10 days. Anorexia lasting from two to five days, fever 105° to 107°F, and depression are common signs. Hemolysis of the red blood cells occurs and varies in degree and duration. An occasional mature animal may develop a transient hemoglobinuria. In calves, there may be a massive hemolysis with hemoglobinemia, hemoglobinuria, and death. The urine is coffee-colored if hemoglobinuria exists. In some herds, cessation of milk production is the only sign of bovine leptospirosis. Milk becomes thickened and yellowish-colored, with an appearance of colostrum, but, occasionally, it is pink in color. The udder is flaccid. Bred cows usually abort 7 to 10 days following a febrile reaction. In herds experiencing a high abortion rate, no clinical signs are noted before abortion. In unvaccinated herds, 25 to 40 percent of the cows may abort. Fetuses at all stages of gestation may be aborted; however, abortion during the latter stages of gestation is most common. Anemia and icterus are common when young calves are affected with bovine leptospirosis.

Porcine Leptospirosis

Leptospirosis of swine in the United States is generally caused by *Leptospira pomona*. It is usually a benign subclinical disease, except for abortion. Infertility is associated with *L. bratislava* and is characterized as a repeat breeder syndrome.

Clinical Signs

The incubation period for porcine leptospirosis is about seven days. Sick animals will have a fever of 105° to 107°F for two to four days. In most cases, there are no clinical signs. Meningitis has been reported on rare occasions. Illnesses including anemia, icterus, and hemoglobinuria, have been encountered in swine infected with *Leptospira pomona*. Abortion is most likely to occur when sows that are more than 59 days into pregnancy are exposed. During the leptospiremic phase, organisms cross the placenta and infect some or all of the fetuses. Abortion usually occurs during the last three weeks of gestation. The sow may abort dead pigs, the sow may "farrow early," with some live pigs that die soon after birth and with some pigs that are mummified, or the sow may farrow at the normal date, with weak pigs that have a poor survival rate and with a possibility of one or more mummified pigs in the litter.

Ovine Leptospirosis

Although leptospirosis rarely occurs in sheep, it is characterized by anemia, hemoglobinuria, and abortion in these animals.

Clinical Signs

Sheep are relatively resistant to leptospira infection. Clinical disease is rarely recognized. It is doubtful that subclinical disease is even common because of the low incidence of antibodies in surveys of sheep serum for leptospirosis. Clinical disease is more likely to occur when sheep are debilitated or have concurrent disease. *Leptospira pomona* has caused abortion, followed by icterus, depression, and death in ewes in the United States. Since there are so few cases of leptospirosis in sheep, the dollar loss is very low.

Equine Leptospirosis

Equine leptospirosis is a mild to moderately severe illness characterized by fever, icterus, depression, and sequelae such as abortion and periodic ophthalmia, also known as "moon blindness." Horses can be venal carriers for a period of three months. Other species of livestock are more likely to be the source of equine infections.

Clinical Signs

The incubation period of equine leptospirosis is 7 to 10 days. Infected animals will run a fever of 104° to 106°F for about three days. Affected horses are depressed and have poor appetites, and their mucous membranes become icteric. Coffee-colored urine is not a symptom. Pregnant mares either abort or have weak-living foals. In natural or experimental infections with *Leptospira pomona*, periodic ophthalmia, or "moon blindness," may occur during the acute illness or up to 24 months later.

Prevention of Leptospirosis in All Species

Purchase clean animals, isolate them, and then retest. In cattle and swine herds, blood sampling should be performed to determine titers to *Leptospira spp.* Control is based on vaccination, sanitation, and treatment. Annual or twice yearly vaccination is the most effective approach to control.

Treatment of Leptospirosis in All Species

Treatment of animals, which should be prescribed by a veterinarian, may include blood transfusions, administration of selected antibiotics, and good care. Antibiotics give fairly good results if cases are treated promptly. It appears that selected antibiotics must be used to eliminate shedders. High levels of the tetracycline drugs (400-500 g per ton) can be used in swine complete feeds for 14 days to help remove the carrier phase. A single streptomycin injection stops leptospiruria in the majority of swine. Cattle can be treated with long-acting oxytetracycline, in two doses ten days apart. In human leptospirosis the M.D. should be consulted relative to treatment.

Bovine Venereal Campylobacteriosis

Bovine venereal campylobacteriosis is an infectious venereal disease that occurs worldwide. It is an important cause of infertility, early embryonic death, and occasional abortion in cattle leading to economic losses from delayed calving and increased culling.

Cause

Venereal campylobacteriosis in cattle is caused by *C. fetus* subspecies *venerealis*, which is a motile, curved. Gram-negative rod that requires special conditions for culture. *C. fetus* subspecies *fetus* is a major cause of abortions in sheep. The transmission of the organism is primarily venereal, with the use of artificial insemination with infected semen and poorly sanitized reproductive instruments being additional routes for transmission. Bulls less than three years of age appear resistant to infection.

Clinical Signs

The main clinical signs are related to infertility in females. The fertilization and early embryonic development are not affected by *C. fetus* sub. *venerealis,* but the inflammatory response in the uterus causes embryonic death. Repeat breeders, extended heat intervals, and extended calving intervals are the main symptoms until the animal clears the infection naturally. The organism is difficult to culture and may require several types of laboratory analyses to prove the existence in the herd.

Treatment and Control

The use of artificial insemination and the cessation of natural breeding are the best way to control bovine venereal campylobacteriosis. Artificial insemination should be used for at least two breeding seasons, as persistently infected cows can be a source of infection for new bulls

Two vaccinations of animals, 2 to 4 weeks apart, at least thirty days before breeding is a part of many control programs. Infection in bulls can be prevented and treated by vaccinating the bulls twice. There are not any antibiotics specifically approved for treating cattle infected with bovine venereal campylobacteriosis.

Bovine Trichomoniasis

Bovine trichomoniasis is a venereal disease manifested by infertility, early abortion, and pyometra.

Cause

Tritrichomonas fetus is the causative agent. It is a pear-shaped flagellate with three anterior flagella. *Tritrichomonas* is the generic name, but, in most clinical literature, the term *trichomonas* is used.

Trichomoniasis is usually introduced into a herd when an infected bull or cow is added. It is generally spread by sexual contact. Nonvenereal transmission has been reported, but it rarely occurs. Although bulls usually show no clinical signs, most cows develop a vaginitis of varying severity within three to nine days after being bred by an infected bull.

Clinical Signs

In the trichomonad infected herd, the most prominent manifestation is infertility. Many infected cows will return to estrus one to five months after breeding. This is suggestive evidence but not diagnostic because other genital infections, most notably vibriosis, result in much the same herd history. Vibriosis and other genital infections may also occur simultaneously with trichomoniasis. In well-managed herds, signs may be noticed and assistance provided early. Frequently, however, the early signs are ignored until the conception rate becomes five services per conception. The normal conception rate should be 1.3 to 1.7 services per conception. As the older cows become resistant to the infection, the conception rate improves, and sterility tends to be limited to heifers and replacement females (Fig. 11-7).

Within 24 hours after being bred by an infected bull, the cow's vulva and vagina may be edematous and inflamed. In most cows, a clinically apparent vaginitis develops within three to nine days after breeding. These signs are not exclusively associated with trichomoniasis, and they frequently are unobserved under field conditions.

Early embryonic abortion causes many of the cows to have delayed estrus. The caretaker is likely to

Fig. 11-7. Bull with trichomoniasis. This animal appeared normal, but spread the disease during the breeding act. (Courtesy, College of Veterinary Medicine, University of Illinois, Urbana)

consider the cow pregnant, only to observe estrus occurring 30 to 90 days after she has been bred. With less than 90 days of development, a fetus will not be observed in most cases. Obvious abortion may occur in a few cows at three to five months of pregnancy. After five to six months, the likelihood of abortion is minimal. Trichomonad abortions tend to occur during the first four months of pregnancy, while the majority of vibrio abortions occur between the fourth and seventh month of pregnancy.

Prevention

Use artificial insemination and bulls free of the infection. If practical, sell infected animals for slaughter or allow 90 days of sexual rest. Exercise great precaution in introducing new animals into the herd, in breeding outside cows, and in taking cows outside the herd for breeding purposes.

Treatment

Generally speaking, bulls infected with trichomoniasis should be slaughtered rather than treated. Successful systemic treatment of bulls has been reported with the use of dimetridazole, ipronidazole, or metronidazole. However, such usage lacks official regulatory approval.

Infected cows should be provided sexual rest for at least 90 days and then served by artificial insemination or by clean bulls.

Epizootic Bovine Abortion (Foothill Abortion)

Epizootic bovine abortion (EBA) is a non-contagious, infectious disease that is manifested by a high incidence of abortion among cattle during late pregnancy. In the United States, it occurs most frequently in cattle that are being pastured in the foothills or on mountainous terrain; hence, the disease is colloquially known as "foothill" abortion.

Cause

A virus causes foothill abortion. It is transmitted by *Ornithodoros coriaceus*, commonly known as the pajaroello tick.

Clinical Signs

In California, the disease is very severe, causing the highest incidence of abortion from mid-summer through the fall. The seasonal abortion rate parallels the seasonal calving program. When females are raised in an area where epizootic bovine abortion commonly occurs, they build up an immunity to the disease. However, females that are introduced to enzootic pastures from EBA-free areas may have a 75 to 90 percent abortion rate during their first pregnancy. Cows rarely abort more than once. This indicates that an immunity develops as a result of one infection. If they leave an enzootic area and return to an enzootic area, they will abort again. Therefore, the immunity can be lost if it is not maintained by constant exposure to the causative agent.

All breeds of cattle are susceptible to epizootic bovine abortion. The only symptom is abortion during the last trimester of gestation. Occasionally, calves are aborted at term. In rare cases, calves are born alive, but they are weak and only survive for a short time.

Prevention

Replacement females should be raised on pastures known to be enzootic EBA-free areas so that they can develop an active immunity against foothill abortion. This is the only control measure recommended. There is no vaccine available for epizootic bovine abortion.

Treatment

There is no treatment for epizootic bovine abortion.

Pyometra

Cause

Pyometra is an inflammation of the uterine mucosa, characterized by accumulations of pus in the sealed uterus. It occurs following post-partum infection, or as a result of a uterine infection such as trichomoniasis that causes complete dissolution of the embryo or fetus. Pyometra is most common in cattle, but it also occurs in horses, sheep, and swine.

Clinical Signs

Anestrus is a common sign of pyometra in cattle. This is due to an incomplete recession of the corpus luteum. This sign may not evoke concern in postcoital cases of pyometra, but it should serve as an indication of reproductive problems in postparturient infections. Pyometra is characterized by the presence of pus in small amounts and up to as much as 3 gallons in a sealed uterus. Diagnosis is made by a rectal examination to differentiate pyometra from pregnancy.

Treatment

Treatment should be entrusted to a veterinarian. The pus may have to be siphoned from the uterus if expulsion is not accomplished by the use of a drug. This should be followed with intrauterine douches of broad spectrum antibiotics to combat the uterine infection. Terramycin, nitrofurazone, and combinations of penicillin and streptomycin saline solutions may be used.

The pus must be removed from the uterus by the siphoning process in mares affected with pyometra. Antibiotic douches should be used to treat the bacterial infection in the uterus. Several treatments may be required.

Bovine Cervicitis, Endometritis, and Metritis

Cervicitis is an inflammation of the cervix, endometritis is an inflammation of the endometrium,

the membrane lining the uterus, and metritis is an inflammation of the uterus. The clinical signs, diagnosis, and treatment of cervicitis, endometritis, and metritis are quite similar; therefore, they will be considered together.

Cause

Reproductive disease in cattle is likely to be a problem in herds in which there is lack of sanitation when assisting with dystocias, treating cows with retained placentas, and performing artificial insemination. In addition, bulls that are infected with vibriosis and trichomoniasis can spread these diseases and also cause metritis and endometritis.

Clinical Signs

There are usually no outward signs of illness, and there is rarely a fever in cows affected by cervicitis, endometritis, or metritis. Occasionally, cows will have a severe uterine discharge that can be seen on the tail and rear legs (Fig. 11-8). Infections of the genital tract may be diagnosed by the clinical signs when they are present. However, manual examination of the reproductive tract and visual examination of the vagina and cervix are usually necessary to diagnose reproductive infections of this nature. The cervix will be enlarged and may have mucopus on or below it in cervicitis. Endometritis is accompanied by the enlargement and thickening of the infected uterine horn. Purulent endometritis is accompanied by a thick, yellow-white exudate in the anterior vagina. Metritis usually occurs in older cows and is more likely to be chronic. Chronic purulent metritis will be accompanied by severe enlargement and thickening of the

Fig. 11-8. Uterine discharge in a cow with metritis.

uterine horns. A purulent exudate is often present in the vagina. There is very little pus found in the uterus of a cow with chronic metritis.

Prevention

Replacement bulls should be thoroughly examined for reproductive disease before they are turned out with the cow herd. Sanitary procedures should be used in removing the afterbirth from cows with retained placentas. Antibiotic boluses should be placed in the uterus of all cows after the retained placenta has been removed.

Treatment

Treatment should be left to the veterinarian. Most cases are treated by the introduction (in solution or tablets) of an antibiotic or sulfa into the uterus. Injections of prostaglandins to induce estrus at 7- to 14-day intervals are very effective in resolving metritis cases and also prevent the possible contamination of the meat and milk supply with antibiotics.

Equine Cervicitis and Metritis

Cervicitis is an inflammation of the cervix, and metritis is an inflammation of the uterus. Endometritis is an inflammation of the endometrium or the membrane lining the uterus. Endometritis occurs in various forms. It may be catarrhal, croupous, fungal, gangrenous, hemorrhagic, or septic. Endometritis is referred to as cervical when it affects the cervix and corporeal when the body of the uterus is affected. Genital infections that cause inflammations in both the cervix and uterus are quite common in barren mares, but recently mares that have reproduced have also been affected.

Cause

Most genital infections in the mare are due to pneumovagina, complications during parturition, post-foaling problems, and poor breeding practices. Pneumovagina leads to cervical and uterine infections when the aspirated air contains fecal material and other contaminants that can infect the mucous linings. Dystocias often cause tissue damage that leads to uterine or cervical infections. Retained placentas, or por-

tion of the placental membranes, that are not expelled within 8 to 10 hours usually result in a uterine infection. Poor breeding management, including unsanitary back-up breeding practices, is the common cause of cervicitis and metritis.

Clinical Signs

There are usually no outward signs of illness in mares affected with equine cervicitis and metritis. A febrile condition is rare. Occasionally, mares will have a uterine discharge that can be seen on the tail and buttocks. Genital infections may be diagnosed on clinical signs, if there are any. Usually, however, it is necessary to insert a sterile swab into the uterus during estrus and culture the material collected in the laboratory. A veterinarian can often tell if the mare has an infection by the way the uterus feels on manual examination.

Treatment

If the mare is a windsucker, a Caslick operation should be performed by a veterinarian. This should be followed by intrauterine treatment with antibiotics. Other mares may require only an intrauterine infusion of antibiotics. The treatment should be administered while the mare is in heat. The mare should be cultured again at the next period following treatment to be sure the antibiotic worked. The mare should not be bred until the disease-causing bacteria have been eliminated from the reproductive tract.

Bovine Vaginitis

Cause

Vaginitis is an inflammation of the vagina. It occurs as a primary or secondary condition. In infertile cows, it is associated with cervicitis, endometritis, and chronic metritis. Primary vaginitis usually is infectious, and specific forms are infectious pustular vulvovaginitis, granular vaginitis, and viral vaginitis. Granular vaginitis seems to predispose, accompany, or be aggravated by other types of vaginitis, and the combination apparently delays conception. Organisms that are known to cause vaginitis are *Haemophilus somnus*, ureoplasmas, mycoplasma, and herpesvirus.

Clinical Signs

A type of vaginitis that is seen frequently in some herds during certain years is characterized by a slight purulent exudate and a mild cervicitis. When the vulvar lips are opened, the mucosa is bright red, and a few strings of mucopus may be present. The inflammation is milder in the posterior part of the vagina, while the anterior part, adjacent to the cervix, is reddened and somewhat edematous. The external opening of the cervix is edematous with congestion and prolapse of the first annular ring. The prolapsed fold is deep red. This type of vaginitis may be viral in origin, and it may appear in virgin heifers. Frequently, the condition will be present in a herd for two years and then disappear, possibly due to development of immunity. Vaginitis due to trauma, retained placenta, and prolapse often delays breeding.

Treatment

In severe cases of vaginitis, sexual rest, and topical applications of anti-bacterial ointments such as acriflavine, bacitracin, and neomycin should be used to control bacterial infections. In addition, sulfa drugs and broad spectrum antibiotics administered intramuscularly may be used in treatment for bovine vaginitis. Bacterial ointments may be beneficial in mild cases of vaginitis. Spontaneous recovery often occurs in two to three weeks.

Equine Vaginitis

Cause

Vaginitis is an inflammation of the vagina. The acute form of vaginitis does not occur frequently in mares, but the chronic form is very common. Tissue damage due to bruising and tearing during foaling, or due to service by a large, rough stallion, or other forms of injury to the vaginal membranes are sources of bacterial infection.

Chronic vaginitis is quite common in "windsucking" mares. This condition, called pneumovagina, occurs in mares whose labia have lost normal tone. The flaccid vulvar lips permit aspiration of air into the vagina and expulsion as the mare walks, trots, or runs. The aspirated air creates a "balloon" effect in the vagina that places additional stress on the atonic muscles and causes the condition to become worse.

Pneumovagina occurs most frequently in old multiparous mares, but it can occur in young mares with an anatomical predisposition to windsucking. Such mares are flat at the croup and high at the tailhead and have small underdeveloped labia and a sunken anus that pulls the lips apart during physical exertion. Pneumovagina occurs more frequently when mares are in heat, although some mares have this problem while in estrus and during the interestrual period.

Clinical Signs

In mild cases of pneumovagina, there are no apparent signs, with the exception of the obviously poor conformation of the vulva, plus the fact that the mare is a poor breeder. If examined during estrus, the cervical secretions will have a frothy appearance which increases substantially after exercise. In extreme cases of pneumovagina, in addition to ballooning of the vagina, some fecal matter can be seen in the vulvar antrum and sometimes in the vagina. The cervical and vaginal mucosas will appear irritated. In some instances, the vaginitis is characterized by a thin discharge which is apparent at the ventral commissure of the vulva. This discharge usually soils the hindlimbs and tail of the mare.

Pneumovagina is conducive to sterility for two main reasons. The first is that, subsequent to irritation of the vaginal mucosa and the presence of airborne or fecal contaminants, a bacterial vaginitis, a cervicitis, and occasionally an endometritis develop. The second reason occurs without the development of infection. The irritation of the vaginal mucosa causes a change in the pH of the genital secretions and, in so doing, inactivates sperm introduced during coitus. Cases have been observed in which mares, obvious windsuckers, were bred successfully after vaginal irrigation with sodium bicarbonate or after the introduction of sperm directly into the uterus after its collection from the dismounting stallion.

Prevention

There are no preventative measures for this type of equine vaginitis.

Treatment

Antibiotic therapy to control bacterial vaginitis and the Caslick operation performed by a veterinarian are recommended for treatment. Antibiotic or sulfa drugs may be administered intramuscularly and as suppositories placed into the vagina in severe cases of vaginitis.

Mastitis

Mastitis is considered to be the most costly disease to the dairy industry, with losses of $200 per cow yearly. Mastitis is inflammation of the udder, with females of any species affected (Figs. 11-9 and 11-10). Mastitis has many adverse effects, both in the cow and on the production of food products made from mastitic milk. Mastitis decreases milk production, damages the milk secretory cells of the udder (Fig. 11-11), causes systemic illness, and occasionally, death of the animal. The milk product shelf life and consumer acceptance are decreased due to increased in-

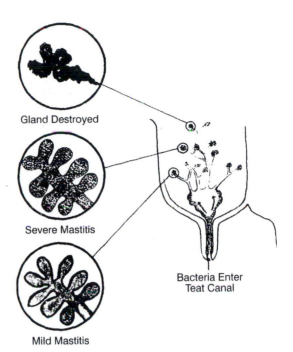

Fig. 11-11. Mastitis destroys cells in the udder. The scar tissue damage is permanent.

flammatory cells (somatic cells), increased whey protein, free fatty acids, sodium chloride, and pH. Large economic losses are seen as a result of culling for mastitis, where genetic potential is lost from the herd.

Cause

Mastitis is caused by bacterial penetration of the udder. It can be related to trauma of the teats, improper milking machine function, and contamination of the environment, which overcomes the natural defenses of the udder. The udder defenses are composed of nonspecific and specific defense mechanisms.

Nonspecific factors of defense are not targeting any specific organism, rather are gland protective for any microbial and fungal organism. Nonspecific factors are the teat orifice, which is elastic connective tissue and works to keep the teat closed against milk leakage and bacterial penetration. The teat canal is lined with keratin that has antimicrobial properties.

Specific defense mechanisms in the udder are immunoglobulins, lymphocytes, and neutrophils. The process of phagocytosis, whereby neutrophils engulf and destroy bacteria, is the most effective mammary defense.

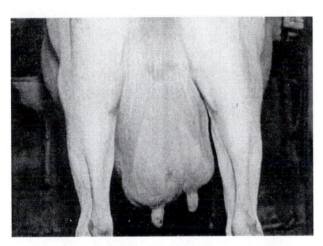

Fig. 11-9. Chronic mastitis. The left rear quarter is infected. (Courtesy, Department of Surgery and Medicine, School of Veterinary Medicine, Kansas State University, Manhattan)

Fig. 11-10. This ewe has mastitis. Note the exaggerated leg movements to avoid contact with the inflamed udder.

CONTAGIOUS AND ENVIRONMENTAL ORGANISMS

Mastitis causing organisms are divided into two groups—contagious and environmental. Contagious organisms are transferred from cow to cow during milking: from infected machines, milkers' hands, and other milking equipment. The environment surrounding the cow, whether drinking, walking, or resting areas, is the source of pathogens in environmental mastitis. Gram-negative organisms are the main infections seen in well-managed, low somatic cell count herds.

Contagious Pathogens

Staphylococcus aureus—*S. aureus* mastitis is the most common contagious mastitis organism and can cause greatly elevated somatic cell counts and lost production. The udder and skin of the teats serve as a reservoir for the infection. Once the bacteria have entered the teat canal, they adhere to the lining of the gland, which prevents the bacteria from being flushed out of the gland during the next milking. The bacteria further produce toxins that permit invasive growth, the survival inside neutrophils, and the formation of abscesses. *S. aureus* predominately causes subclinical mastitis, although acute cases with toxemia and death also occur. *S. aureus* infections have a poor antibiotic response and cure rate; therefore, culling, proper milking hygiene, and teat dipping eliminate the organism.

Streptococcus agalactiae—The main source of infection is the infected udder, although environmental contamination may be a factor in herds with poor hygiene. *Strep. ag.* is susceptible to intramammary antibiotics. The organism can be eliminated from the herd through dry cow antibiotic therapy and teat end disinfection.

Actinomyces pyogenes—Infections are usually associated with teat end damage, and the organism causes a very suppurative mastitis. Proper use of dry cow antibiotic treatments is effective in controlling *A. pyogenes*.

Environmental Pathogens

Streptococcus uberis—*Strep uberis* is a common organism present on the skin of the cow and can be shed in large numbers through the feces. The bacteria multiply in the bedding and infections occur most of-ten in the dry period. Dry cow therapy helps reduce new infection rates.

Coliform pathogens—*Escherichia coli, Klebsiella spp.,* and *Enterobacter aerogenes* are the most common causes of Gram-negative mastitis. Infection results from gross contamination of the teats between milkings or from liner slips during milking a poorly cleaned and dipped teat. Endotoxemia and clinical illness are related to the large numbers of neutrophils destroying the organisms in the udder. Milk cultures of clinical cases are often negative, due to the inflammatory response in the udder.

Mycoplasma species—*Mycoplasma bovis* is the most common cause of bovine mycoplasmal mastitis. The signs are thick purulent milk; a hard, swollen udder early in the course of the disease; and a sharp drop in production. Control of mycoplasma mastitis is through milk culturing and culling of positive animals, as well as proper teat disinfection and hygiene.

Clinical Signs

Mastitis can be divided into two types based on clinical signs—clinical and subclinical mastitis. Clinical mastitis is characterized by one or more of these symptoms: off-color milk; curdy, clumpy, or stringy milk; red, hot, swollen udder; and fever and illness in the animal. Elevated SCC without the visible signs of inflammation characterizes subclinical mastitis. Subclinical mastitis is very costly to the dairy industry, due to milk production losses. SCCs of 500,000 cause a 10 percent production loss, and SCC of one million leads to a 16 percent production loss. The California mastitis test and other individual cowside tests are good for the detection of cows with elevated SCC. Bulk tank somatic cell counts indicate the herd's mastitis prevalence. The current limit for bulk tank SCC is 750,000/ml in the United States, with international pressure to lower the limit to be in line with the saleable limit in other countries. A goal for the well-managed dairy herd is a herd SCC below 200,000 and 85 percent of the individual cows below 200,000.

Prevention

A complete discussion of all the measures important to the control of mastitis is beyond the scope of this text. However, the control of mastitis is focused on three areas—the milking machine and system, the milker, and the cow. Good premilking preparation is the only mastitis prevention tool universally available. This method maximizes oxytocin release for greater milk letdown and lactation yield, as well as maximizes the contact time of the dip for destroying mastitis-causing organisms present on the skin.

MILKING MACHINE AND SYSTEM

1. Ensure that there is adequate vacuum capacity and reserve capacity.
2. The milk and pulsation lines need to be properly sized for the number of units and milk flow.
3. The pulsation ratio should not exceed 60 percent milking and 40 percent resting.
4. The inflations should be replaced according to the manufacturer recommendations, or anytime cracks occur.
5. The air flow and regulator response must be sufficient to recover the vacuum within three seconds when airflow is admitted through milk clusters.
6. The proper hot water temperature, sanitizer, and wash system must be used to control bacteria counts.

MILKER

1. Milk only clean, dry teats. *See "Once Under Method of Milking Cows"*
2. Attach the milking cluster within one and one-half minutes of udder preparation.
3. Minimize air entering the system when attaching the unit.
4. Position the unit for it to function effectively and correct the position to prevent liner slips.
5. Do not machine strip or overmilk.
6. Dip the teats with an approved teat dip immediately following milking.
7. Culture the quarters showing abnormal milk.
8. Use an intramammary antibiotic tailored to the organisms in the herd.

9. Positively identify treated animals, milk them last, and withhold their milk from shipment.
10. Maintain teat and udder skin health through clean bedding, assuring an uncontaminated water supply, and maintaining effective fly control.
11. Provide proper nutrition, especially selenium and vitamin E.
12. Keep mastitis case and treatment records.

COWS

1. Consider docking the tails and flaming/clipping the udders to reduce debris on the udder.
2. Provide adequate lying space, feeding space, and watering space to prevent teat injuries.
3. Consult with the herd veterinarian regarding the use of vaccines for coliform mastitis.

Once Under Method of Milking Cows

1. Knock the dirt off the teats and strip each quarter, observing for abnormal milk.
2. Dip the teats with an approved teat dip.
3. Stimulate and rub the teat dip into the surface of each teat, paying special attention to the crevices around the teat orifice.
4. Dry and clean each teat with an individual towel.
5. Apply the milking units.

Treatment

The treatment for mastitis varies with the severity of the case and the organism involved. See Appendix B for examples of good herd management practices for treating and recording mastitis cases, as well as forms for recording individual animals treated for mastitis in light of antibiotic residue prevention. A mild case of mastitis may respond to oxytocin therapy to induce complete milk-out of the affected quarter and allow the cow's natural defenses to cure the infection.

Moderate cases of mastitis may require intramammary antibiotics formulated specifically for use in the udder of a cow, a horse, a goat, or a sheep. The intramammary antibiotic preparations are packaged in single-use tubes that prevent disease transmission from animal to animal. Multiple-use containers, such as those of injectable penicillin used to treat *Strep ag.* cases of mastitis, are not recommended. Multiple dosing from one container to save treatment costs has induced many cases of non-treatable mastitis in cows. These cases have resulted from contamination of the container, and thus the udder, with yeast organisms. Commercial intramammary antibiotic formulations to treat mastitis cases are available in both prescription and over-the-counter packages. Prescription antibiotic formulations can be purchased only from a veterinarian, who can consult with the owner regarding the proper diagnosis and treatment of clinical cases. Owners are discouraged from using two types of antibiotics in a quarter at one time, as well as from using the drugs in an extra-label method. Milk withdrawal times must be followed to ensure an antibiotic-free milk supply.

Severe cases of mastitis present as animals with high fever; listlessness; anorexia; hard, bloody quarters; and possibly diarrhea and recumbency. A veterinarian should be consulted to diagnose a suspected case of severe mastitis and develop a treatment plan. A cow with coliform mastitis will often be culture-negative due to the severe influx of neutrophils into the udder, killing the bacteria. A cow with coliform mastitis will often not benefit from intramammary antibiotics for this reason, but is in dire need of supportive care to treat the effects of endotoxic shock from killing so many bacteria. Oral and intravenous fluids, anti-inflammatory drugs, and calcium supplements are often used in the treatment of severe cases of mastitis.

Mastitis-Metritis-Agalactia

Mastitis-metritis-agalactia (MMA) is a major problem of sows and gilts at farrowing time. Each component of this syndrome may occur alone or in various combinations. Mastitis is defined as an inflammation of the udder, and metritis as an inflammation of the uterus. Agalactia literally means no milk or interference with milk secretion.

Cause

The organism commonly thought to cause the MMA syndrome is *Escherichia coli*. Other organisms that have been associated with the syndrome include: *Actinobacillus, Actinomyces, Aerobacter, Citrobacter, Clostridia, Corynebacterium, Enterobacter, Klebsiella, Pseudo-* *monas, Mycoplasma, Proteus, Streptococcus, Staphylococcus,* and *Chlamydia*. It is particularly noteworthy that *E. coli* is also the most common organism isolated as the cause of diarrhea in baby pigs during the first three weeks of life. However, the disease is complex, and proof of the cause is often difficult.

Clinical Signs

Some or all of the following signs of disturbance may occur at farrowing, or three to four days later: loss of appetite, fever, shivering, and a foul-smelling vaginal discharge that is yellowish-white in color. The udder sections are hot and inflamed, and the sows may lie on their bellies and refuse to get up. The sows often show a disregard for their litter.

Clinical signs depend to a large extent on the exact cause of trouble and on the primary disease. For example, if hormonal imbalance is the only cause and if agalactia is the only disease, an affected sow would be unlikely to show any sign other than milklessness. On the other hand, if agalactia is secondary to some kind of infection that has caused mastitis or metritis or both, the sow would probably be visibly sick, in addition to not having any milk. The complex is usually encountered in this latter form.

Prevention

Prevention revolves around sound management, good nutrition, superior sanitation, and proper swine husbandry. When sows are kept in confinement, it is especially important to keep the sows from becoming overfat and constipated, and to reduce stress, particularly near the time of farrowing. The first condition can be controlled by limiting the ration, and the addition of 6 to 10 percent molasses or extra bran to the ration will prevent constipation. Also, underfeeding may contribute to the problem.

Treatment

The list of suggested treatments is long, confusing, and not always effective; among them are:

1. Cross-fostering the piglets to a normal mother. This appears to be the most effective solution.

2. Antibiotics or nitrofurans to eliminate infections.

3. Oxytocin, a hormone to cause milk let-down.

4. Relief of constipation by (a) using a warm, soapy enema, (b) feeding a ration high in molasses or wheat bran, or (c) providing morning and evening exercise.

5. Supplemental milk and glucose to keep the baby pigs alive.

6. Oral antibiotic or nitrofuran for the baby pigs if diarrhea occurs.

There is no confirmed evidence that vaccines or antibacterial agents have any beneficial prophylactic effect.

Prolapse of the Uterus

Prolapse of the uterus may occur in any female, but it is most common in cattle and sheep. This condition usually occurs immediately following parturition. It is limited to the few hours following expulsion of the fetus when the cervix is open and the reproductive tract lacks tone.

Cause

The cause of uterine prolapse is not known. It occurs in some cases of dystocia, dry births of fetuses that must be manually removed, and occasionally in cows that have milk fever. Atonic muscles of the reproductive tract due to overfeeding and a lack of exercise are likely to contribute to this problem.

Clinical Signs

A prolapse of the uterus is complete. The large mass of material that is everted hangs almost to the ground when the animal is standing. A uterine prolapse in a cow and ewe is differentiated from a vaginal prolapse by the presence of cotyledons, which are the large, round, intermittently spread attachments for the placenta. The mass tends to become larger the longer it remains outside the body because the arterial blood enters the uterus, and the venous flow is restricted.

Treatment

A uterine prolapse is a medical emergency and a veterinarian should be called immediately. To minimize damage to the uterus, the cow should be kept as quiet as possible until the veterinarian arrives. A sheet or similar wrap can be applied around the uterus to reduce contamination from bedding and fecal material. The veterinarian will administer an epidural injection, wash the uterus with antiseptic solution, and replace it into the normal position.

Prevention

The incidence of uterine prolapse can be reduced through proper dry cow nutrition to avoid milk fever, and the administration of oxytocin during manual delivery to maintain uterine tone.

Vaginal Prolapse

Prolapse of the vagina occurs in all domestic species of females, but it is most common in sheep and cattle close to parturition.

Cause

The cause of vaginal prolapse is believed to be due to a hereditary predisposition and pressure on the anterior wall of the vagina by a large fetus in the uterus and a full rumen and colon. Fleshy cattle and ewes with restricted exercise are much more likely to have vaginal prolapses than range animals. Sheep and cattle that are pastured in hill country and dairy cattle confined to stalls with sloping floors are more likely to have the eversion than animals that can rest in the recumbent position on flat surfaces.

Clinical Signs

The first sign of vaginal prolapse is the presence of a small part of the vaginal mucosa while the cow or ewe is lying down. The pink-colored mass usually disappears when the affected female rises to the standing position. In the advanced stages, the vagina may be completely everted and remain outside the vulva. This displacement of the vagina covers the urethral opening and prevents urination. If the prolapse is not corrected, the affected ewe or cow will eventually die from uremic poisoning, necrosis, or another infection.

Prevention

Females that have a history of vaginal prolapse should be culled from the herd or flock.

Treatment

A local anesthetic should be administered epidurally to allow suturing of the vulva. The mass should be cleaned and massaged to reduce swelling. Occasionally, the cervix and bladder will prolapse with the vagina and care needs to be taken to minimize damage to the cervix and bladder. Once replaced, the vagina is sutured in a pattern to prevent further prolapse, yet allow the animal to urinate. The animal needs to be closely watched for signs of parturition so the sutures can be loosen for delivery.

Urinary Calculi

Urethral blockage by stones is known as urolithiasis, "water belly," and urinary calculi. Frequently, the blockage is complete and is followed by rupture of the urethra or bladder.

Cause

The disease can affect both males and females. However, as a rule it is seen in steers before puberty because they tend to have the smallest urinary passageway. The calculi form in the bladder and then cause problems when they are being voided in the urine. A combination of factors is responsible for the problem. The constituents in cattle feeds seem to be primary causative factors; silica, oxalates, and estrogens in plants in large quantities increase the incidence of urinary calculi. In feedlot cattle and lambs, the stones are composed primarily of calcium, magnesium, and ammonium phosphates. Sorghum feeds and beet pulp increase the incidence of urolithiasis. The balance of calcium and phosphorus is very important in calculi formation. If the calcium level is low, the animals are prone to form calculi. Insufficient salt in the ration or water deficiency can cause the urine to become too concentrated and result in problems. Vitamin A deficiency may contribute to urolithiasis.

Clinical Signs

There must be partial or complete obstruction of the urethra before the animal will show any symptoms (Fig. 11-12). When the urethra becomes blocked, the animal will strain to urinate, switch its tail occasionally, tread with its rear feet, and kick at its belly. Sometimes, the animal will void small amounts of urine intermittently. Occasionally, lambs become bloated and constipated the day before they show other signs of urethral blockage. In feedlot cattle, the steers frequently have a partially ruptured penis and urine leaks into surrounding tissue in the abdomen before the animals are diagnosed. These animals have a very swollen abdomen in the region between the external urethral orifice and the rear legs.

Rectal palpation of the bladder assists in making a diagnosis and prognosis. If the bladder is full or greatly distended, and the animal is demonstrating symptoms of urolithiasis, the chances are good that it is a urinary calculus. If the animal is showing symptoms and the bladder is empty, the bladder ruptures, partially heals, and begins to fill again. Rupture of the bladder is followed by signs of uremia and death if surgery is not performed.

Prevention

Good nutrition and management are very important in the prevention of urolithiasis. Animals should be given plenty of fresh water, a balanced mineral supplement, free access to salt, and enough vitamin A for normal health. If cattle are getting a milled ration, salt

Fig. 11-12.   A wether lamb affected with urolithiasis. Note the straining effort to urinate. (Courtesy, Washington Agricultural Experiment Station)

should be added to the ration. If there is a problem with urinary calculi, the proportion of salt in the ration should be gradually increased until it is up to 4 percent.

Treatment

In feedlot or range steers, surgically removing the portion of the penis that is affected and exteriorizing the end of the penis just below the anus works very well. This treatment will also save about 50 percent of the animals with ruptured bladders, if the veterinarian gets to them shortly after their bladders rupture. In breeding animals, tranquilizers and smooth muscle relaxants may be tried. They seem to be most effective in animals that are passing some urine.

When very valuable bulls are involved, surgical removal of the stones from the penis can be tried. The success depends on both re-establishing the flow of urine and preventing the development of adhesions along the penis.

Pizzle Rot

Pizzle rot is also known as enzootic posthitis, sheath rot, and balanoposthitis. It is inflammation of the prepuce and penis of male sheep. Castrates are more commonly affected with pizzle rot than rams, but it is not uncommon in rams.

Cause

Although the disease is associated with castration and protein-rich diets, it is caused by a bacterium indistinguishable from *Corynebacterium renale*, which acts on urea and initiates the ulceration.

Clinical Signs

A small scab above the external urethral orifice may be the only sign of pizzle rot for long periods of time. If the scab extends to the interior of the prepuce, the preputial opening becomes ulcerated (Fig. 11-13). At such times, sheep will become restless and kick at their bellies, and will only dribble urine as they would with a urinary calculus. Preputial swelling commonly occurs after an infestation of blowfly maggots. Pus and fibrous tissue adhesions may form and interfere with urination and protrusion of the penis. When

Fig. 11-13. Pizzle rot in a wether lamb.

this occurs in rams, it impairs breeding permanently. Rarely does the penis become completely obstructed. When this does occur, the sheep will die from secondary obstructive uremia, toxemia, or septicemia.

Ulcerative vulvitis occurs in ewes which are in the same bands in which the wethers are suffering from pizzle rot; however, no ulcerative vulvitis has been found in cows that are being run with bulls suffering from sheath rot. The lesions in ewes only affect the lips of the vulva and consist of scabs and ulcers. A blowfly strike can be a problem in affected ewes.

Prevention

Wool should be removed from around the prepuce or vulva to avoid urine accumulation. The diet should be restricted to reduce the urea content of the urine. Affected sheep should be segregated, and the preputial area in rams disinfected if the causative bacteria is present in the environment. Preventative disinfecting is recommended for rams at six-month intervals. This should begin when the rams are six months old. The antiseptic should be infused into the prepuce and smeared over the skin around the prepuce. In ewes, the disinfectant should be swabbed onto the vulva and surrounding skin.

Testosterone propionate is highly effective in preventing the disease when it is implanted subcutaneously. Affected animals need simultaneous application of an antiseptic. An implant of 60 to 90 mg is effective for three months and can be repeated throughout the year. However, timing the implants to coincide with periods of maximum incidence is more economical.

Treatment

Treatment involves the application of (1) antiseptics such as 5 to 10 percent copper sulfate ointment, quaternary ammonium compounds, or aluminum silicone; or (2) ointment containing penicillin or bacitracin to the affected area. Before application, any scabs, urine, pus, and debris should be removed. Changing the diet to low-quality pasture or straw and implanting testosterone, the male hormone, may hasten healing.

Broken Penis

Broken penis, also known as hematoma of the penis, is seen most often in bulls, especially beef breeds because of natural breeding under range conditions.

Cause

The cause of broken penis is a misdirection of the penis during thrusts of breeding that creates excessive pressure on the penis where it extends from the sheath. The majority of the tears occur through the tunica albuginea on the top of the penis. This indicates extreme downward bending has occurred.

A bull's penis doesn't elongate during erection. The apparent increase in length is due to uncoiling of the sigmoid flexure. The tunica albuginea in the bull is quite dense. The corpus cavernosum has a fine cavernous pattern and does not change in diameter during erection. The rigid nature and tissue density result in rupture or tearing of the tunica albuginea.

The injury is seen more frequently in younger bulls than in older ones, indicating inexperience or overexcitement on the part of the young bulls.

Clinical Signs

Bleeding through the ruptured tunica albuginea causes a large blood clot or hematoma to form in front of the penis. Eversion of the prepuce may occur. The penis may extend partway out of the sheath because the hematoma prevents normal retraction. Urination is not interfered with. Swelling is usually more prominent on one side of the sheath than the other. The bull may be reluctant to move because of the pain.

An abscess often occurs if the hematoma is not surgically corrected.

Treatment

Surgery is the only treatment, and it must be done within two weeks of the injury.

Phimosis and Paraphimosis

Cause

Phimosis is characterized by the inability of the male animal to protrude the penis out of the sheath. The opening may be too small, the penis may be too large, or there may be a combination of both conditions.

Paraphimosis is characterized by the inability of the male animal to retract the penis into the sheath. The penis may be too large, the opening too small, or there may be a combination of both factors.

Treatment

Phimosis is rare in horses. When phimosis exists in any farm animal, the only treatment is surgery.

Paraphimosis is not rare in stallions. The usual history is that the penis has been kicked when erect. This is a common accident when stallions are permitted to cover mares that have not been properly hobbled.

The treatment is long and tedious but usually successful. It may require two or three weeks. The animal's penis should be showered with cold water for several hours daily. Proteolytic enzymes should be given systemically. The penis should be kept in a suspensory at all other times. When the penis can be returned to the sheath, a retention suture should be placed across the front of the sheath to keep the penis in place, and a gauze pack should be placed over the end of the sheath to prevent irritation of the penis by the sutures.

CHAPTER 12

The Respiratory System

The term *respiration* commonly refers to the inspiration and expiration of air. The process, called breathing, is essential for a second kind of respiration which takes place in every cell in an animal's body. It involves the combination of oxygen with organic compounds in the cells to release energy and to produce carbon dioxide and water. This gaseous exchange in the cells is called "cellular respiration." The circulatory system works with the respiratory system to transport oxygen to and carbon dioxide from every cell in the animal's body.

STRUCTURE AND FUNCTION

The two major functions of the respiratory system are (1) to provide an adequate supply of oxygen for the tissues and (2) to release carbon dioxide from the body. Secondary functions include warming, filtering, and moistening air as it is inspired, eliminating water in the process of expiring, and phonation or the creation of sound. The respiratory system is composed of two lungs and the air passages which lead to them. These passages include the nostrils, nasal cavity, pharynx, larynx, trachea, and bronchi (Fig. 12-1).

Nostrils

The respiratory system begins externally with the nostrils. They are paired openings in the muzzle, or snout in the case of swine, which lead into their respective half of the nasal cavity.

Nasal Cavity

The nasal cavity extends from the nostrils to the pharynx. It is divided into halves by the nasal septum. The nasal cavity contains a series of thin plates of scroll-shaped bones called turbinates. The cavity itself and the turbinate bones are lined with a mucous membrane that is richly supplied with blood vessels. Therefore, inspired air is warmed as it passes through the narrow vascular-lined spaces formed by the turbinate bones. In addition, the surface layer of the mucous membrane which lines the cavity has many microscopic hairs to filter out dust particles in the air.

Pharynx

The pharynx is a short muscular structure with seven openings. They include openings from the nasal cavity, Eustachian tubes from the middle ear, the mouth, the esophagus, and the larynx. The pharynx functions as a part of the digestive and respiratory system, since it is a common passage for both food and air.

Larynx

The larynx is a short tube which connects the pharynx and trachea. It regulates the volume of air in respiration and prevents the inhalation of foreign materials, and it is essential in phonation or voice production. The larynx is situated just below the pharynx and is easily located externally.

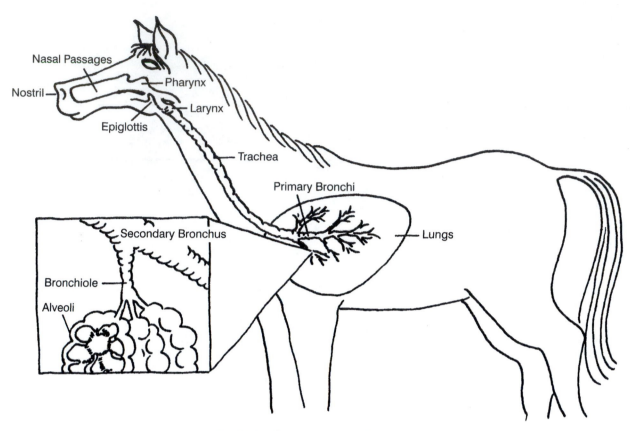

Fig. 12-1.   Respiratory tract of the horse.

The larynx is formed by five large articulating cartilages. The epiglottic cartilage is located just behind the base of the tongue. The epiglottis extends forward over the terminal part of the soft palate as the animal breathes and, thereby, directs air through the larynx and into the trachea. It is closed against the arytenoid cartilages when the animal swallows to prevent ingesta from entering the larynx.

The interior portion of the larynx is divided into two cavities. Each side contains strips of connective tissue which contain the vocal cords. Air passed against these cords causes them to vibrate and produce sound. The range in sounds produced by various species of farm animals is due to the length of their vocal cords, the tension they exert on them, and the volume of air deflected against the vocal cords.

Trachea

The trachea, commonly referred to as the windpipe, extends from the larynx to the juncture of the lungs where it divides into two bronchi, one for each lung. The trachea is kept open permanently by a series of incomplete rings of cartilage embedded in its wall.

Bronchi

The bronchi are the passageways which branch off from the trachea to enter the lungs. The bronchi divide into innumerable smaller tubes called bronchioles which eventually terminate in the alveoli.

Lungs

The lungs are soft, elastic, spongy organs which occupy the major part of the thoracic cavity. They are cone-shaped structures, containing millions of tiny air sacs called the alveoli. Each alveolis is surrounded by a network of capillaries. It is through the walls of the alveoli and their surrounding capillaries that oxygen is supplied to the blood and carbon dioxide is removed (Fig. 12-2).

In normal breathing, about 22 percent of the oxygen inspired goes into the circulatory system and approximately 4 percent of the air expired is carbon dioxide. The following figures show the average composition percentage of air as it is inspired and expired.

	Oxygen	Carbon Dioxide	Nitrogen
Inspired air	20.93	0.03	79.04
Expired air	16.29	4.21	79.50

SIGNS OF DISTURBANCE

There are common signs of ill health that should not be overlooked, even though they are common to disturbances associated with several of the animal systems. Anorexia, the lack of appetite, is a common sign of animal ill health. It is particularly significant in respiratory diseases when accompanied by an elevation in temperature, depression, and a rough hair coat.

Abnormal Respiration

Respiratory disturbances are often manifested by an increased breathing rate. This may be accompanied by shallow inspirations, groaning, and/or labored breathing. Dyspnea, or difficult breathing, is characterized by the animal taking a spraddle-legged stance, with the head and neck extended and nostrils dilated to facilitate breathing (Fig. 12-3). Open-mouth breathing, with the tongue protruding, and excessive abdominal action in breathing are signs of severe respiratory disturbance.

Respirations are noiseless in normal animals at rest. Respiratory noises due to work, exercise, and species' peculiarities, such as the snort of a horse due to excitement, should be distinguished from pathological respiratory sounds. Coughing, sneezing, roarings, wheezing, and groaning sounds are indications of disturbances in the respiratory system.

Nasal Discharge

An abnormal nasal discharge is usually a sign of respiratory disease. It may range from a thin, watery

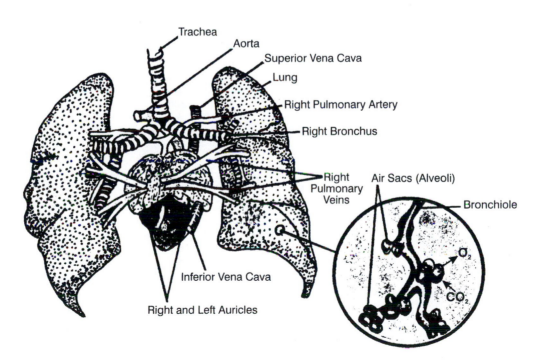

Fig. 12-2.   Organs of respiration and pulmonary circulation. (Courtesy, University of Georgia, Athens)

Fig. 12-3.   A calf showing signs of respiratory infection. (Courtesy, Elanco Products Company)

material to a thick, stringy substance. The color of the nasal discharge is affected by admixtures of pus, blood, and necrotic tissue.

Nasal Distortions

Rhinitis causing atrophy of the turbinate bones in the nasal cavity often results in permanent nasal distortions. Damage to facial nerves, edema, tumors, and severe inflammation of the nose can also cause nasal distortions.

DISTURBANCES OF THE RESPIRATORY SYSTEM

Pasteurellosis

Pasteurellosis is the name used to identify respiratory infections of cattle, sheep, and swine that are caused by pasteurella organisms.

Cause

Pasturella multocida, type A, causes fowl cholera and is a secondary invader in (1) porcine enzootic pneumonia and atrophic rhinitis, (2) shipping fever in cattle, and (3) pneumonia in sheep. *Pasteurella multocida*, type B, causes hemorrhagic septicemia of cattle and buffaloes, and type D is a secondary invader in swine pneumonia.

Pasteurella hemolytica is a secondary invader in (1) shipping fever in cattle, (2) rhinitis in swine, (3) pneumonia in sheep and goats, and (4) septicemia in lambs.

Hemorrhagic Septicemia

Hemorrhagic septicemia is an acute infection of cattle and buffaloes, with most animals dying. The disease has occurred only twice in the United States. The first outbreak occurred in 1922 in Yellowstone National Park and the second in 1967 in Montana. Both outbreaks involved buffalo. In tropical countries, hemorrhagic septicemia is encountered frequently, especially during the rainy season when cattle and buffaloes are used as draft animals.

Cause

Pasteurella multocida is generally believed to be a normal inhabitant of the mucosa of the upper respiratory tract of cattle. It invades tissues and produces a septicemia when resistance of the host animal is lowered by some type of stress.

Clinical Signs

Hemorrhagic septicemia is characterized by a fever of 106° to 107°F and by subcutaneous swellings of the throat, neck, and brisket. Most animals affected with hemorrhagic septicemia die within one to four days.

Septicemic Pasteurellosis in Lambs

Cause

Septicemic pasteurellosis is an acute septicemia of lambs caused by *Pasteurella hemolytica*.

Clinical Signs

Pasteurellosis affects feeder lambs in a satisfactory state of nutrition and frequently occurs after shearing or shipping. It is not uncommon, however, for pasteurellosis to occur in sheep that have not been stressed. There are, in many cases, no symptoms. Apparently healthy lambs are found dead. Those that show clinical signs have temperatures of 106° to 107°F and die in a few hours. The mortality rate may reach 5 percent in some flocks.

Prevention

Immunization with autogenous bacterins is recommended to protect sheep against pasteurella infection. The value of vaccination, however, is controversial. Aureomycin fed in the ration has some value in preventing this disease.

Treatment

Sick animals should be isolated and started on antibiotic therapy as soon as possible. Sulfamethazine and broad spectrum antibiotics are recommended for treatment against pasteurella infections in sheep.

Pasteurellosis in Cattle

Cause

Pasteurellosis in cattle is caused by *Pasteurella hemolytica* and *Pasteurella multocida*. These organisms are frequently isolated from nasal and ocular discharges of cattle with respiratory infections. In such instances, using broad spectrum antibiotics to prevent bacterial pneumonia is wise.

Clinical Signs

The clinical signs of pasteurellosis in cattle include depression, dyspnea, anorexia, and a temperature ranging from 104° to 108°F. A nasal discharge tinged with blood is a characteristic sign of this and other respiratory infections. In addition, affected cattle usually have an increased pulse rate, rapid, shallow respirations, and a characteristic cough. This if followed, in terminal cases, by prostration and death.

Prevention

Cattle should be vaccinated with for infectious bovine rhinotracheitis and parainfluenza$_3$ to prevent pasteurellosis. This is usually more effective than vaccinating with IBR and PI$_3$ vaccines in combination with *Pasteurella hemolytica* and *Pasteurella multocida* bacterins. The basis for this recommendation are (1) vaccinating against four diseases produces more stress than vaccinating against two diseases, (2) viruses are unable to damage cells of the respiratory tract when animals are immunized against viruses of the respiratory tract, and (3) pasteurella bacteria are incapable of serious damage to the cells unless the cell walls have been previously damaged by a virus infection.

Bovine Respiratory Disease Complex

Bovine respiratory disease complex (BRDC) was called shipping fever. It is now more aptly called a complex because of the involvement of bacterial, environmental, and viral factors. The disease is characterized as a respiratory disease composed of upper and/or lower respiratory tract disease.

Cause

Signs of pneumonia are seen soon after significant stress from environmental, physiological, or psychological changes or exposure to new disease organisms. Examples of stress are weaning, commingling of cattle in auctions, shipping, parturition, or thermal stress. Morbidity can be greater than 50 percent in high-risk cattle and mortality up to 10 percent. High-risk cattle are immunocompromised due to parasitism, previous illness, and nutritional deficiencies. Several agents alone, or in combination, can precipitate a pneumonia outbreak. Viral causes of BRDC are infectious bovine rhinotracheitis virus (IBR), bovine viral diarrhea virus (BVDV), bovine respiratory syncytial virus (BRSV), and parainfluenza 3. IBR causes fever and impairs the function of neutrophils, macrophages, and lymphocytes. BVDV is also immunosuppressive. *Pasteurella haemolytica* is found in healthy cattle in the nasopharynx. High-stress environments, viral immunosuppression, or immune system compromise through inadequate dietary energy, protein, and trace elements allows inspired *P. haemolytica*to to colonize the lungs and cause morbidity and death. *Haemophilus somnus* is also a major factor in BRDC. It is very common in feedlots as a factor in arthritis, myocarditis, and thromoembolic meningoencephalitis after a respiratory disease outbreak.

Clinical Signs

The incubation period for shipping fever is 5 to 10 days. In most outbreaks, there is a definite build-up of cases for four to five days, followed by a declining rate so that most cases have developed within 7 to 10 days of the initial onset of disease. The usual morbidity in an outbreak is 20 percent, but if the daily temperature of all animals is determined, the attack rate is closer to 40 percent. Herd mortality rate may be as high as 5 percent. One or two animals may be found dead at the beginning of an outbreak. These peracute cases have a course of 12 to 24 hours. Vague symptoms of prostration, anorexia, and fever may be noticed. Calves with body temperatures as high as 107°F often cannot be recognized as being sick, even by experienced caretakers. The typical acute case has a course of 8 to 12 days and is characterized by a temperature of 104°F or higher. There is a white, viscous nasal discharge and a watery discharge from the eyes (Fig. 12-4). The pulmonary signs include increased respiration and coughing. Diarrhea seldom occurs in uncomplicated shipping fever. Chronic cases of shipping fever will have fast, shallow, or labored breathing, coughing and expiratory grunt, and emaciation.

Infection with, and multiplication of, the parainfluenza₃ virus causes damage to the respiratory

Fig. 12-4. A young calf suffering from shipping fever. (Courtesy, Elanco Products Company)

mucosa. This allows the pasteurella bacteria to multiply and cause the more serious sequelae of the disease—pneumonia. This synergistic action is frequently dependent on an additional stress factor. Maternal antibodies to the parainfluenza₃ virus are passed to the calf in colostrum, resulting in a serum antibody level equal to or greater than that of the dam. The colostral antibody decreases as the calf ages and disappears by weaning time at six to eight months. Antibody action against the virus is again produced when the calf comes in contact with the infectious agent.

Manifestations of shipping fever may be produced during this virus infection if concurrent pasteurella infection and stress of weaning or inclement weather also occur. Infections in the absence of stress do not cause serious disease. Anamnestic antibody responses without signs of disease occur on subsequent exposures to the virus.

Prevention

The prevention of BRDC is through reducing stress and raising the immunity level of the cattle. Suggested means of reducing stress is to castrate, dehorn, and administer the primary vaccine prior to weaning calves. The way to raise the immunity of the animal is through vaccination. Vaccination requires an immunocompetant animal to work. Factors to consider are the age of the animal—a very young calf will not respond to all vaccinations due to colostral immunity. Secondly, the nutritional status of the animal determines the ability to respond to vaccines. Suggested levels of vitamins and trace minerals in the ration are 15 ppm copper, 0.2 ppm selenium, 500 IU/kg vitamin E, and 5000 IU/kg of vitamin A.

The choice of live or killed vaccines depends upon the individual producer's goals and situation. Some live vaccines should not be used in pregnant cows, or calves nursing pregnant cows, as abortions may occur. Killed vaccines must be boostered two to four weeks after the initial dose to achieve maximal immunity. All vaccines must be handled carefully to avoid contamination of the bottle and damage to the vaccine from temperature extremes while working the cattle. *Pasteuralla haemolytica* vaccines have been shown to be effective in reducing morbidity and mortality in feedlot cattle when the vaccines are administered upon arrival.

It is important that the ration provided to animals undergoing stress be palatable and nutritious. More critical to the ration analysis is the water intake. Dry matter intake is limited by the water intake, so recently weaned or shipped cattle must be encouraged to drink. Water troughs should be cleaned daily, and often the sound of the water running will draw the cattle to the water. Dehydration is a significant physiological stress to cattle and the importance of abundant, clean, fresh water that is easily accessible cannot be over emphasized.

Treatment

The treatment of bovine respiratory disease complex is varied, and depends upon several factors. For high-risk cattle, the manager may elect to mass medicate, or treat the entire group of animals with antibiotics before any clinical signs are observed. Both prescription and nonprescription drugs are available to treat cattle with BRDC. Please consult a veterinarian for proper diagnosis and dosing regimen of these drugs. The classes of antibiotics used to treat BRDC are the penicillins, sulfonamides, tetracyclines, cephalosporins, macrolides, florfenicol, aminoglyco-

samides, lincosamides, and fluoroquinolones. Some of these medications are formulated to be administered SQ in a single dose, eliminating every day or twice daily dosing of an animal and increasing the stress.

Ancillary therapy for the treatment of respiratory disease is aimed at reducing inflammation, reducing the febrile response, and stimulating feed intake. Flunixin meglumine is marketed in the United States as a fever reducer, antiendotoxin drug to aid in improving the clinical efficacy of the antimicrobial drug used.

Parainfluenza$_3$

Cause

Parainfluenza$_3$ (PI$_3$) is a virus that affects the respiratory tract of cattle. PI$_3$ is spread by airborne transmission or by contact with infected animals. It also can be spread by contaminated feed, water, equip-. ment, shoes, and clothes. The disease usually is encountered in feedlots shortly after the arrival of infected cattle.

Clinical Signs

Cattle affected with PI$_3$ are gaunt and depressed, and have a snotty nose, a mild cough, and a temperature of 104° to 106°F.

Prevention

Vaccination is recommended to prevent parainfluenza$_3$. Cattle that are vaccinated with both infectious bovine rhinotracheitis vaccine and parainfluenza$_3$ vaccine will tend to have cross immunity to other respiratory viruses. Consequently, they will have some protection from other respiratory infections. For this reason, parainfluenza$_3$ vaccine should be administered to breeding stock as well as to feedlot cattle.

Infectious Bovine Rhinotracheitis

Infectious bovine rhinotracheitis is also known as rednose, infectious pustular vulvovaginitis, and IBR. It is an acute virus infection of cattle that affects the respiratory and reproductive tract.

Cause

Infectious bovine rhinotracheitis is caused by one antigenic type. This herpes virus is easily isolated from the mucosa of the respiratory and reproductive tracts, from the conjunctiva of the eye, and from the brain, when the infection involves these particular organs. The IBR virus can also be found in the aborted fetus of infected dams.

IBR is spread by contact with infected animals or via the air. It also can be spread by contaminated feed, water, equipment, shoes, and clothes. The disease usually is encountered in feedlots within 20 days after the arrival of the infected cattle. Some feedlots have had outbreaks with 50 percent of the cattle affected within 10 days after arrival.

Clinical Signs

RESPIRATORY FORM

The incubation period for the respiratory form of IBR is 2 to 14 days. Typical symptoms are gauntness, depression, a snotty nose, a temperature of 104° to 107°F, and a mild cough. In addition, the nostrils and nose pad may be red and inflamed; thus, the term "rednose."

The nasal discharge becomes white and stringy. Some animals wheeze and breathe from the mouth. Occasionally, there are bubbling sounds in the larynx and trachea, but the lungs sound normal. Diarrhea rarely occurs. The course of the respiratory form of IBR is five to seven days. The mortality rate ranges from 5 to 10 percent. In dairy cattle, respiratory signs are minimal, but there is a severe drop in milk production.

CONJUNCTIVAL FORM

In some outbreaks of respiratory IBR, there are infections of the conjunctiva, ranging from an occasional case to as much as 50 percent of the sick cattle. Outbreaks rarely occur in which only the eye form is seen. The conjunctiva may become very inflamed and swollen, with a pussy discharge emanating from the eye. In extreme cases, the cornea becomes cloudy and ulcers form on it. As a rule, the outbreak will last for three weeks, and the symptoms in one individual will last for one week.

INFECTION OF THE REPRODUCTIVE TRACT

The name given to the reproductive form is infectious pustular vulvovaginitis (IPV). In this form, the virus affects the vulva and vagina of females and the penis and prepuce of bulls. The infection begins by the formation of small pustules, which later become ulcerated. The main problems from this form are interference with breeding and decreased milk production. IPV rarely occurs simultaneously with the other forms.

ABORTION AND NEONATAL DISEASE

Cows may abort three to five weeks after they have had the respiratory form of IBR. The abortion rate may reach 25 percent. Vaccination of cows that are more than $5\frac{1}{2}$ months pregnant can cause up to 60 percent abortions. Fetuses aborted during natural outbreaks are five months of age or older.

Prevention

Vaccination is recommended to prevent IBR (Fig. 12-5). Feedlot cattle should be vaccinated before arrival or immediately after they arrive in the feedlot. If the weather is extremely hot or extremely cold, feedlot cattle should have a booster shot of IBR when they are reimplanted.

Fig. 12-5. Intranasal administration of vaccine to prevent infectious bovine rhinotracheitis and parainfluenza₃. (Courtesy, Jensen-Salsbery Laboratories)

Equine Rhinopneumonitis

Equine rhinopneumonitis is a viral infection of horses that causes abortion, respiratory disease, and encephalomyelitis.

Causes

Equine herpes virus 1 (EHV-1) and 4 (EHV-4) cause rhinopneumonitis. EHV-1 causes abortion, stillbirths, and weak foals that die soon after birth. Respiratory disease in young horses is also caused by EHV-1. Respiratory disease is the only form of EHV-4. The virus is similar to herpes viruses in that latent infections occur. A latent infection is where the animal is infected, but shows no clinical disease. Then, under stressful conditions, the virus can become active and cause disease, exposing more horses. The virus is spread by direct contact with infected horses and through the environment by shared feeders, drinking facilities, grooming equipment, and caretakers.

Clinical signs

RESPIRATORY FORM

The clinical signs of equine rhinopneumonitis are similar to other respiratory diseases in horses. Fever, cough, anorexia, nasal and ocular discharges, swollen lymph nodes, and fatigue are all symptoms. An animal's first exposure to the virus usually produces the more severe signs and later exposure may cause infections not apparent to the trainer and caretaker. Neurologic disease, paralysis, and possibly death may follow EHV-1 respiratory infections, although this is relatively rare.

REPRODUCTIVE FORM

Abortion storms can occur on breeding farms as a consequence of infection with EHV-1. Abortions from EHV-1 usually occur in the last trimester of pregnancy. Mares generally have very mild or unnoticed respiratory infections, and abortions can occur up to three months after the respiratory phase of the disease. Abortions occur suddenly without signs; and the aborted fetuses are not decomposed. Sometimes, live foals are born, but die within a few days from severe pneumonia.

Prevention

RESPIRATORY FORM

Vaccination to protect against equine rhinopneumonitis is similar to equine influenza, in that vaccination of all horses on the farm is very important to control outbreaks of EHV-1 and EHV-4. For low-risk horses that do not travel and are not exposed to other horses, semi-annual vaccinations may suffice. For horses competing and traveling, vaccination every two to four months may be necessary. The vaccination does not entirely prevent infection, but it helps reduce the severity of the disease and lowers viral shedding.

REPRODUCTIVE FORM

The vaccination of all horses on the farm is necessary to prevent outbreaks of EHV-1 abortions. Manufacturers recommend vaccination at five, seven, and nine months of pregnancy, and some veterinarians recommend vaccinating in the third month of pregnancy. Isolation of new arrivals and prompt removal, disinfection, and diagnosis of aborted fetuses are additional steps in controlling the spread of infection.

Treatment

RESPIRATORY FORM

There is not a specific treatment for the respiratory form of equine rhinopneumonitis. Rest in a well-ventilated stall free of dust, and antibiotics to control secondary bacterial infections will enable a complete recovery. Medications to reduce the fever and congestion may hasten recovery. The sick animal must be allowed time to rest in order to fully recover.

REPRODUCTIVE FORM

There is no specific treatment for the abortion, weak and stillborn foals, or the neurologic disease.

Good herd-management practices and prevention are the best methods of treatment.

Tuberculosis

Tuberculosis is a chronic infectious disease caused by acid-fast organisms belonging to the genus, *Mycobacterium*, and characterized by the development of tubercles and abscess formation, with resulting calcification. Human beings, cattle, and swine are highly susceptible to tuberculosis.

Cause

Three types of tubercle bacilli are responsible for tuberculosis in warm-blooded animals. They are (1) *Mycobacterium tuberculosis* (human type), (2) *Mycobacterium bovis* (bovine type), and (3) *Mycobacterium avium* (avian type). (See Table 12-1.)

The *Mycobacterium* organisms are morphologically similar. They are pleomorphic, acid-fast, Gram-positive rods.

TABLE 12-1. Types of Tubercle Bacilli Causing Tuberculosis in Warm-Blooded Animals[1]

Animal	Bovine	Avian	Human
Human	***	—	****
Cow	****	—	—
Horse	—	—	—
Sheep	—	—	—
Pig	****	**	†
Dog	*	—	*
Cat	*	—	*
Chicken	—	****	—

—Not susceptible, or at least occurs so rarely as to be of little or no practical importance.

†Very rarely, if ever, occurs.

*Slight, but occurs occasionally.

**Quite susceptible, but disease usually remains localized.

***Quite susceptible, especially the young.

****Very susceptible, disease may be widespread.

[1]From *Animal Sanitation and Disease Control*, 6th edition, by R. R. Dykstra, The Interstate Printers & Publishers, Inc., Danville, Illinois, 1961.

Bovine animals tend to become infected with *Mycobacterium bovis* organisms by the inhalation of aerosal droplets and dust or by the ingestion of contaminated feed and water. Swine, when running with infected bovines, are infected with tuberculosis by ingesting contaminated feed and water. They also may be infected by ingesting diseased chickens or contaminated feces.

Clinical Signs

Cattle in the United States rarely show symptoms of tuberculosis. Bovine tuberculosis is usually chronic; however, the spread of the disease in a herd may be extremely rapid. Emaciation occurs when the disease is extensive. The respiratory system is the usual center of clinical tuberculosis. The udder is very important because it can be a source of infection for calves and humans. There is evidence that the normal mammary gland may, on occasion, permit the escape of tubercle bacilli from the blood into the milk. The disease is rarely found in the reproductive tract. Cattle are readily infected by bovine bacilli.

The symptoms of tuberculosis in swine are characteristic of chronic respiratory disease. Hogs are usually slaughtered at such an early age that tuberculosis lesions have not reached such proportions as to cause external manifestations of the disease. Hogs are readily infected by avian or bovine bacilli.

Prevention

There is no vaccine recommended for the prevention of tuberculosis in the United States. Control of tuberculosis is based upon a testing program that identifies affected animals. Animals that are positive reactors to the tuberculin test are slaughtered.

AREA TESTING

California is reaccredited during a six-year period by testing for tuberculosis according to the following schedule:

1. Raw milk dairies are tested annually.
2. Commercial dairies are tested every two years (except in areas where annual tests are indicated).
3. Family cows are tested every six years, or at the owner's request, these cows may be scheduled for a test at two-year intervals.
4. Purebred beef are tested every two years.
5. Grade beef are tested when there is a reason to suspect that tuberculosis exists in the herd or that the herd may have been exposed to tuberculosis.
6. All herds with a history of infection within the past 12 years are tested annually.

AGE TESTED

1. In infected herds, all cattle regardless of age are tested.
2. Routine tests are conducted on cattle 24 months of age and older.

INFECTED HERDS

1. Reactors are branded with the letter "T" on the left jaw, appraised, and sent to slaughter within 15 days of appraisal.
2. The entire herd is placed under a hold order.

3. The premises must be cleaned and disinfected.
4. The herd must pass at least two negative tests at 60-day intervals before it can be released from the hold order.
5. A follow-up test is conducted at six months, and then the herd is tested annually.

SUSPECT HERDS

1. Animals sensitive to tuberculin may be classified as suspects in herds not known to be infected with, or exposed to, tuberculosis.
2. Suspects are placed under a hold order and retested at 60- to 90-day intervals until their status is determined. Retests should show that the animals' sensitivity is definitely decreasing, or the animals are classified as reactors.
3. The retest of suspects is based on the theory that most cattle infected with *Mycobacterium bovis* develop the disease progressively and that their sensitivity would be persistent, while animals sensitized by other members of the genus sustain only a transient sensitivity.

Treatment

There is no treatment recommended for tuberculosis in farm animals.

Strangles

Strangles is a bacterial disease that causes fever, nasal discharge, and enlargement of the lymph nodes of the head and neck. Sometimes, the lymph nodes will swell enough to inhibit breathing in the horse; thus, the disease name strangles.

Cause

Strangles is an extremely contagious disease of the upper respiratory tract caused by the bacteria *Streptococcus equi*. Strangles is spread by direct contact with infected horses or through environmental contamination. The bacteria can enter the body through inhalation or ingestion of purulent exudates that contaminate feeders, gates, and corrals. Young horses (up to five years of age) are most susceptible for infection. Strangles occurs most frequently in stables and farms where many horses are kept and new horses are constantly moved into and out of the population.

Clinical Signs

Infected horses become depressed, anorexic, cough, and develop a high fever of 104 to 106° F. A thin, watery nasal discharge develops initially, and then a thick, yellow discharge develops. Lymph nodes of the head, under the mandible and throat area, will swell. After 7 to 14 days, the lymph nodes may abscess open and drain a yellow purulent material. Occasionally, the bacteria may spread beyond the upper respiratory tract and invade organs, such as the spleen, liver, kidneys, and brain. This syndrome is called "bastard strangles." A more rare complication, which is very serious and potentially fatal, is purpura hemorrhagica. Purpura hemorrhagica is an immune system reaction, which damages the horse's blood vessels.

Prevention

Strict quarantine of new arrivals on the premises for up to six weeks allows for the observation of the incoming horses before they have direct contact with the farm population. Vaccination with an intramuscular or intranasal vaccine should be done routinely in areas with an outbreak of strangles or high-risk herds. Vaccination can be done in the face of an outbreak.

Clinically recovered animals may shed the bacteria in nasal secretions for several months after the illness. Recovered horses can be cultured to determine if they are free of the bacteria. All exposed or contact animals should be monitored for fever or other signs of illness for several months—due to the amount of environmental contamination that occurs from ruptured abscesses. Contaminated objects, such as boots, hands, tack, hay stalls, feeders, and soil, should be cleaned and disinfected or discarded.

Treatment

Horses with strangles should be isolated and encouraged to eat soft, palatable feed, because swallowing is painful. The use of antibiotics to treat strangles is controversial, as the antibiotic may slow the course of the disease. Penicillin is effective in killing the *Strep. equi* organism, but it is unable to work in the abscessed lymph nodes. Hot packing of the lymph nodes to hasten rupture and drainage, as well as flushing the abscesses, is necessary to resolve the infection. Horses with bastard strangles and purpura hemorrhagica should be treated individually.

Chronic Obstructive Pulmonary Disease (Heaves)

Chronic obstructive pulmonary disease (COPD) is a lower respiratory disease also known as heaves, broken wind, emphysema, and chronic bronchitis in horses. Labored breathing and a severe expiratory effort characterize COPD.

Cause

The cause of chronic obstructive pulmonary disease in equines is unknown, but it is likely multifactorial, which leads to lower airway inflammation. Being exposed to dusty hay, feed, and bedding, or confinement in a poorly ventilated barn, are contributing factors. The development of COPD may follow a viral respiratory infection, because the viral disease damages the respiratory tract and may allow foreign material to penetrate the lower airway and be contained in the mucous discharges.

Clinical Signs

Horses with the early stages of heaves are alert and nonfebrile and may have an occasional cough during exercise or feeding. The frequency and intensity of the coughing will increase as the disease progresses and the coughing is deep and nonproductive. In severe cases, an increased respiratory rate, flared nostrils, and hypertrophy of the external abdominal oblique muscles, which form a "heave line," may be present. When using a stethoscope, wheezes and crackles may be heard throughout the lung fields. Mucus may be present in the trachea causing wheezing in advanced cases.

Prevention

The prevention of chronic obstructive pulmonary disease in horses is directed towards environmental control and preventing viral respiratory disease. Reversing the clinical signs of COPD through environmental control may take up to a month. Hay, straw bedding, and dusty feedstuffs should be replaced. Soaking the hay for two hours in water may prevent the inhalation of dust and mold spores. In addition, feeding hay from the ground is less likely to expose the lower airway to as much dust as feeding the horse at head level. The horses should be removed from the stalls during cleaning to minimize the exposure to the stirred bedding. The barn aisles should be moistened before sweeping and exercise pens and arenas should be moistened and free of dust. Maintaining a horse outside, even in very cold climates, will usually keep a horse free of clinical signs.

Treatment

The best treatment for COPD is environmental modulation. Several drugs are available to reduce the clinical signs. Corticosteroids are used to control the allergic reaction and bronchodilators are used to relax the muscles of the lower airways and provide relief of the functional obstruction. Other medications may be used to increase the clearance of the mucus from the airways.

Porcine Respiratory Disease Complex

Porcine respiratory disease complex (PRDC) is a costly barrier to efficient and profitable pork production. Although the presence of specific respiratory pathogens, housing types, and climactic conditions vary by region and country, all swine producers experience some degree of respiratory disease. It is difficult to understand the complex factors that contribute to respiratory disease and producers and veterinarians

are constantly challenged to develop cost-effective control and treatment strategies. Viral and bacterial pathogens, as well as internal parasites, can interact with each other to have a profound effect on the severity of disease caused by any one pathogen. One common scenario is that mycoplasma or viral infections disturb the respiratory defense mechanisms, increasing the susceptibility to bacterial pathogens.

Pathogens involved in PRDC (singly or in combination):

1. Porcine cytomegalovirus
2. Pseudorabies virus (Aujeszky's disease)
3. Atrophic rhinitis
4. Swine influenza virus (SIV)
5. Porcine respiratory and reproductive syndrome virus (PRRS)
6. *Mycoplasma hyopneumoniae*
7. *Actinobacillus pleuropneumoniae (APP)*
8. *Pasteurella multocida*
9. *Salmonella cholerasuis*
10. *Streptococcus suis*
11. *Haemophilus parasuis*
12. *Actinobacillus suis*

Clinical Signs

A producer and a veterinarian must have keen skills of observation to determine the normal and abnormal appearances of pigs. Pigs are very excitable and not receptive to submitting to physical examinations. The assessment of an individual pig or a group of pigs is determined by the pig activity, interaction with other pigs, and the posture of the pig. Signs of upper respiratory tract disease are tear staining, conjunctival discharge, nasal discharge, and sneezing. Cough and increased respiration rate and effort indicate disease of the lower respiratory tract. Many times the appetite of the pen of pigs will decrease a few days before an outbreak of respiratory disease. The economic effects of PRDC are related to mortality and morbidity. Morbidity results in decreased feed consumption, decreased average daily gain, longer days to market, and a greater percentage of cull pigs.

Contributing Factors in PRDC

Environmental and management factors have a large influence on the prevalence and severity of respiratory disease. Recent improvements in confinement systems for swine and all-in, all-out production systems have not completely eliminated PRDC due to the possible improper management and lack of routine maintenance of the system. The areas of environmental factors are:

1. Thermal—A temperature below the comfort zone of the pig may cause the pig to use nutrients for heat production and lower the respiratory defense mechanisms. High environmental temperatures allow rapid pathogen buildup, as well as suppress pig appetite.

2. Gases—Oxygen, carbon dioxide, ammonia, carbon monoxide, and hydrogen sulfide are present in hog operations. Excess ammonia concentrations damage the lung clearance mechanisms and allow inhaled particles to deeply penetrate the lungs.

3. Moisture—Excessive moisture chills pigs, improves the survival of microorganisms, and leads to building deterioration.

4. Particles—Dust from feed and animal dander serves as a major carrier of microorganisms.

5. Microorganisms—The concentration of microorganisms in the environment depends upon ventilation rates, humidity, and waste removal.

Management practices can impact respiratory disease by slowing the growth rate of the pigs, increasing the risk or level of exposure to pathogens, and increasing the level of stress, which results in increased susceptibility to disease. Factors that restrict growth rates are inadequate dietary fortification, inadequate floor space, animal movement, and restricted feed and water supply through inadequate feeder space or adjustment.

PSEUDORABIES VIRUS (AUJESZKY'S DISEASE)

Pseudorabies is caused by a herpes virus and is under regulatory control and testing in the United States. It is most commonly associated with reproductive failure in breeding females and death and nervous signs in suckling pigs.

ATROPHIC RHINITIS

Atrophic rhinitis is caused by a complex interaction of infectious agents, with the management and environment factors influencing atrophy of the nasal turbinates. Two organisms are commonly implicated, *Bordetella bronchiseptica* and *Pasteurella multocida*. Both organisms release a toxin, which induces turbinate atrophy and clinical signs of sneezing, tear staining, bleeding from the nostrils, deviation of the snout, and reduced growth rate.

SWINE INFLUENZA VIRUS (SIV)

There are many strains of SIV, which are spread pig to pig by oronasal secretions. In acute outbreaks, all pigs will appear sick within a short period—the incubation period is 1 to 3 days. Fever, inappetence, depression, and a whooping cough are clinical signs in growing pigs, as well as abortions and stillbirths in breeding females.

PORCINE RESPIRATORY and REPRODUCTIVE SYNDROME VIRUS (PRRS)

This disease first appeared in the United States in the 1980s and was referred to as Mystery Swine Disease and Swine Infertility and Respiratory Syndrome. The virus is transmitted primarily pig-to-pig, although airborne transmission over long distances has been proposed. The disease was widespread in Europe and the United States by the early 1990s. In naïve herds, respiratory disease in young pigs and abortion and reproductive failure in breeding swine characterize the disease. In endemically infected herds, the disease is characterized by poor nursery performance and secondary infection by other pathogens.

MYCOPLASMA HYOPNEUMONIAE

M. hyopneumoniae is believed to be the primary initiator of respiratory disease in pigs by harming the respiratory defense mechanisms, which allows the invasion of lung tissues by secondary pathogens, such as *Pasteurella multocida*. Sows transmit infection to the pigs. Then gradual and continual transmission occurs from infected to non-infected pigs in continuous flow operations. Depressed weight gains and feed conversions and a chronic, nonproductive cough are characteristics in an infected herd.

ACTINOBACILLUS PLEUROPNEUMONIAE

Twelve serovars of this organism have been identified and serologic testing is used to diagnose the disease. Very rapid and fatal pneumonia is characteristic in severe outbreaks, which can be triggered by stressing, moving and mixing pigs, or ventilation mishaps. Pigs become febrile, anorexic, stop drinking, and can die in six to eight hours in some cases. Labored breathing, increased respiration rates, and a bloody discharge from the nostrils are additional signs.

STREPTOCOCCUS SUIS

S. suis is a Gram-positive cocci. It infects baby pigs at birth and causes neurologic signs, such as incoordination, paralysis, and tremors, as well as arthritis. Disease outbreaks occur most commonly after weaning and are attributed to stresses, such as weaning, improper ventilation, and a high stocking density.

SALMONELLA CHOLERASUIS

Environmental conditions strongly influence the infection with this Gram-negative organism, as transmission is by fecal-oral routes. The disease causes fever, moist cough, sudden death, and purple extremities and lungs.

Equine Influenza

Cause

Equine influenza is caused by several strains of the equine influenza virus. Equine influenza is spread over long distances and occurs year round for several reasons. First, horses are commonly transported by trailers and planes over all parts of the country and internationally. Secondly, influenza cases are not apparent for several days post infection, while the horse is contagious and spreading virus even though it does not appear sick. Clinical signs may suggest a diagnosis of equine influenza, however laboratory tests are re-

quired for a diagnosis. Paired serum samples, with the first sample taken during the illness and the second sample taken two weeks later, are required for diagnosis. A rise in antibody titers to the virus will indicate exposure to the disease and the antibody reaction to the virus.

Prevention

The best approach to controlling influenza is through prevention. Vaccination with adequate boostering and frequent revaccination will provide protection against infection or reduce the severity of clinical signs if the disease does occur. Frequent boostering is necessary because current vaccines do not provide long-lasting immunity and even horses recovering from natural infection are protected from reinfection for only a few months. Horses with low risk of exposure can be vaccinated semiannually. Animals that are raced, shown, travel extensively, or are exposed to a large number of horses should be vaccinated more frequently (Fig. 12-6).

Clinical Signs

Equine influenza usually presents with a sudden onset of high fever. A harsh, dry cough begins early in the infection and may last for several weeks. The nasal discharge begins clear and watery, then becomes yel-

Fig. 12-6. Protection from respiratory disease through vaccination will keep show horses performing at their peak. (Photo by Springer)

low and thick from secondary bacterial infections. Watery eyes, swelling and stiffness in the legs, colic, anorexia, depression, and enlarged lymph nodes between the mandibles are common.

Treatment

A horse with a mild case of equine influenza will recover in about a week if it is allowed to rest completely. Severely ill horses may require months to recover, especially if they are subject to stress, not allowed to rest, and develop secondary pneumonia. The sick horse should be kept in a well-ventilated stall that is as dust free as possible. Anti-inflammatory drugs may be used to reduce the fever, alleviate the soreness, and encourage the horse to eat. Antibiotics may be required if the fever lasts more than four days, pneumonia develops, or the nasal discharge becomes thick and purulent.

Swine Influenza

Swine influenza is an acute, infectious, highly contagious respiratory disease of hogs that occurs most often during the fall and winter.

Cause

Swine influenza is caused by the influenza virus. The virus usually remains dormant until inclement weather in the fall. Once an outbreak occurs in a herd, the virus spreads rapidly from pig to pig through direct nasopharyngeal contact. The virus is maintained in swine populations by continual passage to young, naïve pigs from older, infected pigs.

Clinical Signs

Swine flu is characterized by sudden onset, rapid transmission through the herd, and sudden recovery, unless complicating secondary infections occur. Swine flu usually follows unusually harsh, cold, damp weather or some change in herd management that creates stress. The period of illness is short, ranging from two to six days. The mortality rate of uncomplicated swine influenza seldom exceeds 5 percent.

Affected hogs suddenly lose their appetite, become weak and lethargic, and run a fever from 104° to 107°F. Respiratory distress is evidenced by rapid, diaphragmatic breathing and a peculiar cough that resembles a hoarse dog's bark. This is often followed by extreme debility, incoordination, depression, and prostration. Bred females infected with swine influenza abort during the course of the disease. Recovery of the sick hogs is abrupt, and the mortality rate is low. The main economic loss is due to loss of weight and an extended recovery period that delays the time that hogs reach market weight.

Prevention

Commercial vaccines are available for swine flu, and they are aimed at reducing morbidity and the associated loss of appetite and weight gain.

Prevention of swine influenza is based on management practices that control lungworms and prevent stress to hogs during the fall and winter months.

Treatment

The virus of swine influenza is not affected by medical preparations. Sulfa drugs and broad spectrum antibiotics are often administered to combat bacterial infections. In most cases, good nursing care, that excludes injections or any excitement of the sick pigs, is the best practice. If the pigs can be kept warm and comfortable, have access to water and be kept free from excitement, recovery is greatly enhanced. Adding anti-bacterials to the drinking water for animals affected by swine influenza may be beneficial.

Atrophic Rhinitis

Atrophic rhinitis is an infectious disease of swine that is characterized by deterioration of the turbinate bones in the nose, slow, inefficient growth rate, and low mortality.

Cause

Atrophic rhinitis is caused by *Bordetella bronchiseptica* and *Pasteurella multocida*. These two bacterial organisms have been isolated from the nasal passages of infected hogs. There is, however, good reason to suspect that other bacterial, protozoal, and viral organisms contribute to the problem of atrophic rhinitis in swine.

Atrophic rhinitis is usually transmitted from a sow to her pigs shortly after birth. It is also spread at weaning when several litters are mixed together. Transmission to susceptible swine can take place at any time. Apparently, the disease does not survive in lots or sheds for more than six weeks. New additions of infected swine to the herd are the major source of infection. Freshly contaminated boots, clothing, wheels, or tires are also potential sources. Rats are believed to harbor infectious organisms that cause atrophic rhinitis.

Clinical Signs

The clinical signs of atrophic rhinitis in swine include sneezing, sniffling, and snorting. A small quantity of clear to purulent mucus material may be seen passing from the nose. In cases where the turbinate bones have been damaged, flakes of blood may accompany the mucus discharge. Nasal hemorrhages often occur when sneezing is hard enough to rupture some of the exposed blood vessels in the nose.

A watery discharge from the eyes is quite common in pigs affected with atrophic rhinitis. The moisture collects dirt and brings attention to the exudate. Most of the affected pigs fail to gain normally, although their appetite remains good. Distortion of the nose in the direction of the most severe turbinate damage occurs in advanced cases of atrophic rhinitis (Fig. 12-7). When both turbinates are atrophied, the nose will not twist. It will appear somewhat shortened and have wrinkles on the skin behind the snout. Death losses from atrophic rhinitis are rare. Pneumonia and other infectious diseases may occur with atrophic rhinitis and, therefore, increase morbidity and the possibility of death to affected pigs.

Prevention

Sows that have atrophic rhinitis can be identified by the examination of the snouts of their pigs at

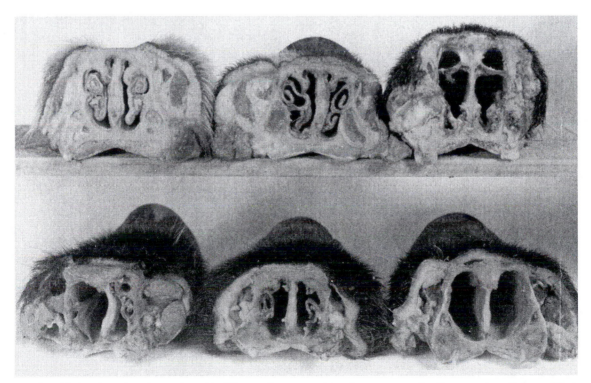

Fig. 12-7.   Atrophic rhinitis in swine. The nasal cross sections show progressive degeneration of the turbinate bones. (Courtesy, Department of Pathology, School of Veterinary Medicine, Kansas State University, Manhattan)

slaughter. If this is impractical, a veterinarian should take nasal swabs of pigs and sows that appear to have atrophic rhinitis. If atrophic rhinitis is confirmed, all sows should be checked by three nasal swabs collected at 7- to 10-day intervals. Infected sows should be removed from the herd. New breeding stock should be purchased from herds that are free from atrophic rhinitis. They should be isolated, swabbed, and found negative of atrophic rhinitis before being admitted to the breeding herd. Rats should be controlled, and all visitors should pass through a footbath before entering the swine unit.

A vaccine is now available to control atrophic rhinitis. Two cc of vaccine should be administered subcutaneously to sows four weeks before farrowing. A second dose is required two weeks prior to farrowing. Sows should be vaccinated during each succeeding gestation with 2 cc of atrophic rhinitis vaccine subcutaneously, two to four weeks before farrowing. Pigs should be vaccinated with 2 cc of atrophic rhinitis vaccine subcutaneously when they are seven days old. A second dose should be administered when they are 28 days old.

Treatment

B. bronchiseptica is sensitive to the sulfonamide drugs. The two most commonly used ones are:

1. Sulfamethiazine medication in the feed at the level of 100 to 450 grams of sulfamethiazine per ton of complete ration.
2. Sodium sulfathiazole administered in the drinking water, at the level of ⅓ to ½ grams per gallon of water.

Young animals should be treated for five weeks, whereas older animals need to be treated only four weeks.

Epistaxis

Epistaxis is a nosebleed. It is usually caused by diseases of the mucosa of the upper respiratory tract.

Epistaxis occurs in all species of farm animals, but horses are the most commonly affected.

Cause

Unilateral hemorrhage is caused by (1) foreign bodies lodged in the nasal passage, (2) nasal granulomas, (3) tumors, (4) severe trauma to one side of the face, and (5) stress from being raced.

Clinical Signs

Nosebleeding from both nostrils is not the same as small amounts of bloody foam in the nostrils which can be caused by equine pneumonia, acute pulmonary edema, or equine infectious anemia. Bilateral epistaxis is occasionally seen in race horses after a hard race. When the nosebleed is accompanied by bleeding from the mouth, it is due to an abscess or tumor in the pharynx, rupture of blood vessels in the lungs associated with rupture of a lung, abscess, or forceful respirations.

In cattle, bilateral epistaxis is seen in infectious bovine rhinotracheitis, malignant catarrhal fever, septic metritis, blackleg, and other systemic infections. Poisoning by nitrates, bracken, sweet clover, or mercurials may also cause bilateral nasal hemorrhage.

When epistaxis is accompanied by bleeding from the mouth, there are frequent swallowing movements, and the animal usually coughs and expels clotted blood.

Prevention

Large doses of vitamin C will help to prevent nosebleeds in race horses.

Treatment

Coagulants are used in the treatment for epistaxis. When large amounts of blood have been lost, a blood transfusion may be necessary.

CHAPTER 13

The Circulatory System

STRUCTURE AND FUNCTION

The circulatory system is made up of the blood, blood vessels, the lymph system, and the heart. (Fig. 13-1). Together, they function as an intricate transportation system to supply all the body tissues with nourishment and to collect and remove waste materials from the cells. Some of the specific functions of the circulatory system are (1) distributing food nutrients, (2) transporting oxygen and carbon dioxide, (3) exchanging oxygen and carbon dioxide, (4) removing waste materials, (5) distributing endocrine secretions, (6) preventing excessive bleeding, (7) combatting infection, and (8) regulating body temperature.

Blood

Blood comprises about 8 percent of the total weight of an animal. It is composed of plasma, blood cells, and platelets. Plasma is the liquid component of blood and accounts for 50 to 60 percent of its volume. Plasma is about 90 percent water. The remaining 10 percent contains small amounts of digested food nutrients. One of these is a protein material called fibrinogen. Fibrinogen is an important ingredient in the clotting of blood. When fibrinogen is separated from plasma, the remaining material is known as blood serum (see Figs. 13-2, 13-3, and 13-4).

Erythrocytes

Erythrocytes are disc-shaped red blood cells that are formed in the red bone marrow and stored in the bloodstream, liver, and spleen. Hemoglobin, an iron-containing protein, is responsible for the red color of the erythrocytes. Red blood cells last only about six weeks in the bloodstream. The liver separates, destroys, and replaces the worn out red blood cells.

Leucocytes

Leucocytes, or white blood cells, are larger in size and much fewer in number than red blood cells. They also differ from red blood cells in that they have a nucleus and are motile. They are capable of passing out of capillary walls in areas of inflammation to engulf and destroy invading organisms. Many leucocytes are formed in the bone marrow, but some are formed in the nodes of the lymph system.

Platelets

The third type of formed bodies in the blood is the platelets. They are small, frail, easily, damaged blood cells. When tissue is damaged, platelets are ruptured, thereby releasing an enzyme which sets in motion a complicated series of events resulting in the formation of fibrinogen.

Blood Vessels

The vascular system is a tubular network comprised of arteries, capillaries, and veins. The blood vessels are divided into two circulatory subsystems within the animal body. The systemic system conveys

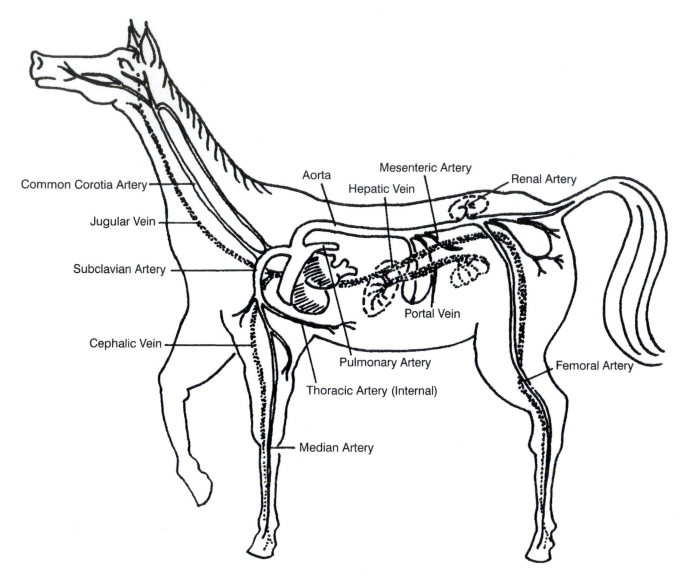

Fig. 13-1.   Circulatory system of the horse.

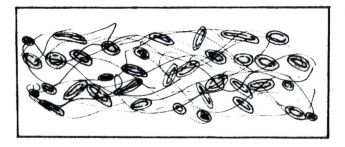

Fig. 13-2.   Blood smear before clotting. (Courtesy, University of Georgia, Athens)

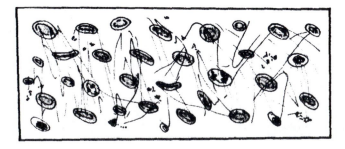

Fig. 13-3.   Fibrin threads form. (Courtesy, University of Georgia, Athens)

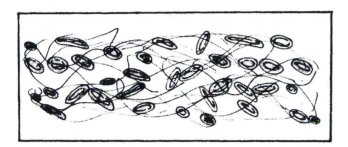

Fig. 13-4. Blood cells trapped. (Courtesy, University of Georgia, Athens)

blood to all the tissues except the lungs. The pulmonary system transports blood to the lungs and back to the heart.

Arteries

Blood is conveyed to all the tissues in the animal body through a branching network of arteries, which have a thick wall composed of muscle and connective tissue. The muscular wall of arteries is quite elastic. It expands as the heart pumps and contracts during the pause between heartbeats. This creates a wave-like motion to force the blood along its course.

Arteries divide into smaller and smaller branches. The smallest arterial branches are called arterioles which are barely visible to the naked eye. The arterioles branch, in turn, and form capillaries.

Capillaries

Capillaries are minute, thin-walled vessels that form networks of connecting tubes between arteries and veins. Food nutrients and oxygen carried in the arterial blood are diffused through the walls of capillaries into the tissues. Carbon dioxide and other waste products of the tissues also pass through the capillary walls.

Veins

Veins form the network system that accepts blood from the capillaries and returns it to the heart. Blood enters the venous system by way of the minute venules and travels through successingly larger veins until it reaches the heart. Veins are relatively thin-walled and inelastic. Movement of blood in the veins is controlled in part by a series of valves which permit the blood to flow forward toward the heart but not away from it. In addition, muscular activity applies pressure on veins, forcing blood along its way.

Lymph System

Lymph is a clear, colorless liquid, except that which drains from the intestine during digestion may contain large quantities of fatty acids, thus giving it a milky appearance. Lymph is derived from blood plasma that filters through capillary walls into the spaces between body cells. Much of the tissue fluid is reabsorbed into the venous capillaries, but the excess material is picked up by capillaries of the lymph system. Once the tissue fluid enters the lymph capillaries, it is called lymph.

The lymph system is a one-way structure of capillaries, vessels, and glands which provides for the return of lymph to the circulatory system. Along the course of the lymph vessels are nodes or lymph glands which filter out infectious organisms and other foreign substances. The nodes also produce lymphocytes which destroy invading bacteria.

Heart

The heart is a powerful, muscular organ located between the lungs in the thorax. The heart is divided into four cavities or chambers (Fig. 13-5). The upper cavities are the right and left auricles and the two lower chambers are the right and left ventricles. Valves between the auricles and ventricles keep the blood flowing in the proper direction.

Blood from all parts of the body, except the lungs, enters the right auricle. The blood then passes through a one-way valve into the right ventricle where it is pumped through the pulmonary artery to the lungs. After passing through capillaries in the lungs where carbon dioxide is released and oxygen is replenished, the blood is moved through the pulmonary vein to the left auricle of the heart. The blood passes through the bicuspid valve into the left ventricle. From the left ventricle, which is the largest and most

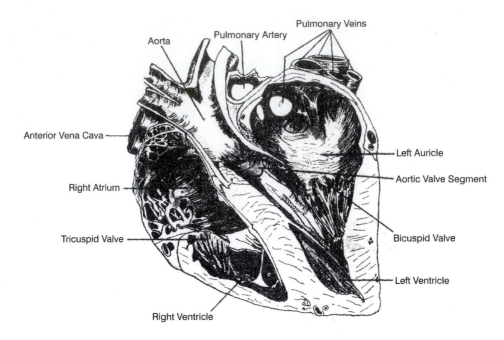

Figs. 13-5.   Cross section of the heart of a horse. (Courtesy, University of Georgia, Athens)

muscular chamber of the heart, blood is pumped to all parts of the body except the lungs.

SIGNS OF DISTURBANCE

The major indications of disturbance to the circulatory system are tachycardia, which is an abnormally rapid pulse rate, distention of peripheral blood vessels, and anemia.

Abnormal Pulse Rate

Drastic extremes from the normal pulse rate are very serious to animal health. Bradycardia, or an abnormally slow pulse rate, is not as common as the opposite condition, tachycardia. Bradycardia may occur in certain diseases that cause pressure on the brain, the comatose condition in parturient paresis, or milk fever, and just prior to the death of an animal.

An increase in pulse rate above the normal may be expected in febrile diseases, severe hemorrhages, and diseases accompanied by pain, excitement, delirium, or mania. A very serious condition exists when the pulse rate exceeds 75 for a horse, 100 for a cow, and 125 for sheep and swine. (Instruction for taking the pulse and normal ranges of pulse rates for domestic animals are found in Chapter 9, under the headings "Sites for Taking Pulse" and "The Pulse Rate."

Distention of Blood Vessels

Signs of disturbance of the vascular system are abnormal pulsation of peripheral arteries and veins and permanent distention of veins. Venous distention occurs when the heart is unable to maintain normal circulation, and the veins become congested with blood. This condition is most easily observed in the jugular veins and mammary veins of cows.

A strong, persistent pulse in the veins is abnormal and indicative of a severe disturbance of the circulatory system.

Other Abnormal Conditions

The following conditions indicate additional disturbances of the circulatory system:

1. Bleeding.
2. Blood in secretions (milk) and excretions (feces and urine).

3. Anemia—a condition in which blood is deficient either in quantity or quality.

4. Edema—an abnormal accumulation of fluids in the intercellular tissue spaces of the body.

5. Hematoma—a tumor containing effused blood.

6. Shock—a circulatory deficiency characterized by a decreased flow of blood.

DISTURBANCES OF THE CIRCULATORY SYSTEM

Anaplasmosis

Anaplasmosis is a non-contagious, infectious disease of cattle, characterized by anemia, due to the destruction of erythrocytes. It occurs occasionally in sheep. Anaplasmosis is a serious problem of cattle in warm, bushy, wet areas of the United States. The incidence of anaplasmosis fluctuates as the population and activity of insect and tick vectors increase and decrease with changing weather.

Cause

Anaplasmosis is caused by a minute parasite, *Anaplasma marginale*, found in the red blood cells of infected cattle. It may be transmitted from infected animals to healthy animals by insects or by surgical instruments.

Three biting insects are known to transmit anaplasmosis mechanically, by carrying infected red blood cells from diseased cattle to healthy cattle. They are horse flies, stable flies, and mosquitoes. Insects must pass the infectious agent from a diseased animal to a healthy animal within a few minutes if inoculation is to take place.

Ticks carry anaplasmosis differently from other insects. The parasites can live in ticks and may be passed through several generations of ticks. People commonly carry anaplasmosis organisms from one animal to another on dehorning saws, castrating knives, vaccinating and bleeding needles, tattoo instruments, and ear notchers. When this type of transmission occurs a large number of cattle in the herd show signs of anaplasmosis at nearly the same time, without a few earlier cases having appeared.

Clinical Signs

The predominant clinical signs are related to acute anemia, and all ages of cattle may be affected. A febrile response where the temperature may reach 106°F is observed early in the course of the disease. Subnormal temperatures may be seen in animals just before death.

Affected animals are weak, depressed, dehydrated, have pale membranes, increased heart rate, and show respiratory distress when excited. The membranes may be yellow and jaundiced, and examination of the blood will show it to be thin and watery. Anaplasmosis is diagnosed based on clinical signs, laboratory values, and necropsy results. A definitive diagnosis is reached when a stained blood smear examined under oil immersion demonstrates the presence of *A. marginale* on the red blood cells. Serologic testing may be beneficial in identifying carrier animals within the herd, as clinically recovered animals remain carrier animals for life.

In the early stages of acute anaplasmosis, the body temperature may rise from 2 to 5 degrees and remain at 103° to 107°F until the animal begins to improve. But, if the infection is overwhelming, the temperature may drop below normal just before death. The heart-beat may double in rate. Breathing may more than double in rate and become labored. As much as 60 to 75 percent of the red blood cells may be destroyed when the animal first appears sick.

Animals affected with anaplasmosis will show sudden weight loss, weakness, and loss of appetite. The muzzle, udder, and visible mucous membranes appear pale or bleached. Constipation is present with dry, hard feces. In addition, a strong jugular "pulse" is present, Infected animals are weak and do not trail the herd; some may die in one to three days after the initial signs of illness. The severity of illness and the probability of death increase with age. In mature cattle, the mortality varies from less than 10 percent to more than 72 percent.

If the acutely infected animal survives the peak of cell destruction and if the formulation of new red cells exceeds the rate of destruction, convalescence usually follows and may continue for several months. The period of convalescence is governed to some extent by the rate of red cell generation, by the presence or absence of secondary complications, and by the treatments given. Blood transfusions are sometimes given to help the recovery of valuable animals. Young animals generally produce red cells faster than do older animals, and they also respond more satisfactorily to blood transfusions.

Aside from the occasional relapse during convalescence, the animal maintains a certain relative resistance to reinfection—that is, resistance to any new introductions of the parasite as long as some degree of infection remains. The resistance depends on the continued presence of the parasite, rather than a lasting supply of the protective antibodies, such as many bacterial and viral pathogens stimulate.

Animals that survive infection become carriers. Although the animals may show no further signs of the disease, their blood is infectious. Such animals generally remain carriers for life, thus posing a threat to other animals. Any carriers freed of the parasite, either naturally or through medication, become susceptible again. Many exposed animals, especially the young, become carriers without experiencing acute anaplasmosis or without having visual symptoms of the disease, since body defenses are able to hold parasitism to a minimum.

Prevention

Anaplasmosis outbreaks occur when there are no control programs, there are carriers and susceptible animals present, and vectors allow transmission of the organism.

Control programs are:

1. Testing and removing carriers, which necessitates a blood sample be obtained from every animal.

2. Testing and clearing carriers, where those animals identified through blood testing are administered tetracycline antibiotics. Oxytetracycline 200 mg/ml formulation can be given to animals four times at three-day intervals at a dose of 9 mg/lb intramuscu-

larly. Chlortetracycline can also be fed at a rate of 5 mg/lb for 60 days, or 0.5 mg/lb for 120 days.

3. Continuous oxytetracycline injections every 28 days throughout the vector season.

Other preventative measures include control of ticks, flies, and mosquitoes and proper disinfection of instruments used in mass procedures such as dehorning, eartagging, castration, and vaccination. Feeding aureomycin at low levels during vector season will control the symptoms of anaplasmosis in susceptible cattle. However, this will not prevent the carrier status in susceptible cattle that are exposed. Cattle fed as little as 150 mg of aureomycin per head per day have been successfully protected against anaplasmosis. Research in Texas indicates that continuous feeding of aureomycin at a level of 0.5 gram per 1,000 pounds of body weight for 120 days during the winter feeding period will eliminate the carrier state of anaplasmosis.

Treatment

A single injection of oxytetracycline stops the increase in infected red blood cells if given early in the course of the disease. This helps reduce the severity of the disease since it is mainly the infected RBCs that are destroyed to produce the anemia.

Acutely ill animals should be isolated, handled quietly, and allowed access to fresh feed and water.

Hemolytic Icterus of Foals

Hemolytic icterus occurs in affected foals during the first three days of their life. It is characterized by depression, jaundice, weakness, and hemoglobinuria. Hemolytic icterus is also called jaundice foals disease. Similar hemolytic diseases occur in calves and pigs. The disease in calves is called neonatal anemia.

Cause

Hemolytic icterus is the result of an incompatibility of the foal's blood to specific antibodies in the colostrum of the mare. The situation is somewhat similar to the Rh factor in humans. It arises when the mare is bred to a stallion with a different blood type, and the foal inherits the blood type of the sire. It usually occurs with the second or third foal from this incompatible blood type mating. Some of the blood cells of the foal enter the dam's circulatory system, and the dam develops antibodies against them. The colostrum, therefore, contains high levels of these antibodies, and when the foal nurses and the antibodies are absorbed, they cause clumping and a breakdown of the foal's red blood cells, resulting in anemia.

Clinical Signs

The signs of hemolytic icterus include general depression, debility, a rapid, weak pulse, rapid respiration, infrequent nursing, and hemoglobinuria. The foal usually becomes weakened to the point that it is unable to stand and nurse. Mucous membranes of the mouth and eye will be jaundiced.

Prevention

A veterinarian can test the blood of the mare late in gestation or the colostrum to determine compatibility with the foal's blood. If there is an incompatibility between dam and foal, the disease may be avoided by preventing the foal from nursing its dam for two to three days. The foal should be bottle fed preferably with colostrum from a non-immunized mare or be allowed to nurse a foster mother until the colostreal period of the natural mother has passed.

Treatment

The anemic foal should be kept as quiet and comfortable as possible. Milk that is compatible to the foal should be supplied until the dam's milk is safe. Antibiotic therapy, including penicillin, aureomycin, terramycin, or streptomycin, may be beneficial. Blood transfusions or a complete exchange of blood, using donors' blood which is compatible to that of the foal, is recommended under some circumstances.

Equine Infectious Anemia

Equine infectious anemia is a viremia of horses, mules, and donkeys that is characterized by depression, debilitation, loss of weight, and swellings on the legs and lower abdomen. It is also referred to as malarial fever, slow fever, swamp fever, and mountain fever. Equine infectious anemia has been reported throughout the United States.

Cause

Infectious anemia is caused by a virus found in blood and tissues of infected equines. The virus remains in the blood of animals that recover from equine infectious anemia. Infected animals shed the virus in discharges from eyes and nose, saliva, urine, manure, mare's milk, and semen.

Clinical Signs

Horses normally develop infectious anemia two to four weeks after they are exposed. However, signs may appear as long as two months after natural exposure.

The animal with equine infectious anemia has a sudden rise in temperature, from a normal of 100° to 105°F, or higher. Fever attacks may be intermittent or continuous. The horse may sweat profusely. Breathing is rapid. The horse appears depressed. It usually loses weight although it continues to eat. The eyes are bloodshot, with a slight watery discharge. Urination is frequent; diarrhea develops in severe cases. Swellings filled with water liquid may form on the legs and lower parts of the body. Weakness causes the horses to develop a wobbly or rolling gait. In some cases, the hindquarters are paralyzed. As the disease progresses, anemia develops. Mucous membranes become pale or yellowish in color, and the pulse weakens. Heart action becomes irregular.

Prevention

There is no vaccine recommended to protect animals from equine infectious anemia. However, the following practices are recommended by the Animal Health Division of the U. S. Department of Agriculture:

1. Disposable syringes and needles should be used. The rule, "one horse—one needle," should be

followed. Other instruments that the caretaker uses in working with animals should be sterilized. All instruments should be thoroughly cleaned after each use and then boiled 15 minutes to sterilize. This will prevent the spread of disease by knives and by dental and surgical equipment.

2. Biting flies, biting lice, and mosquitoes should be controlled in stables and pastures. Stables and immediate surroundings should be kept clean and sanitary at all times. Manure and debris should be removed promptly. The area should be well-drained. A separate tack should be provided for each animal—if bridles, saddles, harnesses, brushes, spurs, whips, currycombs, or bandages must interchanged, each piece should be cleaned thoroughly before reused. For cleaning, a 2 percent trisodium phosphate solution should be used; for conditioning leather items, saddle soap or neat's-foot oil should be used.

3. All new horses, mules, and donkeys should be isolated. Temperatures should be checked daily. New animals should be kept under observation 60 days before they are put with other animals. Mares and stallions that are suspected of being infected with infectious anemia should not be bred. When diagnostic problems occur or when shipping requirements must be met, the Coggins test should be used.

4. At horse shows, county fairs, race tracks, and other places where many animals are brought together, the animals should be quartered in separate, well-ventilated stalls. These stalls should be kept clean and free from flies. Feed and water should be given only in containers reserved for individual animals.

5. Apply the Coggins Test for diagnosing equine infectious anemia. Repeat the test to confirm all positive reactions. Positive reactors are identified with an "A" in a visible brand or lip tattoo, which stands for anemia. Animals so branded are quarantined and cannot be moved except for slaughter or approved research purposes.

6. In order to prevent bringing in the disease, in 1976 the USDA amended the import regulations to require that imported horses pass the Coggins Test to assure that they are free of equine infectious anemia.

Treatment

There is no treatment for equine infectious anemia.

Anemia of Suckling Pigs

Anemia is a condition in which the blood is deficient in either quantity or quality. it is characterized by paleness of the skin and mucous membranes, fatigue, and labored breathing.

Cause

Anemia of suckling pigs is commonly caused by an iron deficiency, but it may also be caused by a deficiency of copper, cobalt, and/or certain vitamins.

The baby pig is born with a total of about 40 mg of iron in the body. With an iron requirement of about 7 mg daily, it is apparent that without supplemental iron, body stores will not last very long. Sow's milk is a good source of all nutrients the baby pig is known to require with the exception of iron.

Iron is a necessary element in the formation of red blood cells and of the compound hemoglobin which is a part of the red blood cell. Hemoglobin is the vehicle by which oxygen is transported from the lungs to the cells of the pig. If iron is lacking in the body of the pig or in its daily diet, it is unable to form hemoglobin; consequently, the cells of the body are unable to function properly due to the lack of oxygen.

Clinical Signs

The signs of anemia in suckling pigs may vary from acute to chronic. The classical picture of baby pig anemia is the sudden death of the largest, healthiest pig in the litter, seemingly with no external signs of trouble. However, the indications of chronic anemia may vary. Pigs may not do well, grow poorly, and appear listless, with the eyelids drooping and the ears and tail hanging limp. The hair coat may be rough, dull, and coarse, and it may stand erect. The pigs will breathe rapidly and appear to have difficulty in obtaining enough air. Anemic pigs often appear fat due to edema under the jaw and around the neck and shoulders.

Anemia lowers the resistance of the pigs to disease. Enteritis is common in anemic pigs and is often considered to be a symptom of anemia. The growth rate may be decreased by borderline cases of anemia in which no outwardly detectable signs of anemia are present in the pigs.

Prevention

Supply dietary sources of iron, copper, cobalt, and certain vitamins (especially folacin, riboflavin, and vitamin B_6). Keep confinement of suckling animals to a minimum and provide dry feeds at an early age.

Anemia in pigs can be prevented by providing supplemental iron in one of the following forms:

1. Inject intramuscularly 100 to 200 mg of iron from iron dextran into baby pigs at 2 to 3 days of age. If pigs remain in confinement and do not have access to creep feed at an early age, a second injection at 2 to 3 weeks of age is desirable. Injection is the method of choice, for it assures that every pig receives its requirement.

2. Orally administer iron dextran in a liquid or a solid preparation. To ensure daily intake by all pigs, it is important to have a preparation that is palatable and readily consumed. Also, placement of the oral preparation at the right location in the creep area is most important.

Edema

Edema is defined as the presence of abnormally large amounts of fluid in the intercellular tissue spaces of the body. It is also called dropsy. Edema is characterized by pitting under pressure. This means that a hollow spot forms and remains for a short time when the finger or hand is used to apply pressure to the edematous enlargement. Edema occurs in connection with infectious diseases such as malignant edema, swine erysipelas, and anthrax; parasitic infestations, such as bottlejaw; and hypersensitivity to a wide variety of foreign proteins that enter the animal's body.

Cause

Edema is usually the result of an impairment of the circulatory system. It may be caused by improper drainage of a lymph, increased capillary pressure, lowered asmotic pressure of the blood, and injury to the capillary wall. Mineral imbalances, protein deficiencies, and allergic reactions to certain drugs, insect bites, plants, parasites, and infectious organisms can also cause edema. Also, edema may be associated with the *E. coli* syndrome.

Clinical Signs

The major sign of edema is the appearance of a painless swelling that pits upon pressure. This accumulation of lymph may appear almost anywhere on or in the body. The swellings are often seen on the lower jaw, legs, abdomen, neck, and udder (Fig. 13-6). The accumulation may increase in size for several days, or it may attain its maximum size in less than 24 hours.

Treatment

Treatment is based upon diagnosis and alleviation of the cause of edema. When the cause of edema has been corrected, the accumulation of fluids will disappear.

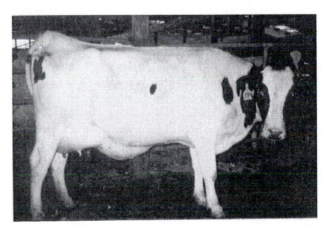

Fig. 13-6. Edema in the navel area of a heifer pre-freshening.

Shock

Shock is a circulatory failure characterized by dilation of the capillaries, low arterial and venous pressure, and a reduction in blood volume. Dilation of the capillaries enables fluids to escape, load up the lymph system, and result in edema.

Cause

Shock can result from severe injury, overexertion, poisoning, surgery, hemorrhage, contact with electricity, sensitivity to antibiotics, serums, etc. In hemorrhagic and traumatic shock, the primary cause is a drastic reduction of blood volume.

Clinical Signs

The clinical signs of shock include pallor, low body temperature, tremors, low blood pressure, and rapid respiration. Coma followed by death of the animal is the usual termination of severe cases of shock that are untreated. Death is more common in animals affected by traumatic shock than those undergoing hemorrhagic shock.

Shock due to overexertion occurs in horses. When a horse is exercised strenuously, it will normally sweat profusely. If the horse has been sweating profusely and suddenly stops sweating, even though the exercise is continued, it may be going into shock. In that case, exercise should be stopped.

Treatment

Treatment is based upon the restoration of blood volume before irreversible circulatory failure occurs. The patient should be kept calm, comfortable, and warm. Blankets to keep the animal comfortable should be used in cold weather. Excessive covering, however, may add to the animal's distress. No coverings are necessary during summer months.

Blood transfusions, a plasma volume expander, such as dextran, or physiologic saline solutions should be administered intravenously.

Hematoma

A hematoma is defined as a tumor containing blood. It is a subdermal effusion of blood in a circumscribed area of the body.

Cause

A hematoma is caused by an injury to any portion of the circulatory system that permits the escape of blood to surrounding areas.

Clinical Signs

The common indication of a hematoma is a hot swollen area on the body. It may be easily confused with an abscess ready to rupture or with edema. Hematomas that affect certain systems or organs may produce a variety of clinical signs. Hematomas within the skull may produce undue excitement, a lack of coordination, convulsions or lethargy, dullness, and stupor. The contrast in clinical signs is a result of damage to different areas of the brain. Hematomas in the respiratory tract may cause dyspnea, painful or labored breathing, and polypnea, rapid, shallow inspirations. Hematoma of the penis often results in the inability of bulls to service cows properly.

CHAPTER 14

The Nervous System

STRUCTURE AND FUNCTION

The nervous system is the governing agency of the body in all domesticated animals. It controls all muscular activities, whether voluntary or involuntary, regulates vital processes such as circulation, respiration, digestion, and egestion, and enables an animal to perceive and adjust to certain stimuli in its environment. The nervous system is made up of two main divisions: (1) the central nervous system, composed of the brain and the spinal cord; and (2) the peripheral nervous system, composed of nerves which extend out from the central nervous system and nerves of the autonomic nervous system (Fig. 14-1).

Central Nervous System

The brain is the central control station for nerve impulses. It is divided into three major parts: the medulla oblongata, the cerebellum, and the cerebrum. The medulla oblongata or brain stem resembles the spinal cord more than it does other parts of the brain. This part of the brain controls many activities, among which are breathing, chewing, swallowing, vomiting, coughing, secreting gastric juices and saliva, and the beating of the heart.

The cerebellum regulates stance, balance, and movement. It receives impulses from the eyes, ears, muscles, joints, and skin and responds automatically. Damage to nerves in this area disturbs the equilibrium and coordination of the animal's body.

The cerebellum is the largest and most active part of the brain. It controls motor movements of the face, jaw, lips, tongue, and other extremities of the body. Memory, sight, smelling, and hearing are also controlled by the cerebrum.

The spinal cord is a continuation of the central nervous system from the medulla oblongata into the sacrum. It is enclosed in a bony canal formed by the neural arches of the vertebrae.

The spinal cord routes impulses that come to it, either to the brain for conscious action or to the lower nerve centers for reflex action.

Peripheral Nervous System

The peripheral nervous system includes all nerve structures beyond the brain and spinal cord. This includes the 12 pairs of cranial nerves (see Table 14-1), the spinal nerves, and the autonomic nervous system. The peripheral network of nerves transmits messages from throughout the body to and from the central nervous system.

The spinal nerves leave the spinal column between each pair of vertebrae. Each nerve has two separate connections with the spinal cord. One of the connections is referred to as the dorsal root, and the other, the ventral root. The dorsal root is sensory, which means that fibers carrying impulses toward the spinal cord enter through the dorsal root. Motor nerves leave the spinal cord through the ventral nerves. Animals whose dorsal roots are severed will have no sensation from stimuli directed to that area, but muscles will continue to function. If the ventral roots are severed, the area served by those nerves will be paralyzed, but sensation will still be present.

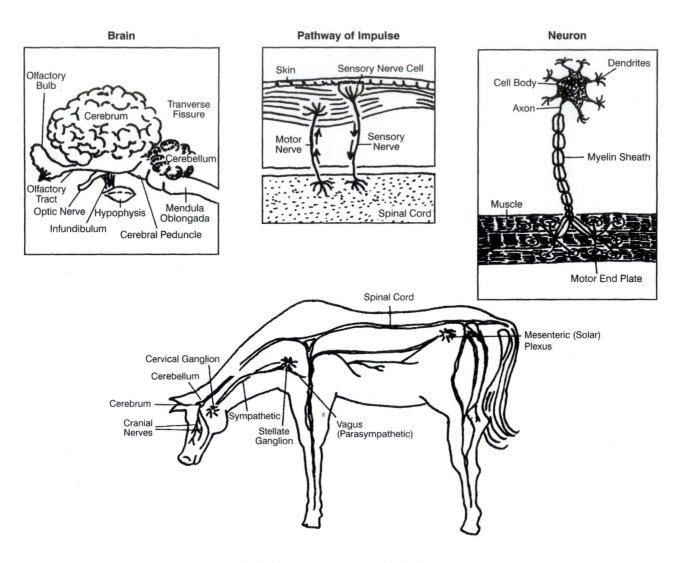

Brain

Olfactory Bulb

Cerebrum

Tranverse Fissure

Cerebellum

Olfactory Tract

Optic Nerve / Hypophysis

Infundibulum

Mendula Oblongada

Cerebral Peduncle

Pathway of Impulse

Skin

Sensory Nerve Cell

Motor Nerve

Sensory Nerve

Spinal Cord

Neuron

Dendrites

Cell Body

Axon

Myelin Sheath

Muscle

Motor End Plate

Spinal Cord

Mesenteric (Solar) Plexus

Cervical Ganglion

Cerebellum

Cerebrum

Cranial Nerves

Sympathetic

Stellate Ganglion

Vagus (Parasympathetic)

Fig. 14-1. Nervous system of the horse.

The autonomic nervous system regulates the internal organs of the body. Its nerves are connected to and regulated by the central nervous system on a subconscious level. Activities controlled by the autonomic nervous system, such as functions of the heart, stomach, intestines, rectum, and bladder, are involuntary and are not consciously controlled.

SIGNS OF DISTURBANCE

Indications of disturbance of the nervous system include abnormal disposition, incoordination, impair-

ment of senses, immotility, and involuntary movement.

Abnormal Disposition

Diseases that affect the central nervous system often manifest themselves by abnormal and drastic changes in disposition. Indications of damage to this system may be categorized into the extreme of derangement and stupor. Rabies is an example of a disease that often incorporates symptoms of derangement and stupor. Pressure applied to the cerebrum by

TABLE 14-1. The Cranial Nerves Whose Names are Derived Partly from the Tissue or Organ Innervated

Name	Function	Tissue or Organ
Olfactory	Sensory	Transmits impulses from nerves in mucous membranes of the nose
Optic	Sensory	Transmits impulses from nerves in the retina of the eye
Oculomotor	Motor	Innervates muscles in the eye
Trochlear	Motor	Innervates muscles in the eye
Trigeminal	Mixed	Transmits impulses from nerves in the face Innervates muscles of mastication
Abducens	Motor	Innervates muscles in the eye
Facial	Mixed	Transmits impulses from nerves in the ear and tongue Innervates muscles in the face
Acoustic	Sensory	Transmits impulses from the inner ear (hearing and equilibrium)
Glossopharyngeal	Mixed	Transmits impulses from nerves in the pharynx and tongue Innervates muscles in the tongue and pharynx
Vagus	Mixed	Transmits impulses from nerves in the pharynx and larynx Innervates muscles in the larynx
Spinal accessory	Motor	Innervates muscles in the shoulder and neck
Hypoglossal	Motor	Innervates muscles in the tongue

tumors, hydrocephalus, and hemorrhage may result in dullness, stupor, and coma. Derangement, delirium, and mania occur with heat stroke, lead poisoning, meningitis, equine encephalomyelitis, and the nervous type of ketosis.

Incoordination

Incoordination occurs with equine encephalomyelitis, certain plant poisonings, milk fever, and acetonemia. Incoordination also results from damage to the inner ear or abnormal pressure on the cerebellum of the brain.

Impairment of Senses

Impairment of senses includes extreme sensitivity to touch, as may be witnessed in salt poisoning, and intense pruritis, which is common in Aujeszky's Disease. Anesthesia, or loss of feeling, and loss of sight are usually due to damage of nerves which innervate those areas of the body.

Immotility

Paresis and paralysis are conditions which adversely affect an animal's motility. Paresis is a slight or incomplete paralysis which prevents an animal from standing. Milk fever, or parturient paresis, is a prime example of this condition of immotility. Other diseases which cause immotility of this nature include ketosis, rabies, equine encephalomyelitis, and listeriosis.

Paralysis is a loss of movement in some part of the body. Tetanus in horses usually results in paralysis of some of the facial muscles. Nerve damage due to dystocias often results in a temporary paralytic condition.

Involuntary Movement

Involuntary movements include spasms, twitching, circling, convulsions, and other signs of damage to the motor nerves of the brain. Tetanic spasms occur with tetanus, grass tetany, and wheat pasture poisoning. Myoclonic congenita is an inherited condition of swine that results in trembling and shaking in baby pigs. Circling as an involuntary movement is associated with the bacterial infection listeriosis. Convulsions occur with many poisonings, tumors, and hemorrhages which apply pressure on the brain.

DISTURBANCES OF THE NERVOUS SYSTEM

Tetanus

Tetanus is an acute, highly infectious, non-contagious diseases that is manifested by tonic contractions of voluntary muscles. It affects human beings, horses, sheep, swine, and cattle. The infectious organism gains entrance through traumatized tissue. Deep wounds that seal over quickly are most frequently involved. Other wounds such as those of castration, docking, parturition, shearing cuts, and openings from dental eruptions and a raw umbilicus are often sites of infection.

Cause

Tetanus is caused by *Clostridium tetani* which is a rod-shaped, anaerobic bacterium that is commonly found in the intestinal tract and feces of most farm animals.

Clinical Signs

The tetanus organism does not leave the region of the wound to get into the bloodstream, as is so frequently the case with other germs; instead, it remains localized, giving off a very powerful poison or toxin which is about 100 times as powerful as strychnine. It passes along the nerves to reach the spinal cord.

The onset of tetanus occurs from a few days to several weeks after the infection takes place. The first noticeable symptom is stiffness, which in 24 hours becomes pronounced. Occasionally, the stiffness affects certain groups of muscles more than others, as in the so-called "lock jaw" in which the muscles of the jaw become set. The ears are stiffly upright, and the tail is held in a semi-rigid extended position (Fig. 14-2). The eyes, upon the least excitement, are retracted into their

Fig. 14-2. Tetanus in a horse. Note the stiff saw-horse stance, the erect ears, and the extended tail. (Courtesy, Dr. Murray Fowler)

Fig. 14-3. A cow with tetanus. (Courtesy, Dr. Murray Fowler)

sockets so that the third eyelid is protruded over the eye (Fig. 14-3).

If the symptoms develop rapidly, the termination of the disease is almost always a fatal one. If the symptoms develop more slowly, or if the stiffness still permits the animal to eat and drink, recovery is more frequent, probably somewhat more than 50 percent.

Prevention

Prevention of the disease consists of a thorough cleansing of wounds as soon a they occur. Docking, castrating, and other management practices should be performed as antiseptically as possible. Wounds with extensive laceration and deep, penetrating wounds are the most serious because their recesses afford hiding places for the germ.

Immunity against tetanus can be obtained through inoculation with either toxoid or antitoxin. Toxoid is an injection of neutralized tetanus toxin to stimulate the animal to build its own antibodies. Antitoxin is a concentrated serum with tetanus toxin antibodies taken from another animal and administered as a preventive measure following wounds, surgery, or parturition. Polyvalent vaccines, which contain up to seven antigens for *Clostridium* spp., are available and highly effective.

In horses, active immunization is achieved through two injections of tetanus toxoid at 2- to 4-week intervals, followed by annual booster injections. If an immunized horse is wounded two months or more following such immunization, it is recommended that the veterinarian administer another toxoid injection at that time. If a horse not previously immunized is wounded, it is recommended that the veterinarian administer antitoxin, which will give passive protection for up to two weeks.

Treatment

Treatment of animals affected by tetanus includes large doses of tetanus antitoxin, sedatives, and nursing strategies to enable the animal to eat and drink when it can. The animal should be housed in a quiet, darkened area and kept as comfortable as possible.

When the site of infection can be located, it should be reopened and thoroughly cleaned. Antibiotics should be administered to combat infection. Injectable procaine penicillin G at 25,000 IU/kg in the muscle twice daily for five days is the antibiotic of choice.

Rabies

Rabies is an acute viral encephalomyelitis that affects all warm-blooded animals. It is commonly called hydrophobia which literally means "fear of water."

The name is a misnomer because affected animals are not afraid of water. They often have difficulty drinking due to paralysis of the tongue and jaw.

Rabies is communicable by means of a bite, by which the saliva-containing virus enters the wound. Rabies occurs most frequently in dogs. Other domesticated animals usually contract the disease as the result of dog bites. Humans are quite susceptible. Cattle are highly susceptible to rabies, horses, sheep, and goats are moderately susceptible, and swine are the least susceptible of the farm animals.

Rabies may occur during any month of the year, but the incidence is usually higher during the spring and fall.

Cause

Rabies is caused by the rabies virus. The causative virus is found in the nerve tissues of the brain and spinal cord, as well as in various secretions such as tears, milk, and saliva. It is important to note that the rabies virus has been found in the saliva at least five days before the infected animal had clinically recognizable symptoms of the disease. This means that the bite of a dog is always to be considered potentially dangerous, and the best procedure is to confine such a dog for daily observation. If no recognizable symptoms appear in the animal within a period of two weeks after it has inflicted the bite, it is safe to assume that there was no rabies at the time of the bite. To destroy a biting dog immediately following the incident is a serious error, because this also destroys the evidence the laboratorian needs to make a diagnosis. To confine the dog until there is clear-cut evidence on which to base a diagnosis is better.

Clinical Signs

The initial clinical signs exhibited from rabies are dependent on the species of animal, the strain of the virus, and the dose of the virus inoculated into the animal.

There are several stages of rabies that affected animals may follow. The initial stage of rabies is usually characterized by a change in disposition. A friendly dog may become shy, attempt to hide, resent attention, and snap or bite when approached. Wild animals often show no fear of people. Nocturnal animals, such as bats or skunks, should be suspected when they appear unafraid of people and are found roaming around aimlessly during the daytime. Rabid dogs assume either a furious or paralytic form of disease following the stage characterized by a change in disposition.

In furious rabies, dogs often bark persistently without cause, move about aimlessly, snap or bite at anything in their path, and attempt to chew or swallow unusual objects. In advanced stages of furious rabies, dogs may have muscle spasms, may drool from the mouth with the tongue hanging out, and mobility may be difficult. The bark is hoarse or it disappears, and the dog becomes too weak and incoordinated to move. Prostration and death follow soon after this occurs. In the paralytic form, dogs become lethargic and prostrate, and they lie quietly until death.

In cattle, depression, a severe drop in milk production, pharyngeal paralysis, and signs of "choke" are the initial signs of rabies. Producers should be aware of the possibility of rabies any time a cow appears to choke, as the virus can penetrate open wounds on the hand of the person examining the mouth of a cow for apparent choke. Cattle, sheep, and goats usually do not develop the furious form of rabies; rather they develop the paralytic form. Cows will die in four to seven days after signs appear. Sheep and goats develop depression, staring eyes, excitation, salivation, and anorexia and rams may exhibit sexual stimulation. The disease rapidly progresses to paralysis, coma, and death.

Prevention

All dogs and cats should be vaccinated with antirabies vaccine. A dog or cat that shows abnormal behavior should not be touched. It should be confined, if possible, and a veterinarian or the local health department called. A rabid animal will show definite symptoms and will die in a few days after the onset of symptoms. For this reason, a 10-day to 2-week quarantine period is necessary to prove an animal rabid. The animal should not be killed unless absolutely necessary—for example, to prevent its escape or to keep it from endangering human health.

Rabies in horses can be prevented through a vaccination program. Stables should be secure and feed kept closed in order not to attract skunks, raccoons, and opossums. Vaccines are available for preventing rabies in cattle in endemic areas. Management goals in a food animal operation should strive to prevent exposure of animals to rabies.

Wild animals or animals that cannot be safely confined or controlled may have to be destroyed. When this is the case, they should not be shot in the head, because the brain must be examined in the laboratory to make a definite diagnosis. The lack of a definite diagnosis might cost exposed persons their lives or subject them to painful and dangerous treatment.

Several new rabies vaccines are available, and others are being developed and tested experimentally. Thus, the choice of a rabies vaccine for animals should be made by the veterinarian.

Treatment

There is no effective treatment for rabies once symptoms have occurred. If a person is bitten by an animal suspected of having rabies, a doctor should be called immediately. All bat bites should be considered rabid until proven otherwise. If a doctor is not immediately available, first aid should consist of scrubbing all wounds to their full depth for 15 to 20 minutes, using soap and changing wash water frequently. Rabies is fatal once symptoms occur. Only one human has ever been reported to have survived.

In humans, the symptoms may be prevented by a series of vaccinations for 14 to 21 consecutive days. Because the treatment is disagreeable and has some risk, positive diagnosis is of utmost importance if humans have been exposed.

Pseudorabies

Pseudorabies is an acute viral infection of pigs and cattle, although sheep are occasionally affected. Rats are also highly susceptible to the pseudorabies virus and probably play an important role in maintaining the infection in swine. Pseudorabies is non-contagious in cattle but highly contagious among swine. It is spread to cattle by bites from infected hogs.

Pseudorabies is commonly called Aujeszky's Disease and "mad itch." Aujeszky was the first researcher to describe the disease. He studied its occurrence in cattle and first thought the disease was rabies. Cattle affected by pseudorabies have an intense, localized pruritis, or "mad itch." Infected swine do not show this symptom.

Cause

An alpha herpes virus causes pseudorabies. Swine are long-term carriers of the virus, as the virus persists in a latent state until stress or declining immunity causes viral replication and shedding again. Pseudorabies is a reportable disease, with state import requirements requiring blood testing of swine. The virus can travel between herds for at least two miles.

Clinical Signs

Cattle affected by pseudorabies almost always have a history of association with swine. Clinical signs include an intense itching at the site of infection. The animals lick, bite, and rub the area with such furor that "mad itch" properly describes the condition. The localized area of pruritis is usually in the hindquarters or vulva. In one or two days, paresis and convulsions develop. Death follows shortly after the animals become prostrate.

Newborn pigs affected by pseudorabies often pass from a normal appearance to a coma and death in 6 to 24 hours. The mortality rate for the litter is very high with young pigs.

Older pigs are likely to show symptoms that indicate involvement of the central nervous system. Such signs may include incoordination, muscle spasms, excitability, prostration, paralysis, and death.

Pseudorabies in older hogs is usually a subclinical infection that goes unnoticed by the caretaker. It may be accompanied by mild fever, depression, and a finicky appetite. In severe cases, older swine may show symptoms quite similar to swine influenza. In these instances, the fever will be high (104° to 106°F), vom-

iting may occur, and nesting or refusal to rise and move away from the shelter, and a loss of appetite are common.

Abortions often occur when bred sows are infected with pseudorabies. Those that do not abort often deliver dead or mummified pigs. Infected breeding females are usually hard to settle.

Prevention

Attenuated and inactivated vaccines that are effective in prevention of the disease are available commercially, but neither kind will prevent infection by field virus, which will be shed by the challenged vaccinates.

Breeding stock should be vaccinated twice per year prior to breeding. Pigs from unvaccinated sows may be vaccinated after 3 days of age. Pigs from immunized sows should be vaccinated when 3 to 8 weeks old.

Treatment

There is no effective treatment against pseudorabies.

Equine Encephalomyelitis

Equine encephalomyelitis is an acute infectious disease of horses and mules that is transmitted by insect vectors. Equine encephalomyelitis is characterized by nervous disorders and high mortality. It usually occurs as individual isolated cases but may become endemic when heavy rains and floods leave suitable breeding places for mosquitoes. Most cases occur during the warm season of the year, when mosquitoes are most active.

Cause

The disease is caused by several distinct viruses. The three most active types in the United States are Eastern equine encephalomyelitis, Western equine encephalomyelitis, and Venezuelan equine encephalomyelitis.

Eastern encephalomyelitis occurs primarily in the east coastal and gulf states. Western equine encephalomyelitis is found most commonly in the central and western states, and Venezuelan virus encephalomyelitis is a threat throughout the United States.

The disease is maintained by an insect-bird reservoir. It is transmitted to mammalian hosts by biting insects, principally mosquitoes of the *Aedes, Anopheles, Culex,* and *Culeseta* species (Fig. 14-4). Birds tend to develop a viremia, and many of them often die. Those that live usually harbor the disease without showing any signs.

Humans and horses are considered "dead end" or accidental hosts. They do not develop a viremia, so the disease is not transmitted from the infected human or horse to other mammalians. With the possible exception of the Venezuelan type, people and horses do not transmit the disease among themselves by contact or by mosquito transmission.

Clinical Signs

Equine encephalomyelitis affects the central nervous system. In horses, this results in nervousness, depression, impaired vision, and walking into objects. They have reduced reflexes, fever, incoordination, and grinding of teeth. Other signs include drowsiness,

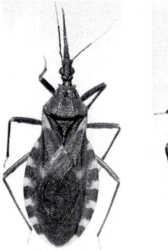

Fig. 14-4. The assassin bug is a natural reservoir of the equine encephalomyelitis virus. The male is pictured on the left; the female on the right. (Courtesy, Department of Entomology, Kansas State University, Manhattan)

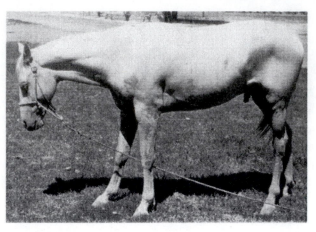

Fig. 14-5. Equine encephalomyelitis in a Palomino gelding. (Courtesy, Dr. Murray Fowler)

sleeping in a standing position, hanging the lower lip, inability to swallow, and prostration (Fig. 14-5). Horses in this condition lie on their sides and flail their feet and legs in a galloping motion. This is followed by paralysis and death. Horses with mild cases recover in a few weeks. Those with severe cases usually die. The mortality rate in horses is around 50 percent with Western and 90 to 100 percent with Eastern and Venezuelan types.

Prevention

Prevention entails vaccination of all horses against the three strains as follows:

1. For Eastern, Western, and Venezuelan strains: trivalent vaccine with EEE, WEE, and VEE, given IM; initial vaccination given one month before mosquito season, and repeated within the year if the mosquito season is quite long; both injections repeated annually as boosters; given to all ages, including foals beginning at 2 to 3 months of age.

2. For Venezuelan strain: attenuated virus cell culture, given IM; one injection only (do not give to pregnant mares); one booster shot repeated annually; including foals beginning at 2 to 3 months of age.

Mosquito control is the best means of protecting people and horses against exposure. Horses can be sprayed with a suitable insecticide or kept in screened stalls. Also, wild or domesticated birds should be discouraged from roosting in the vicinity of horse barns.

Treatment

Treatment is not very effective, because of the rapid course of the disease. Since the Western type progresses more slowly and results in a lower mortality rate than the Eastern and Venezuelan types, it lends itself to more supportive treatment. Good nursing is perhaps the best and most important treatment. The maintenance of fluid and electrolyte balance is recommended. No specific therapeutic agent is known to influence the course of the disease.

Bovine Thromboembolic Meningo Encephalitis

Bovine thromboembolic meningo encephalitis (TEME) is commonly called brain fever. It is a common problem in feedlot cattle in the southwestern part of the United States.

Cause

Bovine thromboembolic meningo encephalitis is caused by a bacterium called *Haemophilus somnus*.

Clinical Signs

Bovine thromboembolic meningo encephalitis is a bacterial infection of the brain; therefore, all clinical signs are characteristic of a malfunctioning brain. The disease tends to strike in certain feedlots. It is usually encountered in cattle weighing less than 400 pounds and usually occurs shortly after animals arrive at the yard. Brain fever is often a problem when the weather is extremely hot or extremely cold.

Affected cattle are drowsy, very depressed, and incoordinated. At the onset of symptoms, affected cattle may not have a fever. They usually have a fever of 105° to 107°F within 24 hours. If sick animals are not treated immediately, their condition deteriorates rapidly and they become weak and ataxic. This is followed by paralysis of the legs, opisthotonos, and finally death.

Prevention

The use of bacterins may reduce morbidity and mortality and decrease the number requiring treatment.

Treatment

Affected animals should be segregated and treated immediately with penicillin and streptomycin or oxytetracycline. Treatment is most effective in the early stages of the disease.

Scrapie

Scrapie is a chronic nervous disorder of sheep and goats that is characterized by a long incubation period, intense pruritis, incoordination, debilitation, paralysis, and death. Sheep between two and four years of age are most commonly affected by scrapie. It seldom appears in animals under 18 months of age. Scrapie is a public health concern because of the assumed relationship to bovine spongiform encephalopathy (BSE).

Cause

The cause and method of spread of scrapie are not fully understood. It has been proven that scrapie is an inoculable disease and that the transmissible agent can be passed in series, indefinitely, from sheep to sheep in filtrates prepared from the tissues of infected sheep. This is characteristic of a virus, but the transmissible agent can withstand boiling and exposure to concentrations of chemical agents which inactivate all known viruses. Moreover, no evidence of the development of antibodies has been obtained and the lesions of nerve cell degeneration do not resemble those usually associated with a virus infection. The agent causing scrapie is thought to be a prion—a protein believed to be spread through birthing fluids.

Scrapie tends to occur more frequently in certain breeds and families of sheep than in others. This has led to the speculation that the susceptibility to scrapie is inherited. Some Suffolk breeders are voluntarily testing their sheep for the chromosome sites for the susceptibility to scrapie.

Clinical Signs

Scrapie is an insidious disease that requires 18 to 24 months for the incubation period. The key signs of scrapie are a progressive weight loss, hind end weakness, recumbency, and death in two to four months. The first signs are nervousness, restlessness, and excitability. Rubbing to alleviate the itching can be observed. Patches of wool will be removed from the tail and rump area as the rubbing continues. Affected animals often grind their teeth. A characteristic "scratch reflex" occurs when the sheep's back is scratched by hand. The infected animals respond by contracting their lips and wagging their tails. There is no loss of appetite, and the temperature remains normal.

Progressive incoordination is apparent in sheep affected by scrapie. There is often a wobbly gait, and the animals may fall when trying to run. Later in the course of the disease, the animals are paralyzed in the hindlegs and are unable to rise. This is followed by convulsions, coma, and death.

Prevention

There is no vaccine to protect sheep against scrapie. Control in the United States is based upon quarantine and slaughter of all sheep and their offspring from an infected flock.

Treatment

There is not treatment for scrapie.

Sunstroke

Sunstroke is an acute disease that is characterized by sudden onset, high temperature, and high mortality. All farm animals are susceptible to sunstroke, but hogs are most commonly affected. This disease should be distinguished from heat exhaustion which is common in working animals and in those being driven or transported for long periods of time. Heat exhaustion is characterized by a gradual onset, normal temperature, depression, and low mortality.

Cause

Sunstroke is caused by exposure to high temperature, high humidity, and poor ventilation. Contrib-

Fig. 14-6. Shade must be provided for animals in hot areas.

uting factors include physical exertion and, in the case of swine, a very inefficient system of heat dissipation.

Clinical Signs

The clinical signs of sunstroke are dramatic. Affected animals will frantically seek shade and water. They salivate profusely, breathe rapidly with the mouth open, and may vomit. The temperature is unusually high. It may reach 110°F. The affected animals become ataxic, prostrate, comatose, and die within a few hours unless the condition is treated promptly. Animals that recover from sunstroke are usually unable to tolerate hot weather and may have permanent mental derangement.

Prevention

The prevention of sunstroke is based upon management practices that keep animals environmentally comfortable and free from undue stress. Hogs must be provided adequate shade, water, and ventilation during hot weather (Fig. 14-6). They should not be moved or handled during the heat of the day. Sows should be bred during the morning or evening, and boars should not be left with sows to aggravate them. New additions should be introduced to the herd during cool periods of the day. (See Fig. 14-7 for livestock weather safety index.)

Treatment

Treatment is ineffective in many cases. Cold water should be applied around the mouths and extremities of the animals' bodies. Cold water should not be poured on the backs of hogs affected by sunstroke. It is advisable to pour water on the floor or ground surrounding the hogs and also around the mouths, feet, legs, and underline. The temperature must be lowered slowly. A sudden drop in body temperature is likely to cause sudden death.

Ice packs placed around the head and cold water enemas can also be used in treating animals affected by sunstroke.

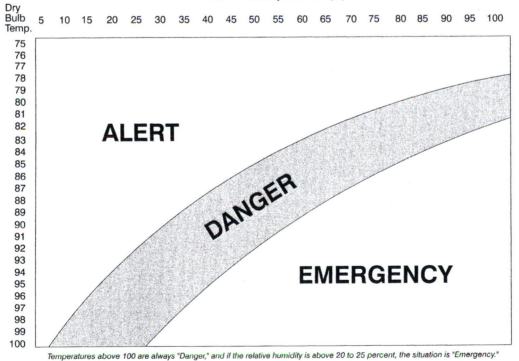

RELATIVE WEATHER SAFETY INDEX

Relative Humidity Intervals (%)

Temperatures above 100 are always "Danger," and if the relative humidity is above 20 to 25 percent, the situation is "Emergency."

Fig. 14-7. Livestock weather safety index. (Courtesy, Livestock Conservation, Inc.)

Equine Protozoal Myeloencephalitis

Equine protozoal myeloencephalitis (EPM) is a neurologic disease of horses that results in weakness, incoordination, and spasticity. EPM is present in the greatest levels in areas inhabited by opossums, such as in the eastern United States. EPM can be an acute or chronic condition and is often difficult to diagnose.

Cause

Equine protozoal myeloencephalitis is caused by a protozoan parasite *Sarcocystis neurona*. Infection occurs when horses ingest the parasite in contaminated hay or feed. The opossum is a carnivore host for the *S. neurona*, and a bird-to-opossum cycle is how the opossum is infected. The birds eat opossum droppings and the opossums eat dead birds continuing the cycle. Horses are infected when opossums have access to feed rooms, pastures, and hay supplies and contaminate the feed through defecation of feces containing the sporocysts.

Clinical signs

The clinical signs of EPM are caused by an inflammation of the nervous system. The classical sign of EPM is focal, asymmetric muscle atrophy. Lameness, respiratory problems, and nearly any neurologic deficit may be apparent. The neurologic signs seen are related to the area of the spinal cord affected, so problems may be seen localized to any muscle group. Weakness and standing wide or ataxia may indicate infection in the brain.

Prevention

The best way to protect horses is to reduce the exposure to wildlife, especially opossums and wild birds. Opossums have a narrow living range, and the presence of dogs may help deter opossums. Other management practices are to repel or euthanize opossums, store grain in sealed containers, minimize mice and rats, and dispose of all uneaten grain, pet food, and human food.

Treatment

The diagnosis of EPM requires analysis of cerebrospinal fluid for antibodies to *S. neurona*. The treatment for EPM is long term, and consists of antibiotics directed against the protozoan. Currently, there are not drugs labeled for the treatment of EPM, although pyrimethamine has been used successfully. A veterinarian should be consulted in the treatment of EPM, as there are special considerations for pregnant mares and vitamin supplementation needs.

PART 3

Miscellaneous Diseases of Farm Animals

CHAPTER 15

Generalized Diseases

ANTHRAX

Anthrax is an acute, febrile, infectious disease of mammals. It is characterized by edema and hemorrhage of subcutaneous tissues, an enlarged, dark-colored spleen, and rapid death of affected animals. Cattle and sheep are more susceptible to anthrax than horses, swine, and humans. Anthrax is commonly called charbon or splenic fever.

Anthrax is endemic to certain areas of the United States. Once an area is contaminated, eradication is virtually impossible because anthrax spores are highly resistant to heat, low temperatures, disinfectants, and prolonged drying. Anthrax may be spread by overflow water or streams from infected areas, by contaminated feed and water, or by dogs, coyotes, and buzzards that have fed on infected carcasses. In addition, biting flies and the misuse of anthrax vaccines can spread this disease.

Cause

Anthrax is caused by a Gram-positive, rod-shaped, spore-forming bacterium called *Bacillus anthracis* The organism becomes encapsulated but does not produce a spore in an unopened carcass. Once the carcass is opened or the bacterium is liberated through body openings, sporulation occurs, and a highly resistant spore is formed.

Clinical Signs

Anthrax may occur in the peracute, acute, subacute, or chronic forms. Clinical signs vary a great deal due to the virulence of the organism and the species of animal affected. The peracute form of anthrax is most common in sheep. Animals are usually found dead with dark-colored, foamy blood exuding from the nose, mouth, and anus. Carcasses fill with gas rapidly, and there is no rigor mortis. When symptoms are observed in animals with peracute anthrax, the clinical signs include incoordination, labored breathing, pounding heartbeat, grinding of teeth, convulsions, and a frothy exudate from the nose and mouth.

The acute and subacute forms of anthrax are most common in cattle, horses, and sheep. The characteristic symptoms are: a rise in body temperature to 105° to 107°F, excitement followed by depression, cessation of milk secretion in lactating females, muscular tremors, labored breathing, and diarrhea that may be hemorrhagic. Horses often show signs of colic and may have blood in both the feces and urine. Edematous swellings may occur along the neck, chest, and abdomen. The final termination of the acute form is usually characterized by convulsions and death in 10 to 36 hours. The subacute form may terminate in death in a week to 10 days, or may result in complete recovery.

The chronic form of anthrax is most common in swine, although they too may be affected by the acute form. In most cases, there is a characteristic swelling in the throat of affected swine that interferes with breathing and swallowing. Hogs may appear to be choking, run temperatures of 107° to 108°F, and have blood-stained, frothy material pass out of their mouths. There is usually general depression, loss of appetite, and vomiting. In many cases, affected swine recover from anthrax but remain carriers of the infectious organism.

Prevention

In infected areas, vaccination should be repeated each year, usually in the spring; and there should be adequate fly control by spraying animals during the insect season.

The nonencapsulated Stern-strain vaccine is used almost exclusively for livestock immunization. Vaccination should be done two to four weeks prior to the season when outbreaks may be expected. Animals should not be vaccinated within 60 days of slaughter.

Prevention of anthrax in people depends on (1) eradication of the disease in animals; (2) elimination of industrial infections (tanneries, woolen mills, and factories utilizing animal hair); and (3) early diagnosis and prompt treatment of infected cases.

Treatment

Treatment for anthrax includes the intravenous administration of immune serum and intramuscular injections of penicillin. Aureomycin and terramycin are also effective when administered in the early stages of the disease. Treatment of animals with the peracute form of anthrax or of those in the advanced stages is usually uneventful.

When an outbreak of anthrax has been confirmed, a quarantine of the premises will be established by the state bureau of animal health. The quarantine will prohibit movement of animals to and from the premises and the sale of milk from affected dairy cattle until the disease is under control. In addition, contaminated milking equipment, barns, feed bunks, etc., should be thoroughly cleaned and disinfected. Bedding and manure should be isolated and treated promptly. Penicillin and antianthrax serum can be used as protective measures for exposed animals.

The carcasses of all animals that have died from anthrax should not be moved when this is possible and should be cremated. Burning will prevent the spread of the disease by flies, dogs, coyotes, vultures, etc. A hot fire is required to cremate animals, especially emaciated ones. Old tires, boards, brush, and other combustible materials that are safe for the person doing the work to use can be placed over and around the dead animal. Road flares ignited and placed inside the tires will start the fire quickly. Large butane burners can also be used to cremate dead animals.

BLACKLEG

Blackleg is an acute, febrile, highly infectious, non-contagious disease of cattle and sheep. Blackleg is characterized by inflammation of the muscles, severe toxemia, and high mortality. It occurs in many parts of the United States and is especially troublesome in the grassland areas of the Midwest, South, and West. Cattle of all ages may be affected by blackleg, but it is most common in young cattle from 4 months to 2 years of age. Blackleg is commonly called symptomatic anthrax, quarter ill, or black quarter.

Cause

Blackleg is caused by a Gram-positive, rod-shaped, spore-forming anaerobe known as *Clostridium chauvoei*. Infection of cattle occurs most often from the ingestion of spores from contaminated pastures. The spores are believed to enter the blood system from the digestive tract and localize in damaged muscle tissue. Blackleg in sheep is, in most cases, a wound infection. Sheep of all ages are susceptible to this disease.

Clinical Signs

Animals infected with blackleg often die suddenly without showing signs of ill health. Visible indications of blackleg include a sudden onset, acute rise in fever to 104° to 106°F, labored breathing, and lameness. Swellings often form in the heavily muscled areas of the neck, loin, and legs. At first, the swellings are hot and painful, but later they become cold, painless, and gaseous. When pressure is applied to swellings, a crackling sensation is imparted, due to the presence of gas beneath the skin. Affected animals often die within 12 to 36 hours. The carcasses bloat quickly and decompose rapidly following death (Fig. 15-1).

Blackleg occurs in all ages of sheep. It usually follows wounds such as those produced in docking, castrating, shearing, and parturition. Swellings that pit on pressure and do not crepitate occur near the site of infection. Anorexia, depression, recumbency, and fever from 105° to 107°F are common signs of blackleg in sheep. Affected animals usually die in 12 to 36 hours. Recoveries are rare.

Prevention

Calves should be vaccinated twice, two weeks apart, between two and six months of age. In high-risk areas, revaccination may be necessary at one year and every five years thereafter.

In "hot areas," vaccinate sheep every two to four weeks, before shearing, castrating, and docking.

Carcasses of animals suspected of dying from blackleg should not be opened. They should be destroyed by burning where the animals died, when this is possible. Suggestions for cremating farm animals are listed in the section dealing with the control of anthrax.

Treatment

Parenteral and multiple local injections of penicillin will sometimes save an animal, provided they are given during the early stages of the disease. But a good immunization program is the key to preventing losses from blackleg.

MALIGNANT EDEMA

Malignant edema is an acute, febrile toxemia of horses, cattle, sheep, swine, and humans. It is a non-contagious disease, characterized by edematous swelling around the wound where the pathogens entered the animal's body. Malignant edema is also called gas gangrene.

Cause

Malignant edema is caused by *Clostridium septicum* and related bacteria. These organisms are commonly found in fertile soils and in the intestinal tracts of ruminants. The organisms gain entrance through the wounds made by castrating, docking, or shearing. In addition, damaged tissue due to abortions, dystocias, or the bites of pigs may also serve as sites for infection.

Clinical Signs

The clinical signs of malignant edema are quite similar to those of blackleg. The affected animals show signs of distress quite suddenly. Anorexia, high fever, labored breathing, and depression are common

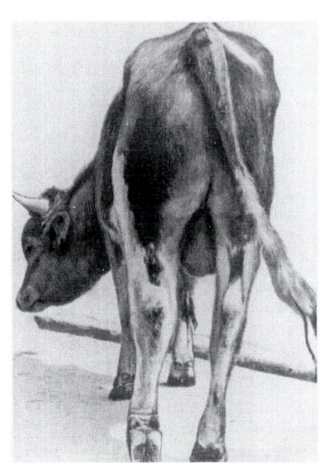

Fig. 15-1.   Young bull with blackleg.

signs. Swelling occurs at the site of infection. The area affected enlarges rapidly with a gelatinous, edematous exudate. The skin above this accumulation pits when pressure is applied. There is no crepitation in the swollen areas, such as is common in blackleg. Fluids that escape from open wounds have a strong putrid odor. Malignant edema is usually fatal in 24 to 72 hours.

Prevention

Since malignant edema is associated with contamination of wounds, the disease can be partially prevented by minimizing wounds and by castrating and dehorning under hygienic conditions.

Vaccination of young cattle with a vaccine containing *C. septicum* (for malignant edema) along with *C. chauvoei* (for blackleg) at the time of the blackleg vaccination(s) will give some protection against malignant edema. Also, antibiotics may be administered four to five days following surgery.

Animals that have died as a result of infection from malignant edema should be cremated where they died, if this is possible. Suggestions for burning dead animals are listed in the section on controlling anthrax.

Treatment

In the early stages of the disease, treatment with massive doses of antibiotics may be effective.

BACILLARY HEMOGLOBINURIA (REDWATER)

Bacillary hemoglobinuria is a peracute infectious disease of cattle, characterized by high temperature, depression, bloody urine, and death in 24 to 36 hours. It is primarily a disease of cattle; sheep are occasionally affected, and, in rare instances, swine are also. Common names for this disease are redwater and infectious hemoglobinuria.

Cause

Bacillary hemoglobinuria is caused by a Gram-positive, rod-shaped, spore-forming anaerobic bacterium called *Clostridium hemolyticum.*

The organism is taken into the digestive system in feed or water and transmitted to the liver where it multiplies and produces a toxin that causes rapid destruction of red blood cells.

Many of the factors favoring the development of this disease are undetermined. The organism is capable of survival in soil and water for many years, particularly in swampy or poorly drained areas. Forage harvested from endemic areas may be infectious. Also, conditions that cause injury to the liver, such as liver fluke disease, seem to trigger redwater disease.

Clinical Signs

The onset of bacillary hemoglobinuria is sudden. Appetite, rumination, and lactation cease abruptly. Affected animals stand off to themselves with their backs arched and abdomens tucked in. There is a disinclination to move, and grunting sounds may be emitted when the animals are forced to walk. The fever rises rapidly to 105° or 106°F in the early stages, but it often drops below normal before death occurs. Respiration is usually shallow and labored. Dyspnea and open mouth breathing are common during the terminal stages.

A characteristic sign of bacillary hemoglobinuria is the dark red, foamy urine. The color results from suspended hemoglobin that is liberated from the red blood cells. Urination is both copious and frequent. This, plus the fact that affected animals take in very little water, results in serious dehydration.

The mortality rate for animals affected with bacillary hemoglobinuria ranges up to 90 percent. Death usually occurs within one to four days after clinical signs appear.

Prevention

Vaccination about two weeks prior to the time of the previous annual outbreak is the best practical control measure. Also, the destruction of snails (by draining stagnant water and using bluestone) may aid in prevention.

Treatment

Treatment is often uneventful due to the peracute nature of bacillary hemoglobinuria. Antiserum from hyperimmunized animals, in conjunction with antibiotics, may be helpful.

BLACK DISEASE

Black disease is an acute infectious disease of sheep and occasionally of cattle. The name *black disease* comes from the dark appearance of the flesh side of the pelt. This discoloration is due to subcutaneous hemorrhages over the back and sides. Black disease is also called infectious necrotic hepatitis. The latter name describes the disease quite accurately. It is an infection that occurs in necrotic areas of the liver. The hepatic necrosis is caused by damage from liver flukes. Black disease always occurs with liver fluke infestation.

Cause

Black disease is caused by a spore-bearing anaerobe, *Clostridium novyi*. Immature liver flukes (*Fasciola hepatica*), which are present in the bile ducts, provide a site for *C. novyi* action.

Clostridium novyi is ingested with feed from contaminated pastures. It migrates to the liver and finds favorable growing conditions in tissues damaged by the invasion of liver flukes. Clostridium novyi produces powerful exotoxins that cause the clinical signs and lesions.

Clinical Signs

The sudden death of several middle-aged sheep in good condition is often the way black disease is manifested. When observed, affected animals appear dull and listless, and they lag behind the flock. This is followed by prostration, coma, and death.

Prevention

Destroy fluke-bearing snails. Vaccinate with a toxoid annually before going on snail-infested pastures (a sheep vaccinated two successive years is very resistant).

Treatment

Treatment is usually impossible or uneventful due to the acute nature of black disease.

SWINE ERYSIPELAS

Erysipelas is an infectious disease affecting swine of all ages, but it is usually most common in young and growing animals. The disease occurs in a variety of forms, sometimes causing severe death loss. However, the greatest economic loss usually results from the chronic form which produces general unthriftiness, rather than from the acute fatal form. The disease may affect other animals and humans. The infection may be of economic importance in turkeys and occasionally sheep.

Cause

Bacteria, *Erysipelothrix rhusiopathiae*, cause swine erysipelas. It is highly resistant to many natural means of eradication. This organism lives and reproduces in alkaline soils and has been known to live in decomposed carcasses up to a year. Swine erysipelas usually spreads from animal to animal by direct contact. It is quite often introduced into a herd through the purchase of breeding stock or feeders.

Clinical Signs

Erysipelas may occur in one of three forms: acute, subacute, or chronic. The acute form of erysipelas is characterized by the sudden death of one or more animals. A body temperature of 104° to 108°F usually accompanies the disease. Some animals may go completely off feed, while others may have a normal appetite. There is usually some vomiting. Blotches

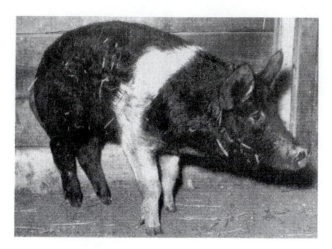

Fig. 15-2. Chronic swine erysipelas. Note the stiff gait and evidence of pain in walking.

often appear anywhere on the skin, varying from pink to dark purple in color. Affected swine squeal loudly when driven and walk with a high arch in their backs due to pain in their joints. The eyes remain clear in hogs affected with erysipelas. Some of the symptoms or signs of hog cholera are similar to those of erysipelas, but sticky eyes is a distinctive difference in the two diseases.

The subacute form of erysipelas shows less severe symptoms. A few skin lesions may appear, but may be easily overlooked. The course of the disease is not as long as the acute form.

The most dramatic physical changes occur in the chronic form, although clinical signs take longer to appear. Affected animals lose portions of their skin, ears, tail, and feet. Skin lesions become dark and firm, and eventually separate from the healing underlying tissue, leaving ugly scars (Fig. 15-2).

The affected pigs become stiff, the joints swell, and the animals evidence pain when they are forced to move. The animals also may have an arched back and show signs of lameness in one or more legs. The lameness may become less as the animals walk. Affected animals may be unthrifty following partial recovery. During a chronic infection, bacterial organisms usually localize in the joints or in the heart valves. The joints of the legs are most frequently affected, causing arthritis. The extent of arthritis varies with the duration of the infection.

Prevention

Prevention of swine erysipelas is best accomplished by sound practices of herd health management, including a program of immunization.

Either killed bacterins or live-culture immunizing strains of low virulence for swine may be used. The formalin-killed, aluminum-hydroxide–absorbed bacterin does not infect other species of animals or humans, and it confers an immunity that, in most instances, will protect the pig from acute forms of the disease until it reaches market age. An oral vaccine of low-virulence is also used. Breeding stock should be revaccinated twice each year.

Treatment

Acute erysipelas may be effectively treated with any of the following antibiotics: penicillin, chlortetracycline, oxytetracycline, or tylosin.

There is no practical treatment for chronic swine erysipelas.

HOG CHOLERA

Note well: Today, hog cholera is of historic interest only, because on January 31, 1978, the U.S. Secretary of Agriculture declared the United States free of hog cholera. It is noteworthy, however, that prior to the National Hog Cholera Eradication Program, hog cholera caused an estimated average annual loss of $2.9 million per year.

Hog cholera is an infectious, highly contagious viremia of swine. It is characterized by generalized hemorrhages, high morbidity, and high mortality. It usually follows an acute course but may become chronic.

Hog cholera is a blood infection. When the virus enters the pig's body, it passes to the bloodstream and develops there. The blood becomes infectious within 12 to 20 hours. The urine and manure usually contain the virus within three days. Secretions of the eyes and nose also become infectious by the third day.

Hog cholera virus enters the animal's body through the mouth, nose, eyes, or through wounds or abrasions of the skin. A susceptible pig gets the disease by contact with infected animals or by contaminated facilities and premises.

Cause

Hog cholera may be caused by any of a number of strains of hog cholera virus and may be complicated by a bacterium, *Salmonella cholerasuis*. The virus will survive in pork products for months and will live for at least six months in pickled, salted, and smoked meats. The ability of the virus to exist outside its host depends on the temperature. Freezing tends to preserve the virus, while heat tends to kill it. In experimentally contaminated water, it lives from two days to seven weeks. The virus, however, is easily killed by sunlight and certain disinfectants.

Cholera is highly contagious. It can be spread from one animal to another in many ways. It may be carried on the tires of a feed truck, on a person's shoes, in old feed sacks, by dogs, by buzzards, or by contact with other cholera-sick animals.

Clinical Signs

In peracute hog cholera, one or more animals may be found dead. Those found alive usually die within a week. The acute course, lasting from one to three weeks, is the most common course in hog cholera. Clinical signs in the first stages include lethargy or mild inactivity, depressed appetite, and nesting. This is followed by more severe signs of distress, including depression, anorexia, high fever, vomiting, and constipation. In a day or two, the constipation changes to a thin, light-colored diarrhea. The temperature often rises to 108°F and then drops suddenly before death. Sick pigs often huddle together and pile on each other in the nest. Shivering and signs of chilling are common. Later, pigs may wander off and lie by themselves. Conjunctivitis is a common sign in hog cholera. A watery discharge is emitted from the eyes in the early stages; later, in the course of the disease, the discharge becomes thick and gummy (Fig. 15-3).

A loss of coordination accompanied by a characteristic scissor-like gait is quite common in hog chol-

Fig. 15-3.   Hogs in advanced stages of hog cholera. (Courtesy, USDA)

era. Affected animals weave and stagger due to weakness in the hindquarters. A purplish discoloration of the skin is easily detected on white or light-skinned hogs. The blotching is most common on the belly, ears, and snout. The final termination of hogs affected with hog cholera is death due to the infection or due to destruction in compliance with the federal hog cholera eradication program.

Prevention

The passage, in 1961, of Public Law 87–209 for hog cholera eradication in the United States marked the beginning of the end of the disease in this country. But this drastic step was not without precedent. Both Canada and Great Britain had used the slaughter method to stamp out hog cholera.

Federal funds for the control program were made available in 1962; and in 1969, the USDA banned all interstate shipments of modified hog cholera vaccine.

Hog cholera biologics are no longer produced or allowed entry into the United States. Today, hyperimmune serum is the only available treatment. It may be effective in the early stages of the disease or for protection of important animals. Control is by eradication or vaccination. Eradication by slaughter of all in-contact pigs and disposal of carcasses has been successful. However, any suspected case of hog cholera should be reported to the proper federal officials.

Treatment

There is no recommended treatment for hog cholera in the United States. All infected or exposed hogs should be destroyed following verification by a state or federal veterinarian.

JOHNE'S DISEASE

Johne's disease, also called paratuberculosis, is a disease primarily of ruminants that has become increasingly important in the United States in the last several years. Diarrhea, progressive weight loss, and emaciation characterize Johne's disease. Johne's is emerging as a significant economically important disease, as herds with a 10 percent infection rate lose as much as $230 per cow annually, mostly from decreased production in infected, subclinical cows.

Cause

Johne's disease is caused by a bacterium *Mycobacterium paratuberculosis*. Infection occurs after ingestion of the organism in contaminated feed, milk, or water. Nursing calves born to infected mothers naturally have a high chance of becoming infected. Johne's disease organisms are shed in very high numbers in the feces of affected animals before clinical signs develop. Fecal contamination is considered the most important mode of disease transmission. Animals up to six months of age are most susceptible to acquiring the infection.

The bacteria enter the tissue of the small intestine where the multiplication of the bacteria causes a reaction in the intestinal tissue. The infection can be diagnosed in cattle through a blood test using the ELISA test, or through fecal culturing for the organism. There is a problem with diagnosis of early infection because heavy fecal shedding does not usually occur until late in the course of the disease, and blood antibody levels are not high early in the disease.

Clinical signs

Affected animals have persistent diarrhea that is unresponsive to treatment and maintain a good appetite. Weight loss can be very rapid and the animals do not usually have a fever. Most of the disease occurs in three- to five-year old cattle due to the long incubation period. Subtle signs of the disease that may be recognized before diarrhea is a rough hair coat, poor milk production, and a lackluster appearance.

Prevention

The prevention of Johne's disease is necessary, as there is no treatment. Due to the fecal–oral transmission, feeding utensils should not come into contact with manure. In dairy herds, a skid loader is often used to scrape manure, as well as deliver feed to cattle, and this practice should be avoided.

Calves should be separated from their dams as soon as possible and fed colostrum from their dam or a tested negative cow. The organism is spread in milk, so pooled colostrum should not be used as it would have the potential to infect a large number of calves at one time. Milk replacer should be fed instead of whole milk or waste milk from treated cows. The calves should be housed away from the cows so they do not have access to cow manure.

Treatment

Long-term therapy with antituberculosis drugs will improve body condition, but it will not cure the infection or prolong the life of the animal.

CHAPTER 16

Metabolic and Deficiency Diseases[1]

MILK FEVER

Milk fever, also called hypocalcemia, parturient paresis, and downer cows, is a nonfebrile disease of dairy cows, beef cows, and sheep and goats. Milk fever is caused by an acute deficiency of calcium, which causes paralysis, circulatory collapse, coma, and death. Milk fever is one of the most common metabolic diseases in dairy cattle with about 6 percent of the U.S. dairy cattle affected annually. Within herd, variation of the incidence of milk fever can range up to 60 percent. Most cases of milk fever occur within 24 hours of calving. Breed, parity, and milk production level are important risk factors in parturient paresis. Jerseys and Guernseys are most susceptible. Holsteins and Brown Swiss are moderately susceptible. The incidence increases with age and milk production, with first-calf heifers rarely developing milk fever. Milk fever is rare in beef cattle, but high-producing doe goats have a similar incidence as dairy cattle. Subclinical hypocalcemia is a very important problem in dairy herds, affecting about 50 percent of dairy cattle. Subclinical milk fever leads to decreased dry matter intake, increased ketosis, increased retained fetal membranes, increased displaced abomasum, decreased fertility, and decreased milk production for the lactation. Management decisions to improve fresh cow calcium intake and metabolism can have long-term payoffs, even in herds without a high incidence of clinical milk fever.

Cause

Hypocalcemia occurs when the initiation of lactation in dairy cows causes a severe outflow of calcium. The calcium dairy cows produce in colostrum is greater than the calcium required prepartum, including the mineralization of the fetal skeleton. In beef cows, milk production is not as high and milk fever may occur as mineralization of the fetal skeleton drains the calcium reserves and overcomes the homeostatic mechanism for maintaining blood calcium levels. The animal's ability to adapt to hypocalcemia is influenced by several factors. The acid-base status, magnesium status, dietary potassium levels, dietary phosphorus, and blood estrogen levels affect blood calcium regulation. Metabolic alkalosis, low blood magnesium, high potassium, and phosphorus are factors that inhibit either the function of the parathyroid glands or the renal synthesis of vitamin D, which are involved in the blood calcium regulation. Feeding dairy cows a diet high in calcium, such as legume forages, is a known factor in milk fever cases. A high-calcium diet does not appear to "trigger" the body to adapt to low-calcium states, as occurs at the initiation of lactation.

Clinical Signs

The signs of milk fever are divided into three stages. Animals in stage I appear mildly excited, anorexic, weak, and hypersensitive. Animals in stage II have flaccid paralysis, lie on their sternum, are depressed, and may show small muscle tremors in the

---

[1]Vitamin and mineral deficiencies are presented in Chapter 2.

Fig. 16-1. Jersey cow with parturient paresis (milk fever), showing characteristic position of head—turned back over shoulder. (Courtesy, Washington State University, Pullman)

triceps muscles. They may have a muffled heart beat and low body temperature with cold extremities. Additional signs are bloat, gastrointestinal atony, dilated pupils, and poor uterine tone and contractility leading to dystocia and uterine prolapse. In stage III, animals are lying on their side and comatose.

Treatment

The treatment of animals in stage I is by the administration of oral or intravenous calcium salts. Oral calcium products are formulated into gels and can be absorbed from the rumen into the bloodstream in fifteen minutes. The advantage of the oral preparations is that a larger dose of calcium is given than in the intravenous form, in order to prevent a relapse of the milk fever. The animal must have a swallow reflex to use the oral gel—in order to prevent aspiration pneumonia. Animals in stage II and III will die unless treated with intravenous preparations of calcium salts. The calcium must be administered slowly, over a period of ten minutes, because the cow can die from rapid calcium administration. Most cattle will stand within thirty minutes of calcium administration.

Prevention

The prevention of milk fever is through good nutritional management of the dry cows and spring-

ing goats. Dietary calcium restriction is probably impractical and could cause additional problems. Forage analysis for calcium, magnesium, phosphorus, and potassium is necessary to create a balanced ration. Acidification of the diet through the addition of anionic salts has become popular as a means of controlling milk fever. Some diets cannot be formulated to correct the anionic balance if they are high in potassium and calcium, such as legume pastures where manure was spread. Some dairies are planting grass hay to feed dry cows, as it is lower in calcium than legume hays. A nutritionist should be consulted to formulate and monitor dry cow feeding programs for beef and dairy operations.

KETOSIS OF CATTLE

Ketosis is a metabolic disease of gestating or lactating cattle. It most frequently occurs within the first six weeks after calving, but it can occur at other times. Ketosis, unlike milk fever, often affects first-calf heifers. It is more common, however, in older cows. Bulls, steers, or dry, open females are not affected by this disease. Ketosis is characterized by the presence of excess amounts of ketone bodies in the blood, urine, and milk. This disease is also called acetonemia and hypoglycemia. It closely parallels pregnancy disease in ewes, except that ewes are most commonly affected during the latter stages of gestation.

Cause

Ketosis is a metabolic disorder of nutritional origin, characterized by hypoglycemia (low blood sugar). If the increased nutrient requirements are not met by more feed during the high-demand periods (in cows, one to six weeks after calving; in ewes, two weeks before lambing), the animal must draw on body fat reserves. If this is done too rapidly, and without adequate carbohydrates in the ration, ketosis follows.

A positive diagnosis can be made by the examination of the animal's urine by means of the Ross test. If the reaction is positive, it indicates the presence of ketone bodies in the urine, which is diagnostic of this ailment.

Clinical Signs

There are no characteristic clinical signs of ketosis in cattle. General symptoms include a loss of appetite, decreased milk flow, a decidedly "cowy" flavor in the milk, rapid loss of condition, and constipation. Occasionally there are nervous developments, so that affected animals run into fences and buildings, though more frequently they become listless and extremely docile. The back may be arched and the gait wobbly or stilted. The temperature is seldom elevated.

Prevention

In cows, the incidence of ketosis can be lessened by (1) avoiding excessively fat cows at calving; (2) increasing the level of concentrates gradually after calving; (3) feeding good-quality hay in preference to high-silage rations after calving, and avoiding abrupt changes in roughage; (4) feeding adequate proteins, minerals, and vitamins; and (5) providing comfort, exercise, and ventilation. In problem herds, feeding ¼ pound daily of propylene glycol or sodium propionate may be helpful.

For sheep and goats, feed more hay and ½ to 1 pound of grain beginning a month before parturition. Good management is important, including exercise, freedom from parasites, and avoiding stress.

Treatment

Treat cattle with ½ to 1 pound of either propylene glycol or sodium propionate daily, with the dose divided into two treatments for five to ten days. Put treatment in grain if cow is eating; otherwise, give as a drench.

Intravenous injection of glucose solution and glucocorticoids (to increase blood sugar levels temporarily) as well as the oral administration of propylene glycol are helpful. Numerous other treatments are sometimes used.

Sheep and goats before parturition may be given 3 to 4 ounces of propylene glycol, administered orally 3 times daily. A cesarean section early in the course of the disease usually leads to recovery and, if near term, the offspring may be saved.

Dairy goats, after kidding may be treated with 6 to 8 ounces of propylene glycol, given orally twice daily. Severe cases may be aided by intravenous injections of 50 percent dextrose solution. Corticosteroid injection may be used in conjunction with either propylene glycol or dextrose solution in does that have kidded.

GRASS TETANY

Grass tetany is a metabolic disorder of cattle and sheep grazing on lush grass pastures. Bred cows and ewes in the latter stages of gestation and post-partum females lactating heavily are most susceptible to grass tetany; however, it can affect cattle and sheep of any age. Grass tetany is characterized by hypomagnesemia, low blood magnesium, and, in many cases, hypocalcemia. High levels of potassium and nitrogen in grass and wheat pastures combine to limit magnesium absorption when grass tetany occurs in cattle or sheep. Grass tetany is also called grass staggers and wheat pasture poisoning.

Cause

Grass tetany is a nutritional disease caused by an inadequate level of magnesium in the blood. It most commonly occurs among lactating animals grazing rapidly growing, lush spring pastures containing less than 0.2 percent magnesium and more than 3 percent potassium and 4 percent nitrogen (25 percent protein). Forage that is high in potassium and nitrogen should have a magnesium content of at least 0.25 percent. Such low magnesium pastures are most commonly encountered during the first two weeks of the pasture season, although somewhat later in the season outbreaks have been reported during rainy and foggy weather. Sometimes tetany is a problem when cattle are allowed to overgraze a field, then move abruptly to a field of new lush growth. Small grain pastures (wheat/rye/oats/barley) are especially troublesome. Also, the disease may occur when animals are fed poor quality hay, straw, or corn stover—feeds that are low in magnesium. It is not common on legume pasture or in animals wintered on legume hay. (Legumes may contain twice the magnesium concentration of grasses grown on the same soil.)

Several factors adversely influence magnesium metabolism in cattle and may "trigger" grass tetany; among them, drastic fluctuations in spring temperatures, prolonged cloudy weather, organic acid content of plants, hormonal status of the animal, level of higher fatty acids in plants, energy intake of the animal, and additional stress—such as a dog chasing animals, parasites, or a cold rain.

Grass tetany is most likely to occur on pasture plants grown on soils that are low in available magnesium and high in available potassium. If calcium is low as well as magnesium, the hazard of tetany is even greater. Many state soil-testing laboratories provide information on the danger of tetany on pastures, and can recommend corrective fertilization or dolomitic liming (which contains magnesium). Also, the historical record of grass tetany in an area or on a specific pasture is important.

Clinical Signs

The severity of grass tetany on the affected animals and the form it takes are based upon the rate that magnesium is depleted from the blood. When the blood level of magnesium drops rapidly, an acute type of grass tetany occurs. In this form, animals may be found dead, or they may appear healthy and suddenly become deranged, vicious, and unduly excited. They often run blindly, stagger, and fall to the ground. This is followed by convulsions, including running motions and colonic spasms, and coma just before the animals die.

In less acute cases, clinical signs begin with undue excitement, incoordination, and loss of appetite. As the condition progresses, viciousness, staggering, and falling develop. Nervousness becomes more apparent with muscular twitching, particularly of the extremities. The animals have an anxious expression and may grind their teeth and salivate profusely. The third eyelid protrudes or flickers as in tetanus. General tetanic contractions of the muscles follow until the animals near a state of prostration. Sudden noises or merely touching the animals will cause a reflex response. This is followed by labored breathing, a pounding heart, and a comatose condition. If the animals are left untreated, convulsions with periods of relaxation will be seen, terminating in death. Six to 10 hours usually is required from the time the first symptoms develop until the animals pass into the comatose condition. If treatment is not begun before the coma, there is little chance of recovery.

The chronic form of grass tetany is more common when the affected animals are fed hay and/or grain in addition to lush pasture. The first signs include a decrease in appetite, reduction in milk flow, and loss in weight. This is followed by signs of disturbances to the nervous system, including nervousness, excitability, and a lack of coordination.

Prevention

Prevention of grass tetany is always preferred to treatment. Prevention consists of providing magnesium daily throughout the high-risk period, because very little of it is stored in the body. *Note well:* Crash feeding programs begun after tetany appears in a herd are usually not adequate to stop the disease. A magnesium supplement should be started 30 days before grass tetany is usually observed in the area in order to get the animals accustomed to it. Since magnesium oxide or sulfate is not very palatable, cattle may not consume sufficient of them.

Meeting the magnesium requirements of beef cows calls for providing 10 grams of magnesium daily for the dry cows, and 20 to 25 grams daily for cows suckling calves. For dairy cows, 30 grams of magnesium per day is recommended. For calves, 4 to 8 grams per day is needed, depending on their ages.

Lactating ewes and does, just after parturition, which is the most tetany-susceptible period, should receive about 3 grams of magnesium per day.

High levels of aluminium, potassium, phosphorus, or calcium decrease the efficiency of magnesium absorption and/or utilization; so, in areas where the levels of these elements are high, the magnesium allowance should be increased to overcome their antagonistic effect.

Normally, animals on pasture during the summer and early fall months receive an adequate supply of magnesium from the grasses on which they feed. However, during the late fall, winter, and spring months, many pastures are magnesium deficient. To prevent grass tetany during these months, cattle, sheep, and goats on pasture should receive a magnesium-rich feed in addition to pasture and/or have ready access to a magnesium mineral supplement.

Treatment

Treatment of tetany cases can be successful if given early and without excessive handling of the affected animals. Chance of recovery is slight if treatment is delayed 8 to 12 hours; so, call the veterinarian immediately.

Under range conditions, 200 cc of a sterile, saturated solution of magnesium sulfate (Epsom Salts) injected under the animal's skin (inject only 50 cc at any one place on the animal) places a high level of magnesium in the blood in 15 minutes.

Some veterinarians use intravenous injections of chloral hydrate or magnesium sulfate to calm excited animals, then follow with a calcium-magnesium gluconate solution. If the animal again goes into convulsions, a second dose of calcium-magnesium gluconate solution may be required. Intravenous injections should be administered slowly (allow about 15 minutes for a 500 cc bottle) by a trained person because there is a danger of heart failure if they are given too rapidly.

An enema of 60 grams (2 ounces) of magnesium chloride ($MgCl_2 + 6H_2O$) in 10 ounces of water is helpful. The enema may be given with an esophageal or oral calf feeder with the probe inserted 10 inches into the anus. Magnesium is absorbed through the walls of the large intestine and the lower bowel.

Oral administration of magnesium to sick animals, in place of intravenous injections or enemas, has not been effective because too much time is required for the magnesium to reach that part of the GI tract where it can be absorbed.

Herd treatment of the animals that are not down may involve adding magnesium sulfate (Epsom Salts) or magnesium acetate or chloride to the drinking water. Some diarrhea may occur, but this is no reason for concern. To be effective, the treated tanks should be the only source of drinking water. *Note well:* Production will be lowered by this treatment due to lowered consumption of water.

Follow-up treatment may involve removing all animals from the tetany-producing pasture and feeding alfalfa hay (plus concentrate if necessary). Additionally, each animal should consume 30 grams of magnesium daily for one to two weeks, preferably through a highly palatable supplement; force-feeding should be resorted to if necessary.

Cattle that get tetany are likely to get it again later in the season or in later years; they are usually the high producers.

Note well: "Downer cows" should be turned daily—and more frequently if possible.

PREGNANCY DISEASE OF EWES

Pregnancy disease or ketosis is the most important metabolic disease of sheep. It usually affects ewes carrying twins or triplets during the last six weeks of gestation. Pregnancy disease is commonly called lambing paralysis, twin lamb disease, acetonemia, and ketosis.

Pregnancy disease may affect as much as 25 percent of the bred ewes in a flock. Ninety percent of those in the advanced stages of the disease are likely to die if treatment is not provided. When pregnancy disease is diagnosed in a flock, all the remaining ewes that are heavy with lamb should be checked by means of the Ross test or by specially designed keto sticks. A positive response to this test indicates the presence of abnormal amounts of ketones in the urine. Curative or preventative treatment should be initiated immediately to control pregnancy disease in such ewes.

Cause

Pregnancy disease is caused by an insufficient intake of carbohydrates, which results in hypoglycemia and ketonemia. In most cases, poor nutrition and the excessive demand on the ewes' bodies to provide nutrients for the developing fetuses render the ewes susceptible to pregnancy disease. About 80 percent of the growth of a fetus occurs during the last trimester of gestation. Ewes carrying twins or triplets need 75 to 80 percent more feed during the last six weeks of pregnancy than they required during the early stages of gestation. This amount of feed may exceed their ingestive capabilities unless concentrates are provided. Pregnancy disease may also affect ewes after they lamb if the feed supplied is inadequate to support milk production.

Stress is believed to influence the susceptibility of ewes to pregnancy disease. Even ewes in good condition may be affected by pregnancy disease when they have undergone stress that interrupts their normal feeding routine during the last part of gestation. This may occur when ewes are transported or driven long distances, held for long periods during shearing, fasted because of inclement weather, unduly excited by attacks of predators, and affected by other disease problems.

Clinical Signs

Pregnancy disease is confined in most cases to ewes that are heavy with lamb during the last four to six weeks of gestation. Affected ewes normally lag behind other ewes, walk with an unsteady gait, or wander aimlessly off by themselves. Other clinical signs include a loss of appetite, an absence of fever, depression, impaired vision, and signs of nerve involvement, such as grinding of the teeth.

In the advanced stages, affected ewes are unable to stand. They lie with their heads to one side, have difficulty breathing, and may have convulsions before they go into a coma and die.

Prevention

The prevention of pregnancy disease in ewes is based upon adequate nutrition, proper exercise, and the absence of unusual stress during the last half of gestation. If ewes become overly fat, management practices should be used to reduce their weight early in the gestation period and to enable them to gain properly during the latter stages.

Ewes with bad teeth or feet that are valuable enough to retain in the flock should be separated and fed accordingly. All ewes should be wormed prior to the breeding season. Good pasture should be provided during the breeding season and gestation period. Additional feed in the form of grain or molasses is usually required during the last trimester of gestation to prevent pregnancy disease. Ewes should be fed to gain about 30 pounds during gestation. Exercise should be required of all ewes, even if it has to be forced.

Treatment

The treatment of pregnancy disease in the latter stages is not often successful. If ewes lamb during this time, recovery is greatly enhanced. Treatment usually includes 50 to 60 cc of a 50 percent glucose solution administered intravenously and 4 ounces of glycerol or propylene glycol given orally twice a day. In addition, affected animals should be housed in comfortable quarters, disturbed as little as possible, and given access to palatable feed and water.

WHITE MUSCLE DISEASE

White muscle disease is a nutritional myopathy or muscular disorder that occurs more commonly in young calves and lambs than in foals and pigs. It is enzootic in those areas of the United States that have selenium deficient soils. White muscle disease is commonly referred to as muscular dystrophy, stiff lamb disease, or calf rheumatism. It is called polymyositis in foals.

Cause

White muscle disease is caused by a selenium deficiency, due to the continuous consumption of a ration containing less than 0.02 p.p.m. selenium.

Clinical Signs

In the acute form of white muscle disease, the young animals may become prostrate and die very suddenly. A lack of thriftiness is often the only sign of a subclinical selenium deficiency.

Lambs are most often affected during the first month of life. The first signs are usually stiffness, disinclination to stand, and difficulty in rising from a recumbent position. There is no fever and there is no loss of appetite, unless muscles of the tongue and throat are affected. Lambs often become so stiff that they are unable to nurse, but they will take milk from a bottle or suckle their mother if they are given assistance. A characteristic stance of lambs affected with white muscle disease includes the feet set wide apart, a weak back, and the shoulder blades thrown forward in a prominent manner (Fig. 16-2).

The signs of white muscle disease in calves usually include stiffness that may resemble acute laminitis or founder, accelerated breathing, and occasionally diarrhea. Calves from birth to four months of age are most commonly affected by white muscle disease.

Foals affected by polymyositis may show sudden signs of depression, rapid breathing, prostration, and death. In less severe cases, they may show stiffness, disinterest in nursing, and disinclination to stand or move about.

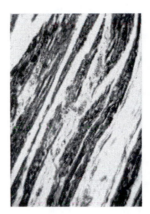 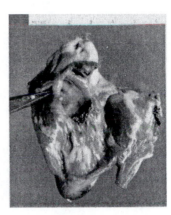

Fig 16-3. Skeletal muscles (left) and the heart (right) affected by white muscle disease. (Courtesy, College of Veterinary Medicine, Michigan State University, East Lansing)

White muscle disease of pigs is most common in pigs from one to four months of age. The stiffness and muscular dysfunction of pigs is quite similar to that seen in other animals affected with white muscle disease.

A post-mortem examination of animals affected with white muscle disease will show pale, whitish streaks in the skeletal muscles that were involved (Fig. 16-3, left). The lesions will be bilaterally symmetrical. The streaks will be found in the same muscle on both the left and right sides of the animals' body. When the heart is affected, lesions in the form of round patches that vary from the size of a pinpoint to 10 mm in diameter may be found (Fig. 16-3, right).

Prevention

White muscle disease can be prevented in lambs and calves, if their dams are fed rations that include feeds with adequate amounts of selenium and vitamin E during the gestation period. The most critical period appears to be during the last trimester of gestation. Selenium may be added to the ration of sheep, cattle, swine, and poultry at the rate of 0.3 p.p.m. to the complete feed.

In areas where selenium deficiency occurs and supplemental feeding is not feasible, subcutaneous injections of sodium selenite to ewes and cows during the last part of gestation and to their offspring at birth are recommended to prevent white muscle disease.

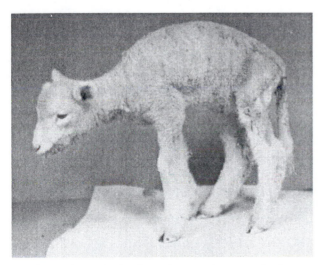

Fig. 16-2. A lamb affected with white muscle disease, commonly called muscular dystrophy. (Courtesy, College of Veterinary Medicine, Michigan State University, East Lansing)

Intramuscular injections of sodium selenite/vitamin E at birth are recommended for prevention of polymyositis and white muscle disease of pigs in endemic areas. In addition, spoiled or damaged feeds that are high in fat should not be fed to pigs.

Treatment

Affected animals should receive early treatment—the intramuscular injection of sodium selenite/vitamin E in aqueous solution at the rate of 0.25 milligrams selenium per pound of body weight. This may be repeated in 2 weeks, but should not exceed 4 doses. *Note well:* Federal law restricts injectable selenium to the order of a licensed veterinarian. Do not use within 30 days of slaughter.

HYPOGLYCEMIA OF NEWBORN PIGS

Hypoglycemia is commonly called baby pig disease or three-day pig disease. The technical name, hypoglycemia, accurately describes the disease. *Hypo* refers to a low level, *glyco* stands for glycogen, and *emia* means in the blood. Together, they describe a condition in which baby pigs have an abnormally low level of blood sugar.

Cause

Hypoglycemia is caused by a sudden drop in blood sugar in newborn pigs. This may occur because of the pigs' inability to effectively manufacture and use the glucose in their bodies the first few days of life, or there may be some abnormality of the colostrum. The more common causes of hypoglycemia, however, include agalactia or the absence of milk, dysgalactia due to mastitis, or weak pigs that are pushed back by bigger pigs in large litters. Inability or refusal to nurse because of damage to the tongue, lips, or gums due to improper and careless techniques used in clipping needle teeth may also contribute to hypoglycemia of newborn pigs. Exposure to cold, damp quarters increases the baby pigs' requirement for glucose to maintain normal body temperature, therefore, compounding the problem.

Clinical Signs

The first signs of hypoglycemia are a lack of coordination, shivering, weakness, and the pigs' hair standing on end. There is a characteristic squeal. The pigs become inactive and become prime prospects to be stepped on or laid on. Later, the pigs go into convulsions. They lie on their sides and go through galloping movements. The forelegs are usually most active. The head is often drawn back, and the pigs may chomp their jaws, creating a frothy material around the mouth. Just before death, a pig's temperature falls below normal, and the heartbeat becomes abnormally slow. Death occurs in untreated animals in 24 to 36 hours.

Prevention

Prevention of hypoglycemia involves the elimination of the cause of the disease. Brood sows with genetic promise of producing ample milk for their pigs should be selected. They should be fed and cared for so that they will be capable of producing sufficient milk, and they should be provided with warm, dry, draft-free quarters for their newborn pigs. When needle teeth are clipped, care should be exercised to prevent damage to the pigs' tongues, gums, or cheeks. Supplemental feeding for pigs whose mothers are incapable of supplying enough milk should be provided by (1) transplanting pigs onto other sows that have recently farrowed, (2) preparing a substitute sow's milk and tube feeding pigs at four- to six-hour intervals, or (3) giving intramuscular or intraperitoneal injections of 5 to 10 cc of 50 percent glucose at four- to six-hour intervals.

Treatment

The treatment for hypoglycemia in newborn pigs includes providing supplemental feeding, placing them in warm, dry quarters, and giving injections of glucose solutions when clinical signs are apparent. Three to 5 cc of 50 percent glucose solution should be given intraperitoneally or intramuscularly every four to six hours. If tube feeding is practiced, 1 tablespoonful of 50 percent glucose added to the substitute sow's milk should be fed every four hours. (See page 122 for procedures recommended for tube feeding orphaned pigs.)

PARAKERATOSIS IN SWINE

Parakeratosis is a nutritional disorder that causes an elephant-hide appearance to the skin of feeder pigs.

Cause

Parakeratosis in swine is caused by a metabolic disturbance that arises from a deficiency of zinc and excess of calcium in the ration. In addition, a deficiency of essential fatty acids in the ration contributes to the cause of the disease.

Clinical Signs

Parakeratosis usually occurs in confinement operations during the fall and winter. Pigs between 7 and 20 weeks of age are most commonly affected. The skin becomes dry and crusty. The crusts thicken and spread along the underline, up on the sides, and around the jowl and ears.

An interesting aspect of the formation of crusts is that they form rather symmetrically on both sides of the pig's body. The crusts formed on one hock may be found in approximately the same spot on the other hock. Parakeratosis differs from mange in that very little itching, rubbing, or scratching occurs.

Prevention

Parakeratosis can be prevented if the proper amounts of calcium, zinc, and fat are included in the ration.

Treatment

The treatment of parakeratosis requires that the calcium content of the ration be reduced to 0.5 to 0.6 percent. In addition, the level of zinc in the ration should be increased by adding 0.4 to 0.5 pound of zinc carbonate or zinc sulfate to a ton of feed.

CHAPTER 17

Localized Diseases of the Skin and Extremities

INFECTIOUS KERATITIS (PINK EYE)

Infectious keratitis is a common infectious disease of cattle and sheep. The reddish-pink color of the infected conjunctiva of affected animals has given rise to the common name, "pink-eye." The infection often includes the transparent outer layer of the eyeball, the cornea. Death from infectious keratitis is rare, but once the disease is established in a herd, severe losses in weight and production can occur. Animals affected with pinkeye may go blind and thus have difficulty locating feed and water (Fig. 17-1). In addition, economic losses accrue due to the expense involved in treating affected animals. Pinkeye occurs most frequently in young cattle that have non-pigmented faces, although animals of any age or color may be affected.

Infectious keratitis is spread by contact between healthy and infected animals. Recovered animals, al-

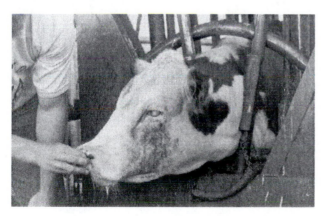

Fig. 17-1. Blindness due to pinkeye.

though normal in appearance, may be carriers of pinkeye. Eye secretions—by contaminating feed, water, and the premises—may be a source of infection. Flies, especially face flies, and other insects are also thought to transmit the disease.

Cause

The most common form of infectious keratitis is caused by a bacterium, *Moraxella bovis*. This organism produces a toxin which irritates and erodes the covering of the eye. Bacterial pinkeye occurs mainly during warm weather. Bright sunlight, wind, and dust may contribute to the cause of the disease. Animals of all ages are susceptible, but the disease is more common in young animals. Well-conditioned cattle are as susceptible as poorly conditioned cattle. Pink-eye of this type is more prevalent in some years than others, even though cattle are kept in the same environment and management remains the same.

The most common virus infection of the eyes of cattle is caused by infectious bovine rhinotracheitis (IBR) virus. It is much less common than bacterial pinkeye. When this organism infects the eyes of cattle, there may or may not be other signs of disease, such as the respiratory infection, vaginitis, or abortion, which is commonly associated with IBR.

IBR conjunctivitis occurs most frequently in the winter, but it may be seen in the summertime also. It affects only cattle. The disease is highly contagious by direct and indirect contact of infected animals with

267

susceptible animals. All breeds of all ages are susceptible, but it is most common in animals under two years of age.

Clinical Signs

Animals affected with pinkeye are easily recognized. The first sign is excessive tear production which spills over the lids and moistens the hair on the face just below the eye. This is followed by a swelling of the eye; evidence of pain, especially when the eye is exposed to bright light such as sunlight; and a change to a pus-type discharge from the eye. Soon a small erosion may be seen on the cornea. The erosion may become an ulcer, and, if neglected, the entire eyeball may rupture through this weakened area. The animals cannot see properly and may bump into objects. If both eyes are affected, the animal may become completely blind (Fig. 17-2).

The first signs of viral infection are redness of the eye and lids and an excess of tears. In contrast to bacterial pinkeye, IBR affects mainly the eyelids and the tissues surrounding the eye. It causes a severe swelling of the lining of the lids. The swelling may be so extensive that the eyelids appear to be inside out. Large amounts of white to yellow pus accumulate in the folds of the swollen lining of the eyelids and run down the face.

The eyeball itself is only slightly affected in most cases of IBR. There will be varying amounts of cloudiness of the eye, and seldom will ulceration or bulging

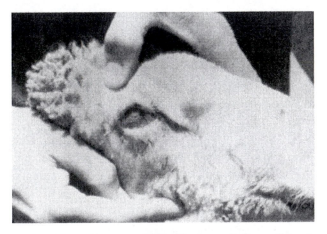

Fig. 17-2. Pinkeye in sheep. (Courtesy, College of Veterinary Medicine, Washington State University, Pullman)

of the eyeball occur. Frequently, there is a watery to purulent nasal discharge. The disease ordinarily runs its course in two to three weeks, but, in a large group of cattle, individuals may be affected over a period of three to four months.

Prevention

The prevention of the pinkeye complex is not easy due to the normal environmental conditions in which cattle are housed. Prevention strategies are aimed at controlling the vectors of transmission of the bacteria. Insecticidal fly tags to control face flies are beneficial, as well as a proper nutritional program. Vaccines have been developed to control pinkeye, but the clinical efficacy is not good for completely preventing cases of the disease.

Treatment

The treatment of pinkeye depends upon the level of occurrence within the herd, and the labor constraints of the cattle operation. Treatment alternatives include topical, subconjunctival, oral, or parenteral administration of antibiotic drugs. The use of dexamethasone injected in the conjunctiva has not been shown to hasten or delay the clinical course of the disease; therefore, it is not recommended. *M. bovis* is susceptible to a wide range of antibiotics, such as cloxacillin, gentamicin, erythromycin, and some sulfa drugs. Topical administration of an antibiotic establishes the highest concentration of the drug in the tear film, but three times daily application is needed to be effective. Topical administration may be feasible in dairy herds with tie-stalls where the cattle can be restrained on a continual basis. An intramuscular injection of long acting, oxytetracycline (20 mg/kg) establishes a therapeutic concentration of the drug in the tear film. Penicillin G (500,000 IU) can be injected into the conjunctiva, but oxytetracycline should not be given subconjunctivally due to the severe ocular reaction to the medicine. The slaughter withdrawal of the antibiotics depends upon the dose rate and administration of a drug. A veterinarian should be consulted in designing appropriate treatment programs.

FOOT ROT OF SHEEP

Foot rot is an infectious disease of sheep that causes heavy economic losses in flocks pastured on low, wet fields. The problems of foot rot are compounded when sheep are corralled in muddy lots and concentrated in small pastures that are poorly drained. The infectious agent lives less than two weeks when separated from the sheep's foot. Foot rot is not readily spread from infected to healthy sheep when the sheep are maintained on dry ground.

Cause

Foot rot is a contagious, infectious disease caused by the organism *Bacterioides nodosus* in conjunction with *Fusobacterium necrophorum*. Possibly other organisms such as *Spirocheata penortha* and *Corynebacterium pyogenes* may be involved. Since the disease can be spread only by infected animals, early diagnosis followed by proper treatment will prevent spread of the disease and will result in considerable savings in time and expense. Walking animals over contaminated areas where infected animals have been is the principal means of spreading the disease, though the incidence is influenced by the weather and temperature.

Clinical Signs

The initial and primary sign of foot rot in sheep is lameness in one or more feet. This may show up within 10 days to 2 weeks after the introduction of the disease. Sheep that have infection in both feet are often seen grazing on their knees or lying on the ground.

The infection starts with a lesion in the soft tissue of the heel or between the claws. It spreads under the sole and inside the horny wall of the hoof. Both claws are usually affected (Fig. 17-3). The horn of the hoof will separate from the tissue beneath it as the infection progresses. In the advanced stages of foot rot, the hoof will be attached only at the coronet. Tissues above the hoof are not usually affected in foot rot of sheep.

Close examination of the feet of sheep affected with foot rot will reveal a dirty gray, cheesy exudate

Fig. 17-3.   Foot rot of sheep.

that has a very disagreeable and distinctive odor. There is a great deal of pain associated with foot rot; therefore, the movement and grazing habits of affected sheep are seriously altered. This results in undernourishment, loss of weight, and increased susceptibility to other infections.

Prevention

Prevention of foot rot includes draining muddy pastures and segregating new animals. Foundation and replacement animals should be purchased from known clean sources. If animals are from a questionable or unknown source, pass through a public market, or are transported by a public conveyance, their hoofs should be trimmed on arrival, and then they should be walked through a foot bath and isolated for one month. Cross infections of foot rot between cattle and sheep do not occur, but cross infections between sheep and goats do occur.

The best preventive measure is a 10 percent zinc sulfate bath, made by mixing 8 pounds of zinc sulfate in 10 gallons of water, and 1 to 2 inches deep. For best results, the foot bath should be placed between the pasture and the water supply.

Treatment

Place infected sheep in a clean, dry pen and treat as follows:

1. Examine every foot of every animal. Trim each foot showing infection, removing enough of the horn of the hoof thoroughly to expose all diseased tissue.

2. Walk all animals through a suitable disinfectant solution and move to clean ground. Visibly affected animals should be kept standing in the solution 5 to 10 minutes. The two most widely used disinfectants are (a) formaldehyde, 10 percent, and (b) copper sulfate, 20 percent. Repeat foot bath at weekly intervals until foot rot disappears. Then continue at two-week intervals for another two months. Two weeks after initial antiseptic treatment, examine feet of each animal a second time, to detect and trim infections overlooked the first time or developed subsequently.

3. After trimming, and treatment in a foot bath, place animals in a clean, dry pasture or lot. One that has not been used for 30 days would be considered clean. An animal with foot rot may spread infection up to three years, but contaminated land loses its ability to infect within three weeks.

Sulfonamide or antibiotic therapy may accompany trimming and foot baths with good results.

FOOT ROT OF CATTLE

Foot rot is an acute or chronic contagious disease of cattle. It is characterized by lameness, due to an inflammation of the tissues within and around the foot. In chronic cases, the infection spreads into one or both of the coffin joints and causes an arthritic condition that lasts for long periods of time. Foot rot of cattle is commonly called foul foot, foul claw, and, correctly termed, infectious pododermatitis. It is completely unrelated to foot rot of sheep.

Cause

Foot rot is a contagious, infectious disease caused by the organism *Bacterioides nodosus* in conjunction with *Fusobacterium necrophorum*. Possibly other organisms such as *Spirocheata penortha* and *Corynebacterium pyogenes* may be involved. Since the disease can be spread only by infected animals, early diagnosis followed by

Fig. 17-4. Concrete pads around water troughs and dry, well-drained holding areas help prevent foot rot.

proper treatment will prevent spread of the disease and will result in considerable savings in time and expense. Walking animals over contaminated areas where infected animals have been is the principal means of spreading the disease, though the incidence is influenced by the weather and temperature.

Clinical Signs

The common signs of foot rot of cattle include lameness, elevated temperature, swollen foot and leg, reduced rate of gain or milk flow, and a characteristic odor from the infected area.

In acute cases, lameness develops suddenly and often affects several of the animal's feet. Inflammation between the toes is visible from both the front and rear of the foot. A moist necrosis can be felt when the foot is handled. Chronic cases of foot rot may affect animals for months before clinical signs appear. Foot rot in this form may, therefore, go undetected and neglected until it is well established. In such cases, irreparable arthritis and tendonitis often occur. The appetite is not noticeably diminished, fever is usually within the normal range, and lameness may be inapparent. Affected cattle may occasionally shake a foot as if they are trying to dislodge an object. If all four feet are affected, the animals may move as if they are walking on eggs. Serious loss of weight and reduction in milk production are often the only indication of chronic foot rot.

Prevention

Prevention of foot rot includes draining muddy pastures and segregating new animals. If animals are from a questionable or unknown source, pass through a public market, or are transported by a public conveyance, their hoofs should be trimmed on arrival, and then they should be walked through a foot bath and isolated for one month. Cross infections of foot rot between cattle and sheep do not occur.

Oral iodides have been beneficial as preventives in cattle in some cases.

Treatment

Systemic and local treatment with antibiotics and sulfonamides is recommended. Other procedures that may speed recovery are cleaning the foot, applying a protective dressing, wiring the claws together, and removing the necrotic interdigital mass. Zinc methionine has been recommended for both treatment and prevention. Walking cattle through a 3 percent formalin foot bath, a 5 percent copper sulfate foot bath, or mixed powdered copper sulfate and lime twice a day decreases the incidence of foot rot.

LAMINITIS

Laminitis is a painful condition of the foot in cattle and horses, with different causes and symptoms. Laminitis is the swelling of the laminae, or sensitive tissues in the foot. Laminitis in cattle has a different etiology and pathogenesis than in horses and will be discussed separately.

Cattle

Cause

The majority of laminitis in dairy cattle arises from nutritional problems. An estimated 70 percent of the cases of acute and chronic laminitis is attributed to feeding problems, with infectious agents, facility design and maintenance, genetics, hygiene, and animal behavior being other contributing factors. Rumen acidosis, which arises from feeding rapidly fermented, high-concentrate rations, is believed to be the major factor in endotoxin release causing damage to the laminae of the foot. The laminae are in close association with the hoof wall and the bone of the foot, so any insult that causes swelling will cause blood leakage from the laminae and pain, ulcer formation, and abscess formation. The modern production methods of concrete confinement for dairy cattle have led to the increased incidence of laminitis because the concrete can erode the hoof wall faster than it grows and long periods of standing on concrete can cause pressure bruising of the laminae.

Clinical Signs

Cows with acute laminitis will present as hesitant to walk, a camped-under stance, crossing of the front legs, or down and reluctant to rise from extreme pain. Other signs to alert the producer that rumen acidosis is present may be a drop in milk production, diarrhea, and dehydration. Chronic sub-acute rumen acidosis leads to chronic changes in the laminae and abnormal growth patterns in the hooves. The hooves may develop horizontal "hardship lines," which can indicate a ration change, such as calving or opening a new silo. The claw of the hoof will begin to flatten and widen out as the quality of the hoof deteriorates (Fig. 17-5). The pedal bone may rotate downward in extreme cases. The combination of the rotation of the pedal bone, the production of softer hoof horn, and the accelerated growth of the hoof on the lateral claw predisposes the foot to sole ulcers, another clinical sign of laminitis. Sole ulcers, sole abscesses, and white line disease cause major production losses annually due to decreased milk production, weight loss, and poor breeding performance in dairy cows.

Prevention

The prevention of laminitis is much more rewarding than the consequences of sole ulcers and abscesses and the treatment. An analysis of the production records for the herd and individual cows can reveal when subclinical rumen acidosis is occurring. Low milk-fat percentages, variations in milk production, low peak milks, thin cows, and the practice of

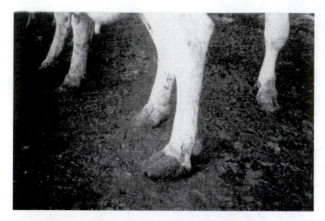

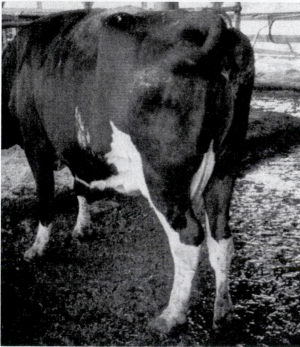

Fig. 17-5. Chronic laminitis in cows evidenced by long, flattened toes (top) and toed-out, hocks-in stance (bottom)

feeding large concentrate meals at one time are signs of a problem with the ration and or cow comfort. A veterinarian and nutritionist should be consulted on a routine basis to analyze forages and formulate a ration balanced for the forage: concentrate ratio. A ratio of 60 percent forage and 40 percent concentrate is the upper recommended limit for proper rumen function to avoid rumen acidosis. The trace mineral availability and formulation is also necessary in strong hoof horn formation. The free stalls or lying areas should be comfortable so cows spend time lying down and relieving pressure from the feet. Preventive hoof trim-

ming to maintain the proper weight-bearing mechanics of the foot is a good herd practice that will increase the longevity of the cow in the herd.

Treatment

Acute cases of laminitis are treated with nonsteroidal, anti-inflammatory drugs to reduce swelling and pain, as well as block endotoxin from the rumen. Sole ulcers should be pared to allow drainage and a block applied to the sound claw to remove weight bearing from the affected claw. Antibiotics are not necessary in treatment of laminitis, sole ulcers and abscesses, and white line disease. These are hemorrhagic diseases causing erosion of the hoof horn exposing the sole. Foot baths for the entire herd and the treatment of hairy foot warts will also treat hoof disease.

Horses

Cause

Laminitis, also called founder, has many contributing factors in horses. The most classic initiating cause is an engorgement of grain, which results in vasoconstrictive substances affecting the laminae in the foot and causing swelling. Many metabolic, physical, and chemical factors can cause laminitis. Poor hoof balance through incorrect or improper hoof trimming and shoeing can lead to bruising of the sole and founder. Pressure founder occurs when one limb is bearing excessive weight due to an injury to the other leg, such as an abscess in one foot leading to founder in the opposite foot. Laminitis can be common following illnesses such as colic, influenza, diarrhea, and other febrile diseases. Endocrine abnormalities, such as hypothyroidism and ovarian dysfunction, can lead to secondary laminitis. A veterinarian should be consulted in cases of founder to determine the cause and prevent future cases.

Clinical Signs

Horses with acute laminitis are reluctant to move and have a "walking on eggshells" appearance when

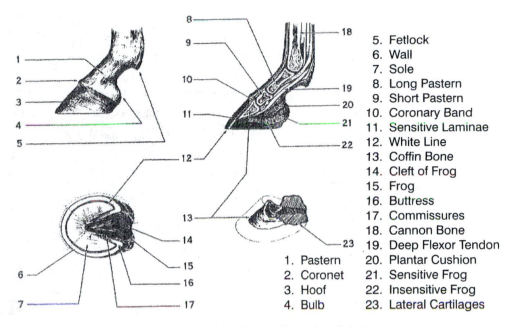

5. Fetlock
6. Wall
7. Sole
8. Long Pastern
9. Short Pastern
10. Coronary Band
11. Sensitive Laminae
12. White Line
13. Coffin Bone
14. Cleft of Frog
15. Frog
16. Buttress
17. Commissures
18. Cannon Bone
19. Deep Flexor Tendon
20. Plantar Cushion
21. Sensitive Frog
22. Insensitive Frog
23. Lateral Cartilages

1. Pastern
2. Coronet
3. Hoof
4. Bulb

Fig. 17-6.   Parts of the foot and lower leg of the horse.

walking. They may stand with the feet camped under or extended to the front if only the front feet are affected. The digital pulses are increased and this is an excellent diagnostic tool in determining if laminitis is the cause for soreness. The digital arteries are on the lateral and medial aspect of the pastern and finger pressure on the arteries can detect if an increase intensity of the pulse is present. The hooves will become warm to the touch later in the course of an acute episode. The loss of integrity of the laminae in the foot can allow the pedal bone to rotate. In extreme cases, the tip of the bone will protrude through the sole of the foot. Horses affected by chronic founder will have overgrown, flattened hooves with horizontal grooves that diverge at the toe. (Fig. 17-7)

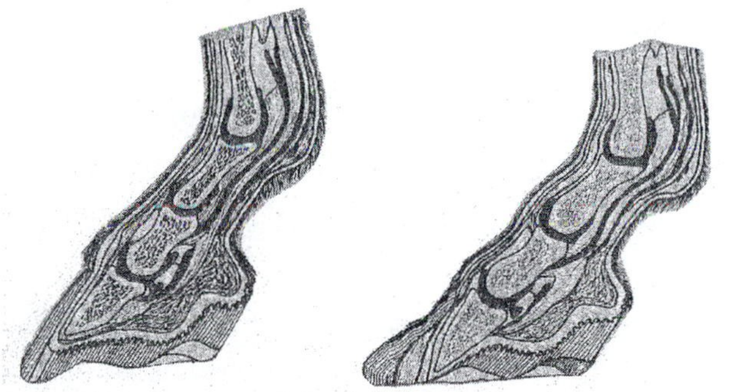

Prevention

Proper nutrition and securing feed supplies to prevent overeating are preventive measures. Proper and regular hoof trimming can prevent biomechanical causes of laminitis. Regular vaccination and deworming programs will lessen the chances of infectious causes of founder.

Treatment

Acute cases of laminitis are treated by applying cold ice water to the feet for 10 to 15 minutes, twice daily, during the initial few days to reduce the swelling of the laminae. Stall rest, with deep bedding to keep the horse comfortable either standing or lying, is needed. The veterinarian, depending upon the severity of the case, may prescribe blood-thinning drugs. Phenylbutazone, 1 gram per 500 lbs., is used to alleviate pain and swelling, and may be used for several weeks on a decreasing dose. A farrier should be consulted to apply special shoes to try to prevent rotation of the pedal bone. Chronic cases of founder require

Fig. 17-7.   Normal foot of a horse on the left. Chronic founder on the right. (Courtesy, Auburn University, Auburn, Ala.)

regular trimming, and often ablation of the anterior hoof and remodeling of the foot.

WARTS

Warts are benign tumors that can affect most animals. In farm animals, warts are a greater problem in cattle, especially calves and yearlings, than in other species. Horses rank second to cattle in susceptibility to warts. Warts are correctly termed infectious papillomas.

Warts are spread among animals of the same species by direct contact with infected animals and by indirect contact with contaminated feed, bunks, needles, water troughs, tattooing equipment, ear notchers, nose tongs, etc. Insects are also believed to transmit warts from infected to healthy animals.

Cause

Warts are caused by a virus that is highly host specific. There is very little chance of natural transmission of infectious papillomas among species. The virus is believed to gain entrance to the animal's body through cuts and abrasions in the skin.

Clinical Signs

Warts affect all ages of cattle, but young animals are more commonly affected than adults. Warts are classified according to the location in which the are found on the body of cattle: cutaneous, genital, and esophageal.

Cutaneous warts occur most frequently on young cattle on the skin around the eyes, mouth, neck, and shoulders. They may spread from the original location to any area of the body. Adult milk cows are most often affected on the teats and udder. These warts are often hard and dark-colored. Occasionally, warts become large and pendulous. In such cases, they sap the strength and stunt the growth of animals. Large warts on the face and neck are often broken open due to rubbing (Fig. 17-8). This can cause anemia and provide a source for bacterial infection. The greatest loss

Fig. 17-8.   Bleeding warts. (Courtesy, Department of Surgery and Medicine, School of Veterinary Medicine, Kansas State University, Manhattan)

from cutaneous warts, however, is due to the damage of hides.

Genital papillomas occur in the sheath or on the penis of males and in the vulva and vagina of the females. Large warts in these areas can interfere with breeding. Esophageal warts occur in the mucous linings of the esophagus and occasionally in the rumen. Large warts in the esophagus can interfere with swallowing and may cause esophageal choke.

Warts often occur on the nose and lips of horses under two years of age. They do not usually cause a problem and generally go away without treatment in a couple of months.

Prevention

The virus that causes warts is usually transmitted to healthy animals by contact from infected animals and materials via cuts or abrasions in the skin. Prevention, therefore, should be based upon eliminating nails, sharp edges on equipment, wires, etc., that can puncture an animal's hide. Equally important is the necessity for sanitation and disinfection of vaccinating needles, dehorning equipment, ear notchers, etc.

When warts are a problem, especially in dairies and feedlots, animals should be vaccinated with a commercial wart vaccine. Two doses are required. They should be administered at two-week intervals.

Treatment

Soften with oil for several days, then tie off the growth with thread or snip it off with sterile scissors and paint the stump with tincture of iodine.

The wart vaccine helps in some cases, or the veterinarian may resort to surgical removal of extremely large warts.

RINGWORM

Ringworm is a chronic dermatomycosis that affects cattle, horses, people, and occasionally swine. Dermatomycosis is a term that describes a fungal infection of the skin. Ringworm is characterized by scaliness, a partial loss of hair, and itching. it is more common in young animals, but animals of any age are susceptible. Ringworm is easily spread by direct contact with infected animals. It is also spread by contact with contaminated brushes, blankets, combs, halters, feed bunks, stalls, etc. Lice may transport viable spores of the fungus over the body of a single host and among hosts when they move from one animal to another.

Cause

Ringworm is caused by *Trychophyton verrucosum* in cattle and *Trychophyton equinum* in horses. Both organisms are infectious to human beings. Their spores are highly resistant to destruction by dehydration and sunlight. They can live several years in barns and sheds. Ringworm, therefore, is most common in animals kept in close quarters, such as in feedlots, and those kept in barns. It is more common in the winter than other seasons of the year.

Clinical Signs

Skin lesions develop on affected animals about three weeks after they are infected with *Trychophyton* spores. The affected skin becomes inflamed, tiny vesi-

Fig. 17-9. Cow with ringworm. Note the raised circular areas. (Courtesy, College of Veterinary Medicine, University of Illinois, Urbana)

cles are formed, and an exudate occurs. This is followed by the formation of scaly, silvery, gray-colored crusts. The patches are generally circular masses, measuring ½ inch to several inches in diameter. They may occur on any part of the body, but they are generally found around the eyes, ears, neck, and tailhead of affected animals. Loss of hair is a common sign of ringworm. This is due to rubbing in response to the itching caused by the infection and to damage to the hair follicle by the invading organism (Fig. 17-9).

Prevention

There is no vaccine to prevent ringworm in farm animals. Prevention is based upon sanitation and disinfection of barns, stables, and equipment. In addition, housing areas for animals should be properly ventilated and kept as dry as possible.

Treatment

Clip the hair from the affected areas, remove scabs with a brush and a mild soap. Paint affected areas with tincture of iodine or salicylic acid and alcohol (1 part in 10) every three days until cleared up.

Certain proprietary remedies available only from veterinarians have proved very effective in treatment.

EXERTIONAL RHABDOMYOLYSIS

Exertional rhabdomyolysis, also called tying-up, Monday morning disease, and azoturia, is a disease in which horses show a stiff gait, muscle cramping, and pain. Rhabdomyolysis is usually seen after mild or moderate exercise. However, the signs can be seen following endurance rides, transportation, pasture turnout, and anesthesia. Three types of exertional rhabdomyolysis have been identified, due to either mithchondrial myopathy, a defect in skeletal muscle excitation–contraction coupling, or a polysaccharide storage myopathy (PSSM) in Quarter horses.

Cause

PSSM is caused from the accumulation of high muscle-glycogen concentrations and an abnormal polysaccharide in skeletal muscle. Horses with PSSM are usually described as heavy-muscled, calm individuals. The precise cause of PSSM in horses is not known, as the enzyme systems for the muscles have not been shown to be deficient. Further research into the uptake of glucose and synthesis of glycogen is needed.

Clinical signs

Classical signs of exertional rhabdomyolysis in Quarter horses usually develop 15 to 30 minutes after light exercise. In mild forms, the horse develops a tucked-up abdomen, muscle twitches in the flank, and stands with the feet extended. The horse may sweat, stop, and refuse to move. Some horses will lie down or buck when they anticipate exercise-producing pain. In severe episodes, the breakdown of muscle tissue can be severe enough to release myoglobin in the bloodstream, which can lead to kidney failure.

Affected horses will have firm, painful muscles of the hind legs and biceps. Serum enzymes greatly increase and in some cases may remain elevated. Muscle biopsies and special staining of the sample is required to diagnose PSSM. Other causes of exertional rhabdomyolysis can be diagnosed from serum analysis and clinical signs.

Prevention

The prevention of exertional rhabdomyolysis, and specifically PSSM, is through establishing regular routines, slowly increasing the exercise, and feeding a diet low in soluble carbohydrates. Horses with PSSM benefit from a total turnout rather than stall rest. A fifteen-minute exercise test on a lunge line can be used to assess the fitness level and to decide whether to increase the workload.

Horses will benefit greatly from a dietary change, which is often hard for the owner to understand. The grain, molasses, and sweet feed should be discontinued or decreased. A high-fat diet, supplied by rice bran or corn oil, is well tolerated in horses. Corn oil can be fed at a level of 1 to 4 cups daily mixed with alfalfa cubes. A balanced vitamin, mineral, and electrolyte supplement must be fed. Dietary and exercise changes are necessary to prevent exertional rhabdomyolysis.

Treatment

Horses acutely affected with exertional rhabdomyolysis can be made more comfortable after receiving intravenous acetylpromazine. Caution should be used, because dehydration can be further exacerbated. Nonsteroidal, anti-inflammatory drugs, such as flunixin meglumine and phenylbutazone, can be used successfully in well-hydrated horses showing signs of cramping. The severely affected horse will need aggressive IV fluid therapy to help flush the myoglobin through the kidneys. Some horses can die from exertional rhabdomyolysis 24 to 48 hours after an attack.

SWINE POX

Swine pox is an acute, infectious viral disease that is characterized by skin eruptions on the underline of pigs. Swine pox is usually confined to young pigs from one to two months of age. Mature hogs are seldom affected.

True swine pox affects only swine. The cow pox virus is also known to affect swine and human beings.

Immunity developed in swine from one type of pox does not protect against the other.

Cause

Swine pox is caused by a virus called *Variola suilla.* The cow pox virus is called *Variola vaccina.* The swine pox virus is spread from infected to healthy animals by the bloodsucking hog louse, *Haematopinus suis.* An infected hog louse can infect swine for several weeks or a month. Other insects may spread the disease mechanically to healthy swine that have open wounds.

Clinical Signs

Inflammation and skin eruptions that scab or crust over are often seen on the soft, thin skin areas of affected pigs. The belly, armpits, and parts of the sides are common sites of infection. Affected pigs may do poorly, due to a reduction in feed intake during the course of the disease. Secondary bacterial infections may occur in the open lesions that are caused by the pigs' rubbing and scratching. In very rare cases, swine pox may become septicemic and cause a loss of appetite, high fever (107° to 108°F), chilling, debilitation, depression, and rapid dehydration. This form usually terminates in death (Fig. 17-10).

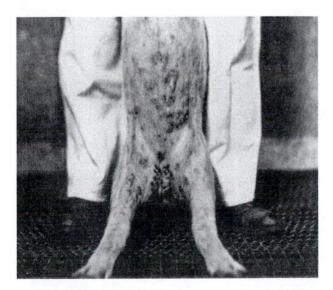

Fig. 17-10. Pig with swine pox. Note the scabs which cover little crater-like spots. (Courtesy, College of Veterinary Medicine, University of Illinois, Urbana)

Prevention

Swine pox can be controlled in most cases when pigs are not subjected to infestations of lice, flies, and mosquitoes that commonly spread the disease. Hogs affected with swine pox should not be purchased or added to a healthy herd of swine.

Treatment

There is no treatment for swine pox.

ABSCESSES

Abscesses are localized collections of pus in the tissues of the body. Abscesses are often described as being "hot" or "cold." Hot abscesses are acute formations of purulent exudate. Cold abscesses usually result from chronic accumulations of pus formed by the disintegration of tissue. Boils, carbuncles, and pustules are types of abscesses.

Causes

Abscesses may be caused by a variety of bacteria, but one group, *streptococci,* causes 85 percent of the abscesses in swine. Other pyogenic organisms include *Escherichia coli, Pasteurella multocida, Spherophorus necrophorus, Staphylococcus aureus,* and *Salmonella typhimurium.* These organisms enter the animal's body through the mouth, respiratory tract, and breaks in the skin due to wounds, injections, castration, docking, dehorning, etc.

Clinical Signs

The clinical signs of abscesses are dependent upon the size and location of the pus-filled enlargements. The signs may vary from none to serious inflammation, prostration, and death. Abscesses along the throat may interfere with eating, drinking, and breathing; those in the vaginal or rectal area may cause difficulty in defecation or in the passage of urine; and abscesses on the udder may cause dams to refuse to permit their offspring to nurse.

Prevention

Clean feed, clean quarters, and aseptic techniques in administering medications, castrating, docking, dehorning, and assisting with dystocias will certainly help to reduce abscesses. Sharp objects that can break the skin of animals should be removed from holding areas. Animals with abscesses that have ruptured should be isolated and treated.

Antibiotics fed in the ration and oral vaccines for swine are recommended when abscesses are a recurring problem. In addition, injections of iron compounds and antibiotics should be administered in the neck rather than in the ham.

Treatment

Abscesses should be lanced and drained when they are "ripe" or soft in the center. This process of "ripening" or bringing the abscess to a head, can be facilitated by the application of poultices.

Before surgery, some of the contents from the center of the abscess should be extracted with a hypodermic syringe and needle to make sure the enlargement is not a hematoma. Pus should be a viscous, smooth, yellowish-colored material. Hematomas should not be excised.

The abscess should be lanced in a manner that assures the best drainage. The pus should be removed, collected, and destroyed. The cavity should be flooded with hydrogen peroxide and packed with gauze soaked in tincture of iodine. Protection should be provided against insects and reinfection by pathogens. This can be accomplished by applying insect repellants to the surrounding area and by providing the animal housing in a clean area. Antiseptics and antibiotics can also be used in therapy. Sodium iodine given intravenously will help the animal clear up the infection.

NAVEL ILL

The term *navel ill* includes those infections that affect newborn farm animals as a result of bacterial or viral entrance through the navel shortly after they are born. Navel ill may be expressed in affected animals by septicemia, enteritits, arthritis, and as abscesses.

Cause

Navel ill in cattle is usually characterized by umbilical abscesses, septicemia, and enteritis. Joint involvement is not as common in calves as it is in lambs and foals. The septicemia or bacteremia forms of navel ill are caused by *Escherichia coli, Listeria monocytogenes, Pasteurella, Streptococcus,* or *Salmonella* species. Enteric forms of navel ill are caused by *Clostridium perfringens,* types A, B, and C, and by pneumoenteritis due to a virus.

Infection of the umbilicus occurs soon after birth as a consequence of failure of passive transfer, animal weakness, and poor hygiene, which exposes the umbilicus to excessive organisms. Failure of passive transfer, in which the animal does not receive or absorb enough quality colostrum, is the most important causative factor. The colostrum is high in antibodies to fight infections and the newborn animal will be very susceptible to septicemia, diarrhea, and pneumonia without adequate colostrum. Inadequate colostrum intake may occur due to leakage of milk from the dam, poor udder conformation, a weak animal from dystocia, or pigs and lambs born after birth mates have nursed the colostrum. *Actinomyces pyogenes* is the most common organism isolated from calves with navel ill. *E. coli, Staph.* spp. and *Srep.* spp. are also found.

Clinical Signs

There is local inflammation around the navel where the infectious agent first enters the body. From the local infection at the navel, the disease may spread to the liver or travel via the urachus to the bladder and result in chronic ill health. The disease may also spread systemically to produce a septicemia. In blood-borne infections, localization usually occurs in the joints and produces a suppurative or non-suppurative arthritis; less frequently, bacteria localize in the eye and cause a total inflammation of the eye. They may localize in the heart valves and cause endocarditis.

Fig. 17-11. Calf with navel ill. Note the swelling in the joints. (Courtesy, College of Veterinary Medicine, University of Illinois, Urbana)

In addition, the meninges is occasionally infected, and meningitis results. Most of the secondary signs take several days to develop (Fig. 17-11).

Dehydration and electrolyte imbalance can occur very rapidly in newborn animals, whether or not diarrhea and vomiting are present. This is probably due to deprivation of fluid intake as much as to loss of fluid. The extreme depression observed in many cases is probably caused by biochemical changes in addition to the effects of bacterial toxins.

Deep palpation of the abdomen of the animal may reveal a round, firm, cylindrical structure extending from the umbilicus to the cranial abdomen where the liver may be abscessed. A weak, thin, and unthrifty appearance may be the sign of an umbilical abscess and close inspection may show the problem.

Prevention

Newborn animals should get plenty of colostrum before they are six hours hold. Animals over 24 hours old are unable to utilize antibodies from the colostrum. A clean area should be provided for the animals to be born in. Open pasture is the cleanest when it is not overstocked. The navel should be dipped in 7 percent iodine, which will cauterize the blood vessels and thereby stop the bleeding. It is the most effective solution for killing any bacteria that are present.

Treatment

Antimicrobial therapy should be given to animals that are sick, depressed, febrile, or draining pus from the umbilicus. Ceftiofur, amoxicillin, ampicillin, penicillin, and florfenicol are drugs with a broad range of antimicrobial activity and are approved for several species of farm animals. Record and identify treated animals to observe the proper drug withdrawal time before slaughter.

When abscessation of the umbilicus has occurred, the abscess needs to be opened and the area flushed with dilute povidone iodine solution for several days. Fly control is necessary to prevent opportunistic infestations with maggots.

POLYARTHRITIS

Polyarthritis is an infectious disease of lambs that is characterized by lameness, retarded growth, and lateral or sternal recumbency. There are two major types of polyarthritis: the suppurative, or the pus-forming type, and the non-suppurative.

Cause

Suppurative arthritis is caused by *Streptococcus* and *Micrococcus* bacterial organisms. The non-suppurative type of arthritis is caused by *Erysipelothrix insidiosa*, *Corynebacterium pseudotuberculosis*, and a virus of the *Psitacosis lymphogranuloma* group. Infectious organisms causing polyarthritis in lambs enter through the fresh umbilical stump and through docking and castration wounds.

Clinical Signs

The suppurative type of arthritis is characterized by lameness in one or more legs due to inflammation and pus in the hock, stifle, elbow, or knee. In some cases, the vertebrae are also affected. Joint enlarge-

Fig. 17-12. A lamb affected with non-suppurative arthritis.

ments are usually quite pronounced due to the accumulations of pus.

Clinical signs of non-suppurative arthritis may appear at any time when the lambs are one week to five months of age. The temperature of sick lambs is usually normal, and the appetite remains good. A loss of condition and retarded growth are due to the lambs' disinclination to endure the pain required to graze or move about. The lameness usually involves all four legs in acute cases (Fig. 17-12). Some lambs apparently recover completely, but others continue to show evidence of retarded growth for several months after the initial signs have disappeared. These lambs develop a chronic lameness in one or more legs. Enlargement of the affected joints may occur in chronic cases of arthritis.

Prevention

Polyarthritis can be prevented or greatly diminished if the umbilical stump of newborn lambs is treated with tincture of iodine. Lambs should be docked and castrated away from permanent lots and corrals. When this is impossible, docking should be performed with a hot iron. Lambs should be moved to pasture immediately after they have been docked and castrated.

Treatment

Affected animals should be isolated so they can be treated and so other lambs will not be exposed. If treatment is begun early, antibiotics are effective. The recommended antibiotics include chlortetracycline, oxytetracycline, penicillin, and tylosin. Advanced cases may not respond to treatment.

Outbreaks of polyarthritis in feedlot lambs can be treated with aureomycin or terramycin fed in the feed at the rate of 200 to 250 mg per head per day for one week.

FLY BITE HYPERSENSITIVITY

Hypersensitivity to insect bites can occur in all species of farm animals. The reaction is caused by an allergic reaction to the feeding pattern of several insects, with clinical signs appearing immediately after a bite or several days later. *Culicoides* spp. are common culprits in causing alopecia and intense pruritis in horses.

Clinical Signs

The clinical signs of fly bite hypersensitivity include intense itching, loss of hair, and thickening of the skin. Hives can also be a symptom of sensitivity to insects. The hives will be large, flat topped, and can occur on all parts of the body.

Prevention

Horses should be kept inside, and Fly Wipe should be used on them frequently to prevent fly bite hypersensitivity.

Treatment

The treatment for fly bite hypersensitivity includes cortisone, antihistamines, and Dia-glo in the feed. Two to four ounces of apple cider, vinegar, or brewer's yeast administered orally on a daily basis tends to provide an odor that repels insects.

PHOTOSENSITIZATION

Photosensitization is a disease that occurs in all species of animals. It is a common problem in cattle, sheep, and swine. The term *photosensitization* means an unusual sensitivity to light. It is a condition whereby an animal becomes hypersensitive to direct sunlight due to the ingestion or injection of photodynamic agents or due to a malfunction of the liver. The former is referred to as a primary photosensitization, and the latter, hepatogenous photosensitization.

Photosensitization should not be confused with sunburn. While both conditions are more severe on light-pigmented animals, they are completely different diseases. Skin lesions develop more rapidly, and the inflammation is usually much more severe in photosensitization.

Cause

Photosensitization is the sensitization of light-colored skin to sunlight. Some feeds, forages, and certain medicines contain substances which may sensitize the skin (primary photosensitization). In other cases, products of metabolism, which normally would be removed from the body, accumulate because of faulty liver function (hepatogenous photosensitization). Primary photosensitization usually occurs in the spring when plants are lush, green, and growing rapidly. St. John's Wort (Klamath weed) and buckwheat are two of the most common sources of photosensitizing substances. Also, rape, kleingrass, kale, trefoil, alfalfa, alsike clover, Swedish clover, lamb's tongue, and plantain have been associated with photosensitization at one time or another.

Clinical Signs

Animals with light hair costs and little pigment in the skin, or a portion of the skin, are most frequently affected with photosensitization. The muzzle, eyes, face, and light areas over the back are usually affected first. Areas of the belly and udder, which are exposed to the sun when the animal lies down, may also be affected.

Fig. 17-13. Photosensitization in sheep. Note the lesions around the eyes and muzzle. (Courtesy, College of Veterinary Medicine, Washington State University, Pullman)

The first signs are usually redness and swelling of the skin. Later, tissue fluids ooze from the affected areas, and crusting of the skin occurs, with resultant matting of the hair. In severe cases, the eyelids and nostrils may be swollen closed. In extreme cases, sloughing of the skin and gangrene result. Photosensitized animals show eye sensitivity to sunlight. They seek shade or turn their backs to the sun.

Where skin lesions are severe and extensive, shock may occur. The temperature is often elevated, and the animal is weak. There may be difficult breathing and blindness. Excitement or depression, excessive salivation, diarrhea, and hemoglobinuria may occur. Pain is often evidenced in sheep and swine by their having a wobbly gait and by their suddenly dropping to the ground with the rear legs extended. In hogs, this is often preceded by an exaggerated sway-back stance that terminates when their belly strikes the ground and their rear limbs are distended. White-skinned pigs are frequently affected by sunburn without the presence of photosensitizing substances. When photosensitization does occur in these pigs, the results are usually rather severe. The swelling around the eyes, muzzle, face, and ears, due to photosensitization in open-faced sheep, gives rise to the common name of "big head." When the udder of a dairy cow is affected, severe mastitis may result from infection of the skin lesions being transmitted to the gland.

Photosensitization in itself is seldom fatal, especially if animals can be protected from the sunlight. Deaths do occur, however, from liver damage, shock, starvation, and secondary bacterial infection of the dead skin.

Prevention

The control of photosensitization is dependent upon keeping animals from having access to photodynamic plants. Though simple in principle, it may prove difficult under range conditions. On ranges where the problem occurs, cattle and sheep may have to be removed during certain periods of the year.

Treatment

Photosensitized animals should be housed or given access to shade. Immediate change of forage is essential to eliminate the source of photosensitization. Support treatment such as fluids, antibiotics, antihistamines, and corticosteroids may assist in returning the animal to normal. Affected parts of the skin may be treated with methylene-blue solution or some other nontoxic dye. The skin will heal when photosensitization is controlled. Even after a severe case, the skin and hair will return to near normal. Animals with hepatogenous photosensitization should be removed from the herd.

SUNBURN

Cause

Sunburn is a dermatitis caused by ultra-violet rays from the sun or from heat lamps. Animals housed inside are not affected by the sun's rays if the light is fil-

tered through a window pane. Sows and pigs may be burned, however, if heat lamps are placed too close to the skin.

Sunburn is a problem in non-pigmented breeds of swine and shorn sheep that are exposed to bright sunlight. Young white-skinned pigs are highly susceptible to sunburn. Light-skinned cattle placed in bright sunshine without a period of acclimation are also likely to sunburn.

Clinical Signs

The severity of clinical signs depends upon the sensitivity of the animals' skin, length of exposure, and brightness of the sun. Inflammation that is extremely sensitive to the affected animals usually occurs 12 to 18 hours following exposure. The skin is bright pink or red in color and hot to the touch. Small vesicles form on the skin, erupt, and cause itching and discomfort. Sunburned animals will seek out shade and will usually show a preference for water over feed. Severely affected pigs often have a high temperature, become weak, and collapse. Very young pigs may be so severely affected that their ears and tails necrose and slough off.

Death is not common in animals affected with sunburn.

Prevention

Protection from bright sunlight should be provided for white-skinned pigs and freshly shorn sheep. All hogs should have access to shade during warm weather. This protection is more essential to prevent sunstroke than sunburn, although the latter is a problem for white-skinned hogs of any age exposed to bright sunlight.

Treatment

Sunburned animals should be placed in a shaded area for a few days. Zinc oxide powder or ointment may be used when necessary to relieve itching.

CHAPTER 18

Plant and Chemical Poisonings

A poison is any substance which through its chemical action kills, injures, or impairs an organism. The effect of most poisons depends upon the amount consumed, the concentration, the rate of absorption and elimination, and, to some extent, the species of the animal involved. Substances which inhibit the action of poisons are called antidotes.

Poisoning of livestock is usually a result of poor management. Chemical poisoning of animals is often the result of carelessness by the animal caretaker in using insecticides, disinfectants, rat poisons, paint, and certain feed additives such as the arsenicals. Poor pasture conditions due to overgrazing or drought reduce the amount of palatable forage and thereby induce hungry animals to eat poisonous plants, however, they are relished by certain animals and eaten in preference to desirable forage. The following management practices are recommended to reduce poisoning of farm animals:

1. Chemical materials such a fertilizers, rodenticides, insecticides, and disinfectants should be stored in areas isolated to all animals.

2. Animals should be kept away from freshly painted buildings, equipment, etc.

3. Chemical containers, papers saturated with paint, and fertilizer bags should be destroyed.

4. Rodent poisons should not be placed in or near feed materials.

5. When possible, infested pastures should be grazed with species-resistant animals.

6. Infested pastures should be grazed when poisonous plants are least toxic. Larkspurs, for example, lose their toxic effect as they grow older.

Cynaogenetic plants are more dangerous when their normal growth has been interrupted by frost, drought, or trampling. They, too, tend to become less toxic as they mature.

7. When poisonous plants occur only in small localized areas of the pasture, fencing or herding animals may prevent consumption.

8. Animals should be fed well before they are released into new pastures or moved through areas infested with poisonous plants.

9. Overgrazing should be prevented, and supplemental forage should be provided when pastures are poor.

10. The population of poisonous plants should be eradicated or reduced by mechanical destruction or by the application of herbicides.

Poisoned animals should be treated as soon as the condition is recognized. Treatment of animals that have eaten poisonous plants is based on four principles:

1. Destruction of poisonous substance within the alimentary tract.

2. Prevention of absorption into the bloodstream.

3. Promotion of excretion and evacuation.

4. Treatment of symptoms.

Poisoned animals should be removed from accessible poisonous plants and provided with feed, water, and shelter. Some animals may need hand feeding and watering. Specific antidotes are available for some poisons; however, others require symptomatic treatment, which includes the use of laxatives and stimulants and good nursing.

POISONOUS PLANTS

Cyanogenetic Plants

Some plants are capable, under certain conditions, of producing hydrocyanic (prussic) acid (HCN) a highly poisonous substance. They are known as cyanogenetic plants. The acid forms when an enzyme or rumen bacteria act on glucocides. Neither the glucocides nor the enzyme is individually poisonous, and, under normal conditions, they do not come in contact with each other. Poisoning usually follows plant damage, such as wilting, freezing, drought, cutting, crushing, etc. Young growing plants usually contain more of the glucocides than mature plants. Drying of plants usually reduces the toxicity, but animals occasionally are poisoned by eating hay from cyanogenetic plants.

Of the domesticated animals, cattle and sheep are most susceptible to the action of HCN. The minimum lethal dose of HCN for these animals is close to 1.052 mg per pound of body weight. Horses and swine are not nearly so susceptible as ruminants because the high concentration of hydrochloric acid (HCl) in their stomachs combines with HCN, as it is liberated, and reduces the toxicity. All animals exhibit a certain degree of tolerance to prussic acid, and it is only when the prussic acid enters the bloodstream, via the stomach, at a greater rate than the animal's tolerance for it, that fatal poisoning results.

Hungry animals of low vigor are more likely to succumb following the consumption of the prussic acid-containing plants because these animals consume larger amounts and lack the vitality to withstand their action. Animals that have consumed these poison-containing plants stagger, fall to the ground, breathe rapidly, and die in a few minutes from respiratory failure. The heart will continue to beat for some time after the cessation of breathing. If the animals' lives are to be saved, the poison must be prevented from forming by the combination of the enzyme and the glucoside. If the poison has actually formed and absorption into the bloodstream has started, a quick-acting antidote must be introduced into the bloodstream to counteract its harmful results.

Sodium nitrite and sodium thiosulfate are specific antidotes for HCN poisoning. Methylene blue has a greater range of treatment for poisonings of this nature than either of the two aforementioned drugs. It, therefore, is more commonly used in the treatment of HCN poisoning. It should be given at the rate of 125 cc per 1,000 pounds of body weight. Solutions should be given intravenously because the poison acts rapidly. Stock should be moved from the field in which poisoning occurred. Animals should be fed hay, straw, or any roughage known to be free of HCN.

Cockle-Bur

Cockle-bur poisoning has been observed in hogs, sheep, and cattle. Hogs and sheep are equally susceptible. The consumption of very young plants in the amount of 1½ percent of the animal's live weight is likely to prove fatal; one-half of this amount may result in the development of symptoms in animals, which ultimately recover. Cattle require twice as much in proportion to their size before serious results are observed. There is good reason to believe that horses also are susceptible to the poisonous action of the cockle-bur. They probably avoid poisoning from the plant because they are more careful feeders. Daily sublethal doses of the cockle-bur plant are not harmful because the poison is not cumulative (Fig. 18-1). It is rapidly eliminated from the body.

Research work has demonstrated that young cockle-burs are poisonous at the time the first pair of leaves is partially developed, or just after germination.

At this stage of growth, the young cockle-burs are very appetizing to hogs and sheep. The stomachs of dead animals, poisoned by cockle-burs, are usually loaded with this material.

Animals that have consumed the young cockle-burs in this stage may vomit. They are depressed and, before death, may have spasms, though some animals die quietly. Death, when fatal doses have been consumed, takes place in from 1½ to 8 hours.

Fig. 18-1. Cockle-bur.

The obvious preventative measure is to keep animals out of all lots in which cockle-burs are likely to make their appearance. After the plants have attained some growth, they are no longer relished and apparently lose their poisonous properties.

There is no specific remedy against cockle-bur poisoning once the poison is absorbed. However, if animals are observed eating these plants, they should be moved to a lot free from cockle-bur plants. Hogs should be induced to drink whole milk or cream as soon as possible. Cattle and sheep should be drenched with mineral oil.

Oleander

Oleander is a hardy ornamental shrub that grows throughout the southern part of the United States. All animals, including humans, may be poisoned by ingesting parts of the oleander plant. All parts of the plant are poisonous, but the seeds contain the greatest amount of toxic material. Oleander produces abdominal pain, vomiting, diarrhea, stimulation of the heart, and constriction of the blood vessels. Poisoned animals tremble, develop a progressive paralysis, become comatose, and die. Animals poisoned by oleander should be kept warm and quiet. They should be given emetics or a gastric lavage to remove the contents of the gastrointestinal tract.

Oak

Oak poisoning is a serious problem for cattle and sheep producers in the southwestern part of the United States. The buds, small leaves, flowers, and stems are palatable and poisonous. As the leaves mature, they become less palatable, and animals usually stop eating them by the time the leaves are three-fourths grown. After leaves become older, animals again will eat them, but seldom in the quantity that they ate the young growth. Acorns, when eaten in quantity, produce signs and lesions similar to oak bud or leaf poisoning.

The symptoms appear in from one to four weeks following the beginning of the oak leaf ration. The outstanding clinical symptoms are constipation, with the passage of blood-stained, dark-colored feces, and unthriftiness. From 2 to 3 percent of the herd usually die, and some of the remaining ones are likely to be permanently stunted.

Due to the severity of the losses from oak poisoning, many cattle are penned and fed for three to four weeks when the oak first buds in the spring. Even so, some of them will later eat sufficient oak to be poisoned.

Calcium hydroxide is an efficient antidote to prevent poisoning by tannic acid. Supplemental feed containing 9 percent hydrated lime has reduced losses in cattle in experimental feeding trials. Feeds containing 10 percent hydrated lime are being tested to reduce or prevent oak poisoning in cattle on the range.

Loco-Weed

Loco-weed poisoning has been generally observed in the semi-arid regions of the plains states and in the eastern Rocky Mountain region.

The weeds involved are the white loco or rattle weed, the purple or woolly loco, and the blue loco. In Texas, the early loco and the garbancillo have also been incriminated. Affected species are horses, sheep, and cattle.

Loco-weeds grow from long roots that may extend 2 to 3 feet into the ground. The white loco is generally found on the hills, while the woolly variety occupies the depressions. These plants are legumes and apparently increase soil fertility.

The outstanding symptoms of "locoed" animals are those of a nervous nature, including an involvement of the senses. Affected animals are easily startled. They run into things and through barbed wire fences. At times, affected animals become very aggressive. They seem to miscalculate distances or have a distorted vision, as evidenced by their exaggerated efforts to pass over very minor obstacles. In the course of time, they show evidence of malnutrition by a loss of weight, and general weakness appears almost at the same time. Death is the final result if the animals are continued on an almost exclusively loco ration.

The loco eating habit seems to become established when good pasture vegetation is scarce. Before the loco eating habit is formed, animals pay little attention to the weed, but once the taste is acquired, they show a decided preference for it.

The actual poisonous factor has not been well established. The barium in the plant has been accused. Others contend that the plant's action is largely the result of its exclusive use as a dietary form, or that it produces a nutritional disturbance. When other wholesome vegetation is available, the grazing of the loco-weed with it does not appear to be harmful.

Animals raised where loco-weeds are common are less likely to eat them than imported livestock are. Susceptibility among animals of the same species is variable. When palatable range forage is scarce, the use of good supplemental feed tends to reduce the amount of loco consumed. Locoed animals should be removed from infested pastures and placed on good feed.

Castorbean

The castorbean fruit contains ricin which is toxic to human beings and to all livestock. The other parts of the plant contain ricin in lesser amounts. Poisoned animals develop nausea and violent purging. There is often blood in the feces. Muscular tremors, general weakness, and emaciation develop in cases of prolonged illness. Children are reported to be especially susceptible to castorbean poisoning.

Water Hemlock

Water hemlock is a perennial herb in the parsley family. The family also includes poison hemlock, celery, carrots, parsley, and dill.

Water hemlock is distributed throughout the United States. It is found in wet meadows and pastures, along streams, and around permanent springs.

The rootstocks of water hemlocks are extremely poisonous. Two to six rootstocks ($\frac{1}{2}$ to 1 pound) can kill a mature cow in less than an hour. Dried rootstocks are as poisonous as fresh ones. All animals, including humans, are poisoned if they eat enough water hemlock. Cattle are most often grazed in areas where water hemlock grows and are more likely to pull up the rootstocks from wet ground. The young leaves, containing cicutoxin, may cause livestock losses early in the growing season. As the leaves grow larger however, the poison becomes less concentrated. Mature leaves rarely cause poisoning.

Water hemlock produces such a rapid death that illness will not be observed in most cases. Animals may be found dead near the source of the plant. Diagnosis is based on finding that the plant has been eaten and perhaps finding parts of the rootstocks in the rumen or paunch.

The first signs of poisoning may be seen from 10 to 60 minutes after the plant has been eaten. The animal walks stiffly and acts uneasy. The eyelids may begin to twitch. In a few minutes, the muscles of the neck begin to twitch and may cause the head to jerk; soon many muscles are twitching, and the animal may appear to be jumping up and down. As the muscles continue to contract, the animal falls down, goes into violent convulsions, and fights for its breath. The legs are flung about wildly, the neck may be arched down between the legs or pulled backwards, and the third eyelid is frequently pulled over the eyeball. The mouth and face become contorted, and violent chewing motions occur, with frothing at the mouth. Periodically, the animal will relax and lie quietly for a moment, but a slight noise or touch will start the convulsions again.

There is no specific antidote for water hemlock poisoning, and, because death usually occurs too quickly for anyone to treat the animal, prevention is the only hope of controlling losses from poisoning. The caretaker must recognize the plant and eliminate it, or keep livestock away from it.

Poison Hemlock

Poison hemlock, a native of Europe, occurs at lower elevations throughout much of the southern part of the United States. Unlike the water hemlocks, it is not semi-aquatic, and, although frequently found in moist ground, it may thrive in fairly dry soil.

Poison hemlock has been recognized as a poisonous plant for many centuries. Its poisonous properties are several alkaloids which are found in the green leaves, stems, and fruit of the plant (Fig. 18-2). Owing to the volatile character of the poisons, this hemlock largely loses its toxicity on drying. Thus, dry hay containing poison hemlock is not dangerous.

The plant is rarely eaten, but, if wholesome forage is lacking, livestock may eat the young foliage, with fatal results. In rare instances, the small seed-like fruit may occur as an impurity in grain in sufficient quantity to poison livestock fed on it. Poison hemlock is not nearly as dangerous as water hemlock, because the animal must eat 1 to 2 percent of its body weight

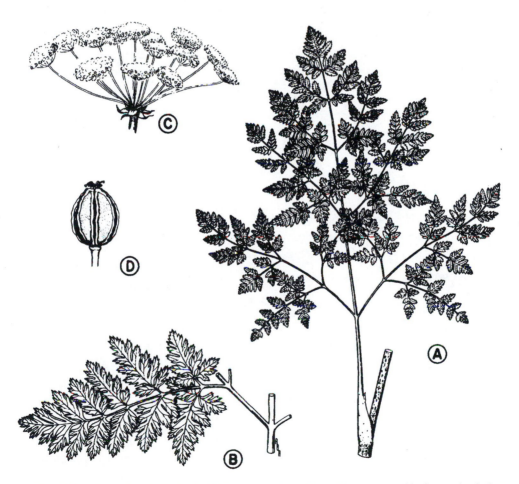

Fig. 18-2.   Poison hemlock. A—leaf, 20 percent of actual size; B—portion of leaf, one-third of actual size; C—flower cluster, actual size; D—fruit, six times the actual size. (Courtesy, University of California-Davis)

to produce poisoning; even then, the animal may recover if given proper care.

Poisoning signs appear two to four hours after the plant has been eaten. Poison hemlock affects the animal's brain. The animal loses strength, its hindlegs buckle beneath it, and it staggers and falls. It may get up again with difficulty, or it may stay down. The front legs soon become paralyzed and the animal appears drowsy and eventually goes into a coma. An animal may be down and paralyzed for a day or two and still recover. Death can occur in 5 to 10 hours after these signs have been noted. Autopsies reveal nothing significant.

There is no specific antidote for hemlock poisoning, but keeping prostrate animals shaded or warm and providing feed and water will save most of them. Laxatives may help to remove the material from the rumen. Drenching an animal in a coma may produce a fatal pneumonia.

Pine Needles

Abortions in cattle grazing in the yellow ponderosa pine areas of the western part of the United States have been reported by cattle producers for over 60 years. Their observations have suggested abortions are caused by consumption of pine needles by cows in the last trimester of gestation. The problems are apparently seasonal, with the concentration of abortions in the late winter or early spring.

Research at South Dakota State University has determined that the substance in the needles that causes cows to abort is present in the needles year-round. Researchers suggest two reasons why abortions are noticed only in the winter and only in cows in the late stages of pregnancy. First, unborn calves that dies in the early stages of gestation are resorbed by the mother cow. Second, the problem generally occurs during January through April because the scarcity of feed at this time of year forces cows to graze on the pine needles.

In most cases where abortions have occurred due to pine needle poisoning, cows have had problems with retained placentas. There are also reports of losses of cows due to peritonitis and other secondary diseases associated with the abortion syndrome.

An antitoxin has not been developed to counteract the substance in pine needles that causes abortion. At this time, the only recommendation for controlling this disease is to keep bred cows and heifers off ranges that are populated with ponderosa pine trees.

White Snakeroot

Synonyms of white snakeroot poisoning are trembles and milk sickness. There are several species and varieties of the genus *Eupatorium* which are poisonous to cattle, sheep, goats, swine, and horses. Human beings may be affected by drinking milk from cows grazing on white snakeroot plants.

Signs of white snakeroot poisoning are trembling, especially after exercise, depression, weakness, and inactivity. Stiff movements with frequent stumbling and falling may also occur. The animals are often constipated, and they may have blood in the feces. Breathing is laborious, and the breath odor is pungent.

The toxin of white snakeroot is an unsaturated alcohol called *tremetol*. It is found primarily in green tissue and decreases as the plant dries. The poison is cumulative and is transmitted in milk.

Animals should not be permitted to graze pastures where white snakeroot grows unless there is adequate palatable feed. Milk from affected animals should not be used for animal or human consumption.

Poisoned animals should be removed from infested pastures as soon as possible. Young animals should be given milk from cows that have not had access to white snakeroot. The use of purgatives, stimulants, and laxative feeds improves the chances of recovery. Drugs should be given by stomach tube or injection, as some animals have throat paralysis.

Localized areas of white snakeroot should be isolated by fencing, or the plants should be pulled and burned. White snakeroot is susceptible to herbicidal sprays. The best control has been obtained with amine formulations.

Crotalaria

Crotalaria is a legume used as a cover crop in the southern, central, and eastern parts of the United States. Some species of crotalaria, especially *C. spectabilis*, are toxic to farm animals. The poisonous factor is the alkaloid monocrotaline. Green, wilted, frosted, or dried plants are equally poisonous. Crotalaria seeds are often harvested with corn. They are more toxic than the vegetative parts of the plant. The poison is cumulative. Death often occurs several months after a sufficient quantity of the plant has been consumed. Clinical symptoms are frequently not observed until the last week or two of life. On autopsy, affected animals have an enlarged, hardened liver. In acute cases, other symptoms are nervousness, excitability, blood in the feces, loss of appetite, weakness, prostration, and death.

Controlling this problem of poisoning is based on keeping animals away from *C. spectabilis*. This species should not be included in cover crop plantings.

Black Henbane

Black henbane, *Hyoscyamus niger*, is about 3 feet high. It has dull yellow flowers, short, purple-veined stalks, and egg-shaped, serrated wavy leaves. It grows from the New England states to the Rocky Mountain region. Its alkaloids, hyoscyamine and hyoscine, are valuable in medicine as depressants. Poisoning in animals is indicated by loss of muscular power, stupor, and death from paralysis of the respiratory centers. Treatment is symptomatic.

Jimson Weed

There are several species of jimson weed, *Datura strammonium*, which are poisonous to cattle, horses, sheep, and swine. The plant has a strong offensive odor. It has purplish-white, funnel-shaped flowers. Jimson weed is common in the western and southwestern parts of the United States. The symptoms of poisoning are very sudden, and they include pain, vomiting, dizziness, convulsions, and death. Treat-ment is usually to no avail because of the extreme rapidity of the poison's development.

Sweet Clover

Sweet clover is a valuable forage and soil-building legume. Under certain conditions, however, consumption of the cured sweet clover plant is followed by harmful results, manifested largely by a loss of the clotting power of the blood. As a consequence, the blood gathers in considerable volume in different localized parts of the body, especially beneath the skin, so as to form soft swellings of varying sizes (Fig. 18-3). Animals affected with this condition that show no signs of swellings are likely to have serious bleeding following injuries or surgical operations, such as dehorning and castration.

Evidence indicates that the disease is due to a chemical poison known as dicumarol. This chemical is formed in the heating or spoiling process that occurs when hay is baled with too high a moisture content. The same reaction is possible as silage made from sweet clover is cured. A tobacco-like odor of the hay is an indication to discard it. Even the feeding of apparently clean, bright sweet clover hay is hazardous. Sheep and cattle are most susceptible to sweet clover poisonings. Sheep have reportedly died from eating green sweet clover.

Poisoned animals should be kept quiet and given blood transfusions. Intravenous or intraperitoneal injections of hemostatic solutions should be given to speed blood coagulation. Administration of vitamin

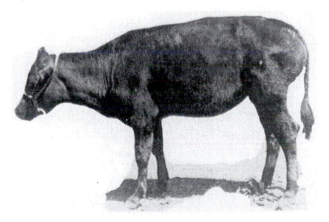

Fig. 18-3. Sweet clover disease. Note swellings over the animal's body. (Courtesy, North Dakota Agricultural Experiment Station, Bulletin No. 250)

K also will speed coagulation. Small or alternate feeding of alfalfa hay will help prevent losses.

Fern

The bracken or brake fern is poisonous to horses and cattle. It is a perennial, growing from 1 to 4 feet high. It has a scaly underground stem, and the leaves or fronds are three-parted or branched, which in turn are composed of many leaflets (Fig. 18-4).

Cattle and horses will eat this fern in the late summer when green vegetation is scarce. Both cattle and horses show a loss of appetite and great depression. Bloody discharges from the mouth and nose are common, and intestinal bleeding may occur in cattle. Death usually occurs in 12 to 72 hours.

Dry fern fronds in hay should be removed, or the hay should not be offered as feed to horses and cattle. Hogs seem to suffer no ill effects from eating the roots. The plant should be cut two or three times during the summer. Ferns grow only in acid soil; therefore, an application of lime to the roots will usually devitalize it or kill it.

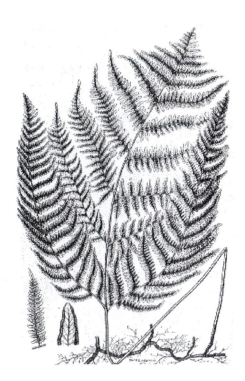

Fig. 18-4.   Bracken fern.

Larkspur

Larkspur is also known as "poison weed" or "cow poison." There are two general sorts: (1) the tall larkspurs—robust plants with coarse, erect, hollow stems, usually between 3 and 6 feet tall; and (2) the low larkspurs—smaller plants with slender stems, usually about 1 to 2 feet tall. The flowers are usually some shade of blue or blue-violet. A few species have red flowers. One of the most distinctive features is the spur that projects from the back of the flower.

Several poisonous alkaloids have been isolated from different species of larkspur, one of the most powerful being delphinine. The greatest stock losses occur early in the season before the plants bloom, while the young stems and leaves are actively growing. The young leaves are the most poisonous. As the plants mature, potency decreases. Research indicates that leaves from fruiting plants are only $\frac{1}{16}$ as poisonous as leaves from young plants. Fully mature plants are generally reputed to be relatively harmless, but livestock losses from tall larkspurs have been reported as late as September and October. Therefore, regarding larkspurs as being dangerous at all times would seem to be wise. The seeds are even more poisonous than the leaves, but they are probably rarely eaten in sufficient quantity to cause trouble. The low larkspurs are reportedly poisonous during the whole life of the plants, but since the tops have usually dried up and died by the early summer, they are dangerous mainly in the early spring.

Larkspurs cause the heaviest losses in cattle. Under range conditions, very few horses are lost, and sheep, with their greater tolerance, are probably only very rarely poisoned. However, it seems likely that large quantities or unusually potent plant material will poison most domestic animals.

The symptoms of poisoning are the same for all larkspurs. The alkaloids act on the nervous system, and they affect heart action, respiration, and muscular activity. Muscular twitching and a staggering gait may be observed, but, more commonly, the animals simply fall suddenly. After some struggling, they may rise to their feet, walk away, and show no further symptoms. More severely poisoned cases may fall repeatedly, and finally be unable to rise. Colicky pains and bloating may sometimes occur. Retching efforts to relieve paunch pressure may lead to vomiting, with the added danger of inhaling paunch contents. Increased salivation and swallowing are observed in some animals. Finally, the prostrated, but conscious, animals die from respiratory failure. Death or recovery usually occurs within 24 hours. No diagnostic changes are found at autopsy.

There is no simple effective treatment for larkspur poisoning. Management practices designed to keep the animals from eating sufficient quantities to cause damage are much more effective than the treatment of affected animals. Veterinary service is recommended when valuable animals are found in the early stages of the disease.

In general, poisonous plants do not grow in large numbers when the range is in good condition and supports an abundance of good forage species. Likewise, poisonous plants are apparently less palatable than most common forage species and are seldom eaten when other plants are abundant. Improper use of range, therefore, may result in an increase in the percentage of poisonous plants present, and may increase the opportunity for those plants to be eaten by the grazing animal.

Animals that are in a good state of nutrition seldom select poisonous plants for grazing. The provision of supplemental feed early in the grazing season, when the growth of range grasses is short, or later when the forage has become less palatable or has deteriorated following early rainfall, will help prevent losses from poisonous plants.

Nightshade

There are several specie of nightshades, *Solanum*, that are poisonous. Horsenettle, black nightshade, and silverleaf nightshade are the most common. Horses, cattle, sheep, and swine are susceptible to nightshade poisoning.

The signs of nightshade poisoning vary with the species of *Solanum*, but they usually include labored breathing, with an expiratory grunt, salivation, and nasal discharge. The temperature may be normal to slightly above normal. In subacute cases, a yellow dis-

coloration of the skin may be observed in lightly pigmented areas. Other signs observed are weakness and incoordination, trembling of the muscles of the hindlegs, anemia, and accelerated heart rate. Affected animals may bloat. They are often found dead with evidence of excessive salivation.

The leaves and fruit of the toxic species of Solanum contain alkaloids, especially solanine. Of the species included, silverleaf nightshade is the most poisonous and frequently causes extensive losses in cattle. Cattle will eat the green plant when there is an adequate supply of fruit, which is as toxic as the green fruit. In addition to the species listed, other plants in the nightshade family are known to be poisonous. The foliage of tomato and potato plants and silage made from potato plants have poisoned animals. When the potato tuber is exposed to light, it turns green. The green portion contains the poisonous element and should not be eaten.

Hungry animals should not be allowed to graze areas where there is an abundance of nightshade. Animals should not be fed from the ground in areas where there is a large quantity of the ripe fruit of nightshade or horsenettle. Most of the cattle that survive for 24 hours after eating nightshade will recover if they are placed in the shade, fed, watered, and kept clean.

Groundsel

Several species of groundsels or ragworts are poisonous to cattle, horses, and swine. Sheep are seldom poisoned by these plants. The species that are most commonly involved in animal poisonings include *Senecio longilobus*, or woolly groundsel; *Senecio riddellii*, or reddell groundsel; and *Senecio jacobaea*, commonly called ragwort. The toxic materials in these plants are a number of pyrollizidine alkaloids that inhibit cell division in the liver and cause hepatic cirrhosis.

Groundsel poisoning, seneciosis, occurs after animals have consumed these plants daily for a month or longer. The poison is cumulative, and the prognosis is usually unfavorable after clinical signs of poisoning appear. Groundsel is poisonous at any time of the year and at any stage of development. The leaves are more toxic than the stems, and young leaves are more toxic than older growth.

There is usually a time lapse between ingestion of groundsels and the first signs of intoxication. The initial signs of seneciosis may be confused with parasitism, malnutrition, or infectious diseases. They include anorexia, emaciation, incoordination, and constipation. The advanced stages of the disease are characterized by continuous, aimless walking, the sudden appearance of nervous disturbances, and tenesmus (a prolonged painful straining to void feces), which often results in a prolapse of the rectum. Constipation may change to liquid, bile-stained feces, but the amount voided remains small. In some cases, poisoned animals become quiet and lethargic; but, most often, affected animals move about continuously in a disorganized, ataxic manner. Such animals are likely to become vicious and attack any moving object that is near. Death may occur suddenly after the appearance of the advanced stage of seneciosis.

The treatment for seneciosis is not likely to be successful after clinical signs have appeared. All animals should be removed from pastures contaminated with groundsels, if possible, when seneciosis is diagnosed. Supplemental feeding of low-protein, high-carbohydrate rations is recommended as a good nursing practice for affected animals that have not progressed to the advanced stage of this disease. Groundsels, like most poisonous plants, are normally unpalatable to animals. Supplemental feeding during droughts or when pasture grasses are inaccessible to animals due to snow and ice is a good management practice to prevent plant poisoning in farm animals.

Tarweed

The tarweed or fiddleneck, as it is commonly called, is a contaminant of grain fields and pastures throughout the western part of the United States. The technical name for tarweed or fiddleneck is *Amsinckia intermedia*. It is an annual with rough, hairy stems and leaves. The flower stem is characteristically curled, and the light, yellow-colored flowers are arranged on only one side of the stem. The seeds, called nutlets, are covered with a hard, wrinkled shell. They range in color from dark brown to black and are about one-third the side of a kernel of wheat. Tarweed seeds are poisonous. *Amsinckia intermedia* contains a pyrollizidine alkaloid that causes liver damage in much the same way as groundsel poisoning.

Tarweed or fiddleneck causes hepatic cirrhosis of cattle, horses, and swine. This disease is called "Walla Walla hard liver disease" in swine, due to the dense fibrotic livers found in affected pigs and the common occurrence of tarweed poisoning in Walla Walla County in Washington.

The clinical signs of tarweed poisoning include anorexia, emaciation, incoordination, and constipation. The advanced stages of the disease are characterized by continuous, aimless walking, the sudden appearance of nervous disturbances, and tenesmus (a prolonged painful straining to void feces), which often results in a prolapse of the rectum. Constipation may change to liquid, bile-stained feces, but the amount voided remains small. In some cases, poisoned animals become quiet and lethargic; but, most often, affected animals move about continuously in a disorganized, ataxic manner. Such animals are likely to become vicious and attack any moving object that is near. Death may occur suddenly after the appearance of the advanced stage of tarweed poisoning.

Pigs are often poisoned by tarweed seeds that are harvested with grain. In one random sample, 1 pound of tarweed seed was found in 450 pounds of mill screenings. Hogs consuming the mill screenings were unthrifty, with rough hair coats, pot bellies, and pale icteric mucous membranes, but had normal temperatures. An autopsy revealed the livers of affected pigs were hard, atrophied, and distorted from the normal shape, and the peritoneums had an accumulation of fluid. When the screenings were discontinued as a part of the ration, no additional cases developed.

There is no know treatment for tarweed or fiddleneck poisoning. Prevention is based on management practices that encourage animals to eat plants other than tarweed and on the removal of seeds from feeds that are fed to livestock.

Yellow Star Thistle

Yellow star thistle, *Centaurea solstitalis*, causes a disease commonly called the chewing disease in horses. The poison is cumulative. It affects the central nervous system of horses that ingest large quantities of yellow star thistle for a prolonged period of time.

Yellow star thistle poisoning is characterized by rapid onset and by awkward tongue and lip movements that culminate with the animals' inability to eat or drink properly. Affected horses commonly engage in persistent chewing movements when there is no feed in their mouths.

There is no treatment for yellow star thistle poisoning.

Molds

Mold poisoning is frequently referred to in literature, but, except in rare and specific instances, the importance of molds and smuts as sources of danger to animals consuming them is greatly overestimated. Certainly, animals almost daily consume with impunity "smutty" corn, or other grains contaminated with stinking smut. However, the fact that many molds or smuts have a decided food value should not be overlooked.

Probably the best rule to remember with regard to moldy feeds is that they are generally classed as unwholesome and therefore are undesirable for animal consumption. Also, their possibly dangerous nature depends on the degree of moldiness. Judgment must be exercised by the feeder in this situation. Cattle are usually less susceptible to the possibly harmful effects of moldy feed than are horses. If the material in question appears very undesirable following a physical examination, it should first be fed to a test animals for 10 days or 2 weeks. Drying by direct sun exposure will also render moldy feeds safer for animal consumption.

Molds reduce the quality of feed. As previously stated, moldy feed is rarely toxic, but it should be used with caution. Economic consideration may dictate that attempts be made to utilize toxic moldy feed for livestock feed. The following suggestions and comments will help to avoid losses:

1. Recommended practices for harvest and storage should be followed to avoid mold growth.

2. Moldy feed should be destroyed, if the quantity and value of the feed are small.

3. Feeding moldy feedstuffs to any breeding animals should be avoided.

4. If signs of estogenic stimulation occur, such as prolapse, the feed should be withheld from animals of any age.

5. Feed unpalatable to younger stock may be acceptable and fed to older growing-finishing stock. Also, older animals are often less susceptible to the effects of mold toxins.

6. Unpalatable moldy corn may be fed to pigs after it has been washed. Previously unpalatable corn that has been rained on may become palatable. Also, some of the toxins are destroyed by sunlight.

7. The odor or taste of unpalatable feed can sometimes be masked with well-liked feed, such as liquid molasses (50 to 100 pounds per ton of complete feed).

8. The feeding value of questionable feeds should be determined by test feeding a small group of animals for at least two weeks.

9. If moldy feed causes problems and no other salvage or utilization can be made, the affected feed can be diluted with good feed and watched closely for signs of toxicity. A general recommendation is that moldy feed should not be used for more than 25 percent of the total feed (3 parts good feed to 1 part moldy feed). Feeding to older animals is preferred.

Ergot

Ergotism is an ailment due to the consumption of ergot or rye smut and of other smuts replacing the grains of grasses. It grows on all plants included under the general name of grasses, especially rye. It has the appearance of blackened kernels resembling oats in size, though somewhat longer and more decidedly curved in its long direction. It is used in both human and animal medicine because it possesses the valuable properties of stimulating involuntary muscles to contract, and it contracts the very small peripheral arterioles. It is this latter property that makes it harmful to

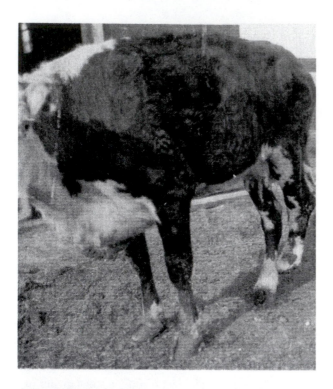

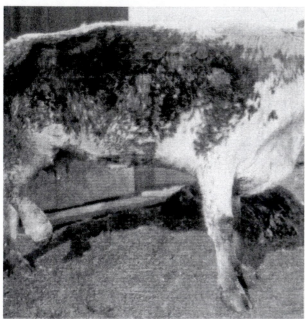

Fig. 18-5. Sloughing of the feet due to prolonged consumption of ergotized grain and grasses. (Courtesy, Kansas State University, Manhattan)

animals, particularly cattle and hogs, when it is consumed daily with grain or forage. It causes the small blood vessels in the extremities of the body to contract so that nutrition is interfered with and such parts die (they undergo dry gangrene).

Ergotism is divided into two different disease syndromes—acute, characterized by nervous signs, and chronic, characterized by gangrene. Signs of nervous ergotism are extreme nervousness, increased heart rate, muscular trembling, and frequent urination. There may be a loss of the five senses, followed by ataxia, prostration, convulsions and death.

Gangrenous ergotism is more often caused by *Claviceps purpurea* growing on some of the cereal grasses. Gangrene affects parts of the body having the poorest blood supply: feet, legs, tail, and ears, any or all of which may drop off (Fig. 18-5). Frequently, cattle are able to walk without hooves, apparently without pain. Abortion is thought to be a sign of chronic ergotism. Although research workers usually have failed to produce abortion with ergot, abortions are frequently associated with gangrenous ergotism in the field. Gangrene and abortions have been observed in herds following outbreaks of the nervous form of abortion.

Grass containing ergot should be mowed, or cattle should be moved. Animals with gangrene often have to be destroyed for humane reasons. Animals with nervous ergotism usually will recover if removed from the ergot-infested pasture, given good feed and water, and not unduly disturbed.

POISONOUS CHEMICALS

Nitrate

The term *nitrate poisoning* is commonly used, but in reality it is a misnomer. Nitrates are not very toxic. They are, however, readily converted into *nitrites* which are quite poisonous.

Probably most of the conversion of nitrate to nitrite takes place in the animal's gastrointestinal tract, although some field studies indicate that nitrite may already be present in the plants before they are eaten. Nitrite converts the hemoglobin in red blood cells to methemoglobin, which cannot transport needed oxygen from the lungs to the body tissues. Thus, animals affected with nitrate poisoning show general symptoms of oxygen deficiency.

Nitrate poisoning is often associated with careless handling of nitrate fertilizers. Empty fertilizer bags should be destroyed; fertilizer spreaders should be filled and cleaned in areas inaccessible to animals, and grazing of freshly fertilized fields should be restricted until the nitrates have had time to disperse into the soil.

Contaminated water is also a common cause of nitrate poisoning. Ponds that take runoff water from barnyards or fields which are heavily fertilized may have concentrations of nitrates or nitrites sufficient to poison animals. Seepage of nitrate materials into wells is another way water supplies can be polluted and cause nitrate poisoning.

There are a large number of plants known to produce nitrate poisoning. Oats are probably the worst offender of the cereal grains. Other field crops that have been known to cause nitrate poisoning are barley, wheat, rye, sorghum, corn, Sudan, alfalfa, millet, soybeans, rape, and fescue. Corn has been reported to have as much as 25 percent nitrate on a dry weight basis. Some grasses and weeds also accumulate nitrates. Examples include fiddleneck pigweed, sunflower, lamb's-quarter, Russian thistle, ragweed, witchgrass, nightshade, fescue grass, johnson grass, smartweed, Sudangrass, puncture vine, Canada thistle, bull thistle, bindweed, sourdock, stinging nettle, elderberry, goldenrod, and sweet potato vines. Vegetables accumulating large quantities of nitrate are beets, turnips, kale, radishes, mangels, Swiss chard, lettuce, celery, squash, parsnips, cucumbers, and spinach. Because they seldom eat sufficient quantities of these vegetables, human beings are not usually poisoned by them; however, when these vegetables are fed to animals in quantity they often cause nitrate poisoning.

The poisonous effects of nitrates vary with different species of animals and with the ration that the animals are eating. Because of these varying conditions, the amount of nitrate reported by different workers to produce poisoning varies widely. Under most conditions, 1.5 percent of the ration in nitrate on a dry weight basis produces acute poisoning and death, while lesser amounts will produce abortion.

One-half percent of the ration over an extended period will cause lowered milk production and weight gains.

Cattle, sheep, swine, and horses are all susceptible to nitrate poisoning. Cattle are more frequently poisoned than other animals. Horses apparently have more tolerance to this disease than cattle, sheep, or hogs.

The symptoms most frequently observed in nitrate poisoning are depression, weakness, rapid pulse, and respiration which is often very noisy and labored. Mucous membranes become dark in color. The recumbent animal may show convulsive movements of the legs. Death is due to asphyxia and, in acute cases, may occur within a few hours after the plant has been eaten. The animal's blood is often dark and sometimes chocolate brown. The fourth stomach of ruminants and the intestine are sometimes congested due to the direct irritating action of high levels of nitrates.

Cows that are not fatally poisoned may abort dead calves.

The following practices are recommended to prevent nitrate poisoning:

1. Animals should be kept away from nitrate fertilizers.

2. Stock water should be tested for nitrates and nitrites if pollution from barnyards or fields is occurring.

3. Weeds in pastures should be controlled.

4. Plants should be tested for nitrates and nitrites before animals are pastured if there is reason to suspect excess nitrates.

Animals poisoned by nitrates should be handled as little as possible since they are suffering from anoxia. The preferred treatment is methylene blue, administered intravenously. It is usually administered in a 1 to 4 percent solution, which also contains 5 percent dextrose, at the rate of 1 gram of methylene blue per 250 pounds of animal weight. The stock should be removed from the field in which they were poisoned and given straw, hay, or roughages known to be low in nitrates.

Lead

Lead poisoning occurs most often in cattle and calves, but all animals, including humans, are susceptible. Lead poisoning results from animals' licking freshly painted surfaces, paint cans, paint brushes, or discarded storage batteries. Drop cloths and papers saturated with paint are also common sources of lead poisoning. The toxic properties of lead are not reduced with age; therefore, lead-based materials of all ages should be kept a safe distance from all animals.

The symptoms of lead poisoning vary according to the amount of toxic material ingested, the age of the animal, and the species involved. The first signs of acute lead poisoning are salivation, slobbering, colic, loss of appetite, and a reduction of milk in lactating animals. The central nervous system is often affected. Animals may walk in circles and run into or over objects because of blindness. There may be seizures similar to those of epilepsy, pronounced trembling, and chomping of the jaws. Death may occur suddenly, during or following such an attack. The animals may show signs of paralysis, be unable to get up, and eventually become comatose before death.

The source of lead poisoning should be identified and removed from the area in which farm animals are housed or pastured.

Arsenic

Arsenic poisoning has been one of the major causes of intoxications in farm animals. The incidence of arsenic poisoning has been greatly reduced in the past few years as arsenicals have been replaced with improved insecticides and rodenticides. Arsenicals are, however, used at the present time as feed additives for swine, in some anthelmentics for sheep, and in some dusts or sprays for plants.

One or more animals will usually die suddenly. Other animals will show signs of trembling, staggering, weakness, incoordination, and prostration. Animals such as swine, cattle, and sheep which are capable of vomiting, may show this sign of distress. In some cases there is severe diarrhea tinged with blood and gastrointestinal mucosa.

Salt

Sodium and chlorine are required mineral elements in the diets of all farm animals. They are commonly expressed as salt requirements. Salt poisoning may occur in livestock and dairy cattle when excessive amounts are consumed. Excessive consumption of salt usually occurs when animals have been deprived of salt for a long period of time and then allowed access to an abundant supply. Symptoms of excessive salt intake may be profuse diarrhea in a group of animals that are bright, alert, and do not appear ill. The history is usually one of gross mixing errors of salt or engorgement of salt after a deprivation period. The ingestion of brine solutions is also a common means of salt poisoning.

The salt requirements for cattle, sheep, and swine are usually adequate when rations contain 0.25 to 0.5 percent salt. In addition, salt is often provided free choice to cattle and sheep. Therefore, intake is likely to exceed minimum requirements and will vary considerably.

Swine can suffer from salt poisoning even if the salt level in the ration is within acceptable limits. Salt poisoning occurs from a lack of water due to electrical failure or frozen water lines. The dehydration leads to staggering and coma in hogs, and often has a high mortality rate.

Range cattlemen use salt in a mixture with feed concentrates to control the intake of protein or grain supplements that are fed from a self-feeder. This type of feeding has been satisfactory, although the amount of salt consumed by the animals greatly exceeds their requirements. Salt poisoning may occur with this system of feeding on ranges where the water supply is limited or when the salt and protein supplement mixtures are fed at too great a distance from the water supply.

Symptoms first noticed in salt poisoning are hypersensitivity to touch, loss of appetite, marked redness and dryness of the mucous membranes of the mouth, and loss of coordination. Paralysis and death may follow.

The autopsy usually shows a distention of the rumen, slight congestion of the mucous membrane of the stomach, and marked edema along the lining of the first portion of the small intestines.

Prevention of salt poisoning is based upon the following practices:

1. Animals should not be allowed access to brine solutions because salt is more toxic in solution than in dry form. A cover should be provided and containers should be drained through holes in the bottom so that salt boxes can be kept dry.

2. Salt in swine rations should not exceed 0.5 percent.

3. Cattle and sheep should be provided with free access to salt at all times.

4. An ample supply of water should be provided for animals at all times.

The treatment of animals suffering from salt poisoning consists of providing small amounts of water at frequent intervals. Severely affected animals are often unable to drink. In such cases, water should be administered with a stomach tube. Mineral oil is also beneficial in protecting the mucous linings of the gastrointestinal tract.

Selenium

Selenium is a trace element required to prevent muscular dystrophy and to stimulate growth in farm animals. It is extremely toxic to all animals at levels beyond the maximum requirements.

Poisoning is caused by the ingestion of plants that have absorbed selenium from the soil. Seleniferous soils, derived from shale formations, have been found in many areas of the western states, principally in the region from the Rocky Mountains to South Dakota and Kansas.

Chronic selenium poisoning results when an animal ingests feed containing 5 to 40 p.p.m. of selenium for a few weeks to several months.

Alkali disease was the name formally applied to chronic selenium poisoning before the exact cause was determined. Symptoms of selenium poisoning are loss of hair from the mane and tail of horses, from the switch of cattle, and from the body of hogs. Rough horns, long and deformed hooves, and later sloughing of the hooves occur in all affected animals. Reproductive

performance may also be affected. Animals may be sterile or slow breeders.

The only practical means of preventing selenium poisoning in livestock is to avoid the use of seleniferous forage and grain. The feeding of a ration high in protein is reported to give some protection to sheep and laboratory animals against the toxic effects of selenium.

Acute selenium poisoning may result in death within a few hours or may follow a progressive sequence of ataxia, loss of vision, and anorexia. In the terminal stages of acute poisoning, the animals may display evidence of abdominal pain by grunting and grinding the teeth. The muscles used in swallowing are often paralyzed, resulting in excessive salivation. Breathing is difficult, and the animals' eyelids are so swollen and inflamed that the animals are almost blind. An animal's temperature usually remains normal during the early stages. Just before death the temperature usually drops below normal.

Molybdenum

Molybdenum poisoning often occurs in cattle and sheep that graze areas where molybdenum is present in excessive amounts in the soil.

Small amounts of the element appear to be beneficial in the growth of nitrogen-fixing bacteria found in the roots of leguminous plants. When excessive amounts are present in the soil, alfalfa and the clovers generally take up much greater quantities of molybdenum than the grasses in the same area and are therefore responsible for most of the poisoning that occurs.

Experimentally, feeding calves 2 to 3.5 mg of sodium molybdate per pound of body weight has produced poisoning. The presence of excess molybdenum in the animals' bodies interferes with copper metabolism, which results in copper deficiency.

Animals vary considerably in susceptibility to molybdenum poisoning. Young animals are more susceptible than older animals. Non-ruminants, such as horses and swine, are more resistant to molybdenum toxicity than ruminants, such as cattle and sheep.

When the molybdenum content of forage is approximately 10 p.p.m., it may take one to seven months of continuous feeding to produce the disease, while symptoms of molybdenum poisoning may appear 24 hours to a week after cattle are put on spring pastures, the forage of which contains 20 to 100 p.p.m. of molybdenum.

The clinical symptoms of chronic molybdenum poisoning are profuse diarrhea, emaciation, swollen vulva, marked anemia, general weakness and stiffness, fading of the hair coat, and occasionally death from severe and prolonged diarrhea. Male animals that escape death from severe molybdenum poisoning are usually sterile. Females often fail to come into heat or to settle. Breeding performance of affected females may be improved by proper treatment, which includes injections of coumol and the addition of cobalt and copper to the drinking water. Specific amounts of each of these materials should be determined by a veterinarian. Watering equipment used in conjunction with the cobalt and copper treatment must be non-metallic.

Mercury

Mercury compounds are highly toxic. Exposure to mercurial compounds is usually the result of carelessness on the part of the animal caretaker. Ointments and antiseptics containing mercury are corrosive and, therefore, highly irritating to the gastrointestinal tract if they are ingested. Animals are occasionally poisoned by mercurial compounds used as fungicides for seed treatment. Carelessness resulting in the admixture of treated seed with livestock feed can cause mercury poisoning.

The symptoms of the acute form of mercury poisoning are dramatic. The animal displays signs of great pain, nausea, and bloody diarrhea. This is followed by extreme weakness, irregular respiration, subnormal temperature, and death from shock. If death is prolonged, trembling will be observed, followed by convulsions, excessive salivation, and finally death due to exhaustion.

PART 4

Parasitology

CHAPTER 19

Introduction to Parasitology

Parasitology is a branch of biology dealing with organisms that live on or in other organisms. The original meaning of the word *parasite* was: "situated along side of." The term referred to being near and possibly sharing food, but it had no reference to pathogenicity. That definition, though honorable in the broad meaning of the word *parasite*, has very little in common with current usage by livestock producers. A parasite, in practical husbandry terms, is an organism that lives at the expense of its host. *Ectoparasites* are parasites that live on the exterior parts of their hosts. *Endoparasites* live within the body of the animals they parasitize. Endoparasites treated in this section will be limited to roundworms and flatworms. (Bacteria, fungi, viruses, and protozoan parasites were discussed in earlier chapters.)

The expense that parasites extract from their hosts costs livestock and poultry producers in the United States 10 to 12 billion dollars each year.

Economic losses from parasitism are greatest in young animals (Fig. 19-1). While complete elimination of parasites is practically impossible, their damage to farm animals can be drastically reduced by sound management programs and the wise use of insecticides and anthelmintics. Prevention and correction of parasitism should be based upon a knowledge of the factors that affect the survival of parasites and their eggs in the environment. A better understanding of

Fig. 19-1. A group of heavily parasitized lambs.

these factors will enable livestock producers to contend with parasitism more effectively.

DAMAGE FROM PARASITES

Ectoparasites adversely affect their hosts in one or more of the following ways: (1) by damaging and irritating the skin; (2) by creating excitement, restlessness, and nervousness by their presence; (3) by transmitting infectious organisms; and (4) by sucking blood.

Endoparasites damage their hosts in the following ways: (1) by absorbing feed intended for their host; (2) by sucking blood and lymph from the animal; (3) by damaging tissues which may cause internal bleeding; (4) by creating mechanical obstructions in the bile ducts, intestines, or circulatory system; (5) by producing toxic substances; and (6) by providing entry for infectious organisms through damage to internal organs and tissues.

The mass migration of parasites through the liver, lungs, blood vessels, and in the abdominal and chest cavities can lead to very serious consequences. Damage to the stomach and intestines often causes the walls to thicken, ulcerate, fill with fluid, and thereby drastically reduce the animal's ability to digest feed. The results of this internal damage and disruption of an animal's normal body functions by parasites are often manifested by poor appetite, dehydration, unthriftiness, persistent diarrhea, progressive weight loss, rough hair coat, slow gains, emaciation, weakness, and anemia.

DIAGNOSIS OF PARASITISM

Control of internal parasites should logically begin before obvious symptoms appear, because these visible signs indicate that a great deal of tissue damage has already taken place. If animals are not gaining properly and appear unthrifty, parasites could be the cause. However, many other diseases cause similar symptoms, so an accurate diagnosis is essential before treatment.

The only accurate method available for the diagnosis of parasitism in live animals is the microscopic examination of feces for the presence of parasite eggs. The fecal examination will show the kinds of internal parasites that are present, and to some extent, the degree of infestation. When possible, post-mortem examinations should be performed to secure more accurate information about the kinds and degree of parasitism than is possible with fecal examinations.

Fecal Examination

Samples of fecal material can be examined for parasite eggs by the direct smear method. This technique is fast and easy, but unfortunately the parasite eggs are not concentrated to the extent they are in the flotation method.

Tap water or a physiologic saline solution can be used as the diluent. A drop or two of the dilution fluid should be placed on a glass slide and a small amount of fresh fecal material thoroughly mixed in it. The mixture should be thin enough to be transparent. A coverslip should be placed over the solution and the solution examined with a microscope. The presence of eggs is a positive diagnosis of parasitism. If no eggs are found, a more refined technique, the flotation method of fecal examination, should be used.

The Flotation Method

The flotation method of examining feces is based on the following facts: Worm eggs are about the same weight as water; when they are washed out of the feces and placed in a solution much heavier than water, the eggs float to the top, where they can be observed with a microscope.

A flotation kit can be purchased quite inexpensively from pharmaceutical companies. The McMaster's kit contains the required glassware with calibrations designed to simplify measurement procedures. It contains complete instructions for conducting the fecal examination.

The first step involves securing or preparing a solution with a specific gravity sufficiently greater than water to mix with the fecal sample, thus enabling the worm eggs to rise. There are several materials, including salt, zinc sulfate, magnesium sulfate, sodium nitrate, and sugar, which can be used. If salt solutions are used, the fecal material can be mixed directly into the solution, instead of first being mixed with water, as must be done with the sugar solution. Salt solutions have an advantage over sugar in floating eggs of flukes and tapeworms; however, they crystallize more rapidly and distort parasitic forms more readily than sugar.

A suitable sugar solution can be made by mixing 1 pound of sugar (454 grams) in 340 ml of water and by adding 10 ml of 5 percent phenol as a preservative.

Two grams of fresh, uncontaminated feces should be mixed with 30 ml of water. The fecal material should be thoroughly mixed into the water and then filtered through a fine screen or several layers of gauze. A test tube should be filled half full with the filtered fecal material and then enough of the sugar solution added to finish filling the test tube. This solution should be placed in a centrifuge at a low speed (1,000 to 2,000 rpm) for approximately five minutes. Then, one drop of the centrifuged material should be placed on a slide and examined under the 10X (16 mm) objective of the microscope.

In cases where a centrifuge is not available, a slide can be placed over a test tube that is filled completely full of the mixed sugar solution and the strained fecal material. This should enable the slide to be in contact with the mixture. The slide should be left in place for a couple of hours to permit the worm eggs to rise. So that the presence of parasite eggs can be determined, the slide should be lifted straight up and examined with a microscope.

When a McMaster slide is used, a well-mixed sample of the sugar and fecal solution should be withdrawn in a large bore pipette and transferred to the slide. The McMaster counting chamber should be filled. The preparation should be allowed to stand for two or three minutes to allow the eggs to rise to the top. Then, the slide should be placed on the microscope and the eggs within the grid counted. The number of eggs counted should be multiplied by 300 to determine the number of eggs per gram of feces.

Diarrhea is common in heavily parasitized animals. This results in a decrease in the concentration of worm eggs. A few eggs in the feces of a scouring animal probably means that there is more parasitism to

deal with than a moderate number in normal fecal material.

LIFE CYCLES

If it were not for the tremendous hazards that parasites encounter in reaching maturity, there would be no contest between the state of health and disease. The parasites would win the battle quite easily. Animal caretakers must know the life habits of these enemies and attack them when they are most vulnerable. The life cycles of most parasites can be grouped into the following types:

1. The first type of life cycle limits the parasitic phase to the outer surface of the host. Ectoparasites such as flies, ticks, keds, mites, and mosquitoes would fit into this category. (See Fig. 19-2.)

2. The second type of life cycle includes those ectoparasites that penetrate the animal body or cavities and damage tissue by migration or irritation. These includes the larval stages of the heel fly and nasal botfly. (See Fig. 19-3.)

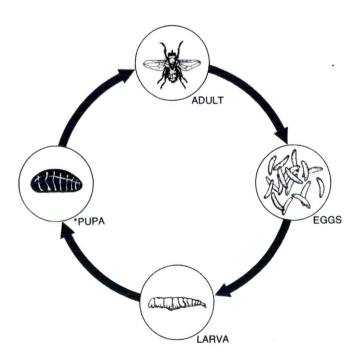

*Lice, mites, and ticks pass from larva to nymph and then to adult.

Fig. 19-2.   Life cycle of external parasites, type 1.

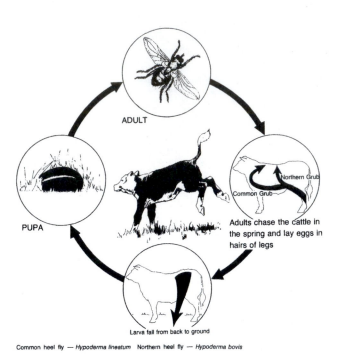

Common heel fly — *Hypoderma lineatum*   Northern heel fly — *Hypoderma bovis*

Fig. 19-3. Life cycle of external parasites with larval migration, type 2.

3. The third type of life cycle includes endoparasites that follow a rather direct route. The eggs are ingested and passed into the stomach or intestines where they hatch and mature. The adults remain in the digestive system during their entire life. They lay eggs which are passed in the feces to complete the cycle. Examples include *Haemonchus contortus*, *Ostertagia* species, *Trichostrongylus* species, and *Nematodius* species. (See Fig. 19-4.)

4. The fourth type includes those parasites that follow a direct route through the primary host but require the assistance of an intermediary host to complete their life cycle. Tapeworms are examples of this type of cycle. (See Fig. 19-5.)

5. The fifth type of life cycle includes those parasites that follow a route through the host that includes migration from the digestive tract through other parts of the body. Examples include ascarids and lungworms of sheep and cattle. (See Fig. 19-6.)

6. The sixth type includes those parasites that combine migration from the digestive system to other organs in the primary host's body and also require as-

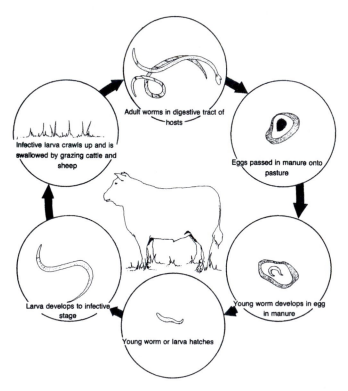

Fig. 19-4. Life cycle—direct route of internal parasites, type 3.

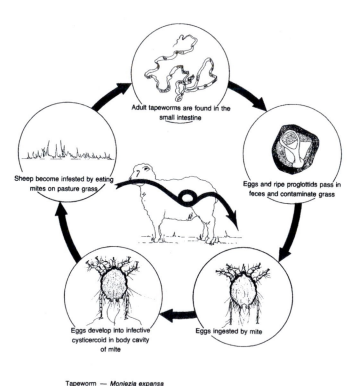

Tapeworm — *Moniezia expansa*

Fig. 19-5. Life cycle—internal parasites with an intermediate host, type 4.

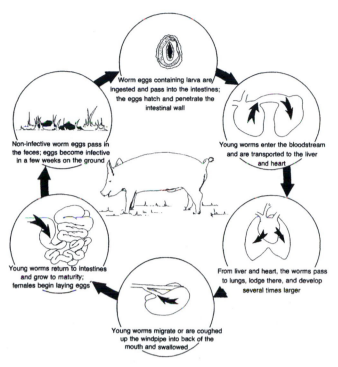

Large roundworm of swine — *Ascaris suis*

Fig. 19-6.   Life cycle—internal parasites with larval migration, type 5.

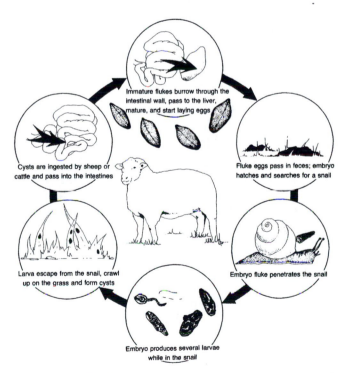

Liver fluke — *Fasciola hepatica*

Fig. 19-7.   Life cycle—internal parasites with larval migration and intermediate hosts, type 6.

sistance of an intermediary host to complete their life cycle. Examples include liver flukes and the lungworms of swine. (See Fig. 19-7.)

CONTROL MEASURES

Parasites of all types are more prevalent in those areas of the United States where temperature and moisture are both high. In addition to climatic conditions, poor management practices are conducive to parasitism. Unsanitary conditions around holding areas, overstocking and overgrazing pastures, improper nutrition, and pasturing young and older animals together are conducive to parasite problems.

Proper Nutrition

Good nutrition is one of the most important factors in a parasite control program. Research at the University of Georgia has demonstrated that supplemental feeding of calves on pasture is advantageous in reducing the danger of parasitism. (see Table 19-1). Yearling calves on pasture that were fed a supplemental ration of corn had higher daily gains, better carcass grades, and fewer nematode parasites than the control group.

Research also indicates that worm counts decrease as the quality of pasture increases. The Georgia station reported that a corn supplement significantly inhibited development of eggs and larvae of cattle nematodes in feces (see Table 19-2). Therefore, corn had a two-fold effect on parasitism. It increased the nutritional level to provide the host with the immunological tools to fight the invasion and reduced the seeding of pastures with infective larvae.

Preventing Overstocking

A concentration of animals on improved pastures, especially irrigated pastures, is an accepted practice. Increasing stocking rates naturally results in a build up in the worm population. The major problem arises, however, when overstocking results in overgrazing. As animals graze closer to the ground and the quality of pasture decreases, the problem of parasit-

TABLE 19-1. Effects of Two Types of Winter Pastures and Supplemental Feeding on Parasitism and Weight Gains of Yearling Calves[1]

| | Type of Pasture | | | |
| | Temporary | | Fescue | |
	No Corn	Corn	No Corn	Corn
Average daily gain, lb	2.56	2.88	3.09	3.33
Carcass grade	7.00	7.38	7.00	7.57
Average number worms	50,126	16,216	30,345	14,439

[1]Courtesy, University of Georgia, Research Report No. 166.

TABLE 19-2. Effects of Supplemental Feeding and Three Types of Winter Pasture on Parasitism and Weight Gains of Yearling Calves (Two-Year Average)[1]

| | Type of Pasture | | | | | |
| | Fescue | | Temporary | | Crimson Clover | |
	No Corn	Corn	No Corn	Corn	No Corn	Corn
Average daily gain, lb	1.39	2.29	2.65	2.66	2.39	2.35
Average number worms	75,250	21,000	13,500	3,500	24,250	7,500

[1]Courtesy, University of Georgia, Research Report No. 166.

ism increases. This is especially true during the warm, rainy season.

In extremely hot, dry weather, closely grazed pastures may contribute to the destruction of eggs and larvae of parasites due to a lack of protection from the sun. This in no way supports the practice of overgrazing. Research at the Georgia station clearly has shown that animals on an overgrazed pasture have more worms when slaughtered, lower daily gains, and carcasses that grade lower than those from a moderately grazed or undergrazed pasture.

Pasture Rotation

Pasture rotation has more benefit for animals grazed on permanent pastures than those grazed on temporary pastures. Evidently, plowing and discing the land for sowing the temporary pasture each year destroys a high percentage of the larval population. Permanent pastures should be divided into three or four sections and grazed on a monthly rotational basis. This practice,

of course, has more merit in warm, humid areas than in hot, dry parts of the country.

Preventing Mixed Grazing

Parasitism is more injurious to younger animals than to those reaching maturity. Older animals tend to develop some resistance to the parasites to which they are exposed. Young animals are, therefore, more susceptible and should not be pastured with older ones.

Sanitation

Many ectoparasites depend on manure piles and filth to hatch their eggs. In addition, the eggs and larvae of internal parasites are passed in the feces. It makes good sense, therefore, to remove manure and to keep it dried out or submerged in water where animals are concentrated in dry lots or holding pens. Earth mounds to improve drainage away from corrals on flat land will help to keep lots dry. Water troughs should be checked frequently to prevent overflowing, and feed bunks should be protected so that manure, pets, and people do not contaminate them.

Low areas in pastures should be drained and permitted to dry before animals are turned in to graze. Water sources should be maintained as sanitary as possible. Irrigated pastures should not drain into ponds that are used for stock water.

TREATMENT

When control through prevention is not practical or sufficient, animals and their surroundings should be treated for parasites. This includes spraying insecticides, using back rubbers or dust bags, and administering medications for internal parasites. Treatments are covered in Chapter 20, Parasites and Their Control.

RECOMMENDED PRACTICES FOR USING PESTICIDES

Indiscriminate use of insecticides, anthelmintics, antibiotics, and feed additives is not only a potential danger for the animals treated and the applicator, but also may result in economic losses due to the rejection of carcasses and milk at the market place. Serious violations of the Food, Drug, and Cosmetic Act could lead to litigation in the courts.

Livestock producers should *always read the label* on all materials they plan to use on their animals and follow the instructions to the letter. A record of detailed information should be kept to assure compliance with withdrawal requirements, to evaluate treatment practices, and to be used as a reference, should residue problems arise at a later date (see Fig. 19-8 and Fig. 19-9 for sample forms).

The following precautions are recommended for the use of insecticides on farm animals and in and around livestock housing areas. The applicator should:

1. Use the correct chemical on the specified livestock.

2. Use the specified amount, no more.

3. Use chemicals at the proper time.

4. Stop applications before sale or use to prevent excess residues.

5. Wear proper protective clothing and use an approved type of respirator for the chemical being applied.

6. Keep fresh filters and check respirators for proper function before each use.

7. Bathe with soap and water and change clothes following use of poisonous chemicals.

8. Read, understand, and follow precautions on container labels.

9. Destroy or decontaminate containers.

The following precautions are recommended for the use of feed additives and drugs in production schemes involving livestock. The applicator should:

1. Read drug labels carefully. Changes are frequently made in drug labels each year. These changes affect the way drugs should be used in treating animals.

2. Use drugs only in the animals species indicated on the label. Drugs meant for one kind of animal can cause adverse drug reactions or illegal drug residues in another species.

3. Always administer the proper amount of drug for the kind and size of animal being treated. Overdosing can cause drug residue violations.

4. Calculate the time of pre-slaughter drug withdrawal accurately. Remember, the time of withdrawal begins with the *last* drug administration.

5. Always use the correct route of drug administration. Giving oral drugs by injection can cause loss of drug effectiveness. Administering injectable drugs incorrectly can lead to adverse reactions, reduced effectiveness, illegal drug residues, and possibly the death of a fine animal.

6. Avoid "double-dosing" animals. Using the same drug in the feed supply and then by injection can cause illegal residues.

7. Keep an accurate record of the drugs used and properly identify animals receiving the drugs. Sending an animal to market too soon after it has been treated or shipping a treated animal because it wasn't properly identified can be a costly mistake. Good drug

LIVESTOCK INSECTICIDE USE RECORD

Producer's Name _____   Type of Livestock _____
                                             (poultry, dairy, beef, hogs, etc.)

Address _____   Year _____

_____   County Extension Office Phone _____

  This form provides a simple method of recording pesticide use on livestock, poultry, and in and around livestock housing. This information is valuable to establish proper intervals between treatments and between last use and marketing of meat, milk, or eggs. This record may also be useful if residue problems arise with products or meat. It can also be valuable in evaluating pest control programs.

KEEP THIS RECORD FOR REFERENCE.

Date of application	Name of chemical used	Rate of application	Percentage of active ingredients	Earliest date eggs, milk, meats, etc., may be marketed	Method of application (dust or spray)	Date marketed	Recommendation made by

Fig. 19-8.

LIVESTOCK FEED ADDITIVE AND DRUG USE RECORD

Producer's Name _____   Type of livestock _____
                                                                  (poultry, dairy, beef, hogs, etc.)

Address _____   Year _____

_____   County Extension Office Phone _____

This form provides a simple method of recording feed additive and drug use in livestock and poultry production. This information is valuable to establish proper intervals between treatments and between last use and marketing of meat, milk, eggs, etc. This record may also be useful if residue problems arise with products. It can also be valuable in evaluating certain practices.

KEEP THIS RECORD FOR REFERENCE.

Date used	Name of drug or feed additive used	Amount/ton of feed or animal	Date of last use	Earliest date meat, milk, eggs, etc., may be marketed	Date marketed	Recommendation made by	Purchased from

Fig. 19-9.

use records also help when professional animal health care is required. Veterinarians need to know what kinds of drugs and how much have been given before they can treat animals effectively and safely.

8. When injecting animals, select needles and injection sites with care. The wrong needle size, the wrong injection sites, or an inaccurate drug amount per site can result in tissue damage, reduced drug effectiveness, or illegal drug residues.

9. Have a reliable source of drug-free feed for animals to eat during withdrawal periods—feed which contains drugs can also cause illegal residues. Clean storage bins and feed troughs thoroughly before putting the withdrawal feed in them.

Chapter 20

Parasites and Their Control[1]

Animals are attacked by a wide variety of internal and external parasites, the prevention and control of which is one of the quickest, cheapest, and most dependable methods of increasing production with no extra animals, no additional feed, and little more labor. This is important, for, after all, the farmer or rancher bears the brunt of reduced meat, milk, and wool production, wasted feed, and damaged hides. It is hoped that the discussion that follows may be helpful in (1) preventing the propagation of parasites, and (2) causing the destruction of parasites through the use of the most effective wormer or insecticide.

ENDOPARASITES

All organisms that live inside a host and damage that animal are called endoparasites. These may include bacteria, viruses, fungi, protozoa, and helminths or worms. This chapter, however, will be limited to those endoparasites of two phyla of helminths that affect farm animals—Nematoda, which includes a vast number of roundworms, and Platyhelminthes, which includes flatworms. Two types of flatworms,

cestodes, known as tapeworms, and trematodes, commonly called flukes, are parasitic to farm animals.

Knowing what internal parasites are present within an animal is the first requisite to the choice of the proper drug, or anthelmintic. Since no one drug is appropriate or economical for all conditions, the next requisite is to select the right one; the one which, when used according to directions, will be most effective and produce a minimum of side effects on the animal treated. So, coupled with knowledge of the kind of parasites present, an individual assessment of each animal is necessary. Among the factors to consider are age, pregnancy, other illnesses and medications, and the method by which the drug is to be administered. Some drugs characteristically put animals off performance for several days after treatment, whereas others often have less tendency to do so. Some drugs are unnecessarily harsh or expensive for the problem at hand, whereas a safe inexpensive alternative would be equally suitable.

Each livestock establishment should, in cooperation with the local veterinarian and/or other advisor, evolve with a parasite control program and schedule. It is recommended that several different wormers be used, and that they should be rotated. Also, a schedule of treatments should be prepared based on knowledge of the life cycles of the various parasites.

Table 20-1 lists the common chemical compounds for the control of internal parasites, by species. Although this is a valuable guide, it is recognized that wormers are constantly being improved, and that new ones are becoming available. So, the producer should consult the local veterinarian relative to the choice of drug to use on the animals at the time.

---

[1]The material presented in this section is based on factual information believed to be accurate, but it is not guaranteed. Where the instructions and precautions herein are in disagreement with those of competent local authorities or reputable manufacturers, always follow the latter two.

The use of trade names of wormers and insecticides in this section does not imply endorsement, nor is any criticism implied of similar products not named; rather, it is recognition of the fact that farmers and ranchers, and those who counsel with them, are generally more familiar with the trade names than the generic names.

Note well: Cognizance is taken that, from time to time, FDA (1) bans the use of some old drugs, and (2) approves the use of new drugs.

TABLE 20-1. Recommended Compounds for Control of Internal Parasites, by Animal Species[1]

Wormer	Trade Name—Manufacturer	Cattle			Sheep									Swine								Horses								Wormer
		Coccidiosis	Gastrointestinal	Lungworm	Brown Stomach	Coccidiosis	Cooperias	Hookworm	Lungworm	Nodular Worm	Stomach Worm	Trichostrongyles	Whipworm	Ascarids	Coccidiosis	Kidney Worm	Lungworm	Nodular Worm	Stomach Worm	Threadworm	Whipworm	Ascarids	Bots	Pinworm	Stomach Worm	Strongyles, Large	Strongyles, Small	Tapeworm	Threadworm	
Albendazole	Valbazen	X	X	X		X	X	X	X	X	X																			Albendazole
Amprolium[2]	Amprole (Merck)	X				X									X															Amprolium[2]
Cambendazole	Camvet																					X		X		X			X	Cambendazole
Carbon disulfide																						X	X		X					Carbon disulfide
Coumaphos	Co-Ral, Baymix (Bayvet Corp.)		X		X		X	X			X	X	X																	Coumaphos
Dichlorvos	Atgard, Equigard, Equiger													X				X	X		X	X	X	X	X	X	X			Dichlorvos
Dithiazanine iodide and piperazine citrate	Dizan (Elanco)																					X		X			X			Dithiazanine iodide and piperazine citrate
Doramectin	Dectomax (Pfizer cattle and swine product)	X	X	X	X	X	X	X	X	X	X	X	X	X	X	X	X	X	X	X	X	X	X	X	X	X	X		X	Doramectin
Eprinectin	Eprinex (Merial cattle product only)	X	X	X	X	X	X	X	X	X	X	X	X	X	X	X	X	X	X	X	X	X	X	X	X	X	X		X	Eprinectin
Fenbendazole	Safe Guard (Hoescht)	X	X											X		X	X	X		X										Fenbendazole
Haloxon	Luxon, Loxon	X		X		X	X		X	X	X																			Haloxon
Hygromycin B	Hygromix (Elanco)													X				X		X										Hygromycin B
Ivermectin	Ivomec (Merck)		X	X					X	X	X			X		X	X	X		X			X			X	X			Ivermectin
Lead arsenate	Bi-forma (Texas Pheno Co.)																											X		Lead arsenate
Levamisole	Tramisol (Am Cyanamid)	X	X	X		X	X	X	X	X	X	X		X		X	X	X	X	X		X				X	X	X		Levamisole
Levamisole-piperazine																								X	X				X	Levamisole-piperazine
Mebendazole	Telmin (Pitman-Moore)																					X		X		X	X			Mebendazole
Morantel			X																											Morantel
Moxidectin	Cydectin (Ft. Dodge cattle product only), Quest (Ft. Dodge equine product only)	X	X	X	X	X	X	X	X	X	X	X	X	X	X	X	X	X	X	X	X	X	X	X	X	X	X		X	Moxidectin
Oxyfendazole		X	X																											Oxyfendazole
Parbendazole	(Helmatac)	X			X		X	X		X																				Parbendazole
Phenothiazine	(Du Pont)			X		X	X			X	X															X	X			Phenothiazine
Phenothiazine, low level	Pheno-Sweet (Farnam)																									X	X			Phenothiazine, low level
Phenothiazine-piperazine	Pheno-Pip (Haver-Lockhart)																					X				X	X			Phenothiazine-piperazine
Phenothiazine-trichlorfon	Equiverm (Texas Pheno Co.)																					X	X	X		X	X			Phenothiazine-trichlorfon
Piperazine	Wonder Wormer (Farnam)													X				X						X					X			Piperazine
Piperazine-carbon disulfide	Parvex (Upjohn)																	X						X					X			Piperazine-carbon disulfide
Piperazine-carbon disulfide-phenothiazine	Parvex-Pheno (Upjohn)																	X						X				X	X			Piperazine-carbon disulfide-phenothiazine
Pyrantel tartrate	Banmith (Pfixer)													X				X						X		X		X	X			Pyrantel tartrate
Thiabendazole	Thibenzole (Merck), Equizole (Merck)	X					X	X	X	X	X					X				X			X		X		X	X		X	Thiabendazole
Trichlorfon	Anthon, Bot-X (Farnam), Dyrex (Ft. Dodge)												X									X	X	X						Trichlorfon

[1] The products listed have 90% efficacy or more. This list is not complete. Inclusion of trade names does not imply endorsement.

[2] In the U.S., it is permitted in feed only as an aid in the control of coccidiosis of chickens and turkeys.

ECTOPARASITES

Ectoparasites of most importance to livestock and dairy producers are flies, lice, ticks, and mites. They rob animal owners of millions of dollars each year. Ectoparasites cause irritation, annoyance, energy loss, blood loss, tissue damage, and lowered vitality to their hosts. This results in stunted growth, decreased milk and meat production, loss of edible meat, and damage to hides. Some of these pests, such as lice, horse flies, and blowflies, may become so serious on individual animals that death occurs. In addition, flies and ticks can transmit anaplasmosis, pinkeye, and anthrax from infected animals to healthy ones.

Recommendations for the control of external parasites change quite frequently because of the development of new insecticides and modifications of state and federal regulations. The wrong insecticide not only could injure the animal but also could produce illegal residues in meat.

Before buying an insecticide, individuals should check for current registration for use on livestock and dairy animals. Farm chemical dealers, veterinarians, and local extension offices have access to the list of approved insecticides.

Table 20-2 lists the common chemical compounds for the control of ectoparasites, with attention to the formula available and the species of pest controlled.

PARASITES AND THEIR CONTROL

A summary of the common animal parasites and their control follows, with the parasites listed alphabetically be species under: Cattle Parasites, Sheep and Goat Parasites, Swine Parasites, and Horse Parasites.

Few dosages for the control of internal parasites are given; instead, users are admonished to *follow the directions on the label*. Also, few insecticides are suggested for the control of external parasites because of (1) the diversity of environments and management practices under which they occur, (2) the varying restrictions on the use of insecticides from area to area, and (3) the fact that registered users of insecticides change from time to time. Information about what is available and registered for use in a specific area can be obtained from the veterinarian, county agent, extension entomologist, or agricultural consultant.

The insecticide recommendations given for horses are for animals not used for human food. Where horses are to be slaughtered for human food, the tolerance levels and withdrawal periods given on the manufacturer's label should be followed with care.

Cattle Parasites

Blow Fly

The blow fly group consists of a number of species of flies all of which breed in animal flesh.

TABLE 20-2. Products for External Parasite Control

Product	Brand Name	Formula	Species Controlled
Amitraz	Taktic	liquid	ticks, mites, lice
Coumaphos	Co-Ral	powder, dust	horn flies, lice, ticks, grubs
Dichlorvos	Ravap	liquid	horn flies, lice, ticks
Fenthion	Lysoff, Spotton	pour-on	lice, flies, grubs
Permethrin	various	ear tag, pour-on	flies, mites, ticks, lice
Cyfluthrin	Cylence	ear tag, pour-on	flies, lice, ticks
Permectrin	various	liquid, pour-on	flies, lice, ticks, mites
Diazinon	various	ear tag	horn flies, ear ticks
Doramectin	Dectomax	injectable, pour-on	grubs, mites, lice
Ivermectin	Ivomec	injectable	grubs, mites, sucking lice
Ivermectin	Ivomec Pour-On	pour-on	horn flies, grubs, mites, lice
Moxidectin	Cydectin	pour-on	horn flies, lice, grubs

DISTRIBUTION AND LOSSES

Blow flies are widespread, but present the greatest problem in the Pacific Northwest, and in the southern and southwestern states.

SPECIES AFFECTED

All farm animals are affected.

SYMPTOMS AND SIGNS

Affected animals have infested wounds and soiled hair. Maggots spread over the body, feeding on the skin surface, producing severe irritation, and destroying the ability of the skin to function. Infected animals rapidly become weak, fevered and unthrifty.

TREATMENT

When animals become infested with blow fly maggots their wounds should be treated directly with coumaphos (Co-ral) spray bomb, used according to manufacturer's directions on the label.

PREVENTION AND CONTROL

Eliminate the blow fly by destroying dead animals by burning or deep burial.

REMARKS

Damage inflicted is similar to that caused by screwworms, and the treatment for both types of maggots is the same.

Cattle Tick Fever

Cattle tick fever is an infectious protozoan disease of cattle caused by a one-celled protozoa called *Babesia bigemina*, which depends upon the tick for transmission and survival.

DISTRIBUTION AND LOSSES

Cattle tick fever is confined to the Gulf Coast area.

Death occurs in about 10 percent of the chronic and 90 percent of the acute cases. Infected young animals are stunted, mature animals are emaciated, and the milk flow of infected dairy animals is greatly reduced.

Death losses are much higher in countries where anaplasmosis also occurs in cattle.

SPECIES AFFECTED

Cattle, especially adults, are susceptible.

SYMPTOMS AND SIGNS

High temperature, rapid breathing, enlarged spleen, engorged liver, pale and yellow membranes, and red to black urine are all symptoms of cattle tick fever.

TREATMENT

Successful treatment of sick animals depends upon early recognition of the disease and prompt treatment. Agents traditionally used include trypan blue, trypaflavine, and quinuronium sulfate.

PREVENTION AND CONTROL

Avoid contact with the cattle fever tick, the only natural agent by which cattle tick fever is transmitted. Treat animals at regular intervals with a suitable insecticide.

REMARKS

The protozoa invade the red blood cells of cattle. The parasite is transmitted to cattle by ticks and is carried over in ticks by egg transmission.

Although immune, recovered animals are permanent carriers of the disease.

In infected areas, native cattle are either immune or only slightly affected. With the exception of a few possible areas along the Gulf Coast, the cattle fever tick has been eradicated—thus controlling tick fever.

Coccidiosis

Coccidiosis is a parasitic disease caused by protozoan organisms known as coccidia.

DISTRIBUTION AND LOSSES

Losses occur worldwide.

There is lowered gain and production in infected animals, along with some death losses. It is most severe in calves.

SPECIES AFFECTED

Cattle, sheep, goats, pet stock, and poultry are susceptible.

Each class of animals harbors its own species of coccidia; thus, there is no cross infection between animals.

SYMPTOMS AND SIGNS

Diarrhea and bloody feces, and pronounced unthriftiness and weakness are symptoms of coccidiosis.

TREATMENT

Amprolium (Amprole) and a decoquinate are effective in treating coccidiosis. They are approved for use in beef and dairy calves, chickens, and turkeys.

PREVENTION AND CONTROL

Avoid feed and water contaminated with the protozoa that causes the disease. Segregate affected animals. Remove and properly dispose of manure and contaminated bedding daily. Drain low, wet areas. Keep animals in a sunny, dry place.

Milk replacers, minerals, and feeds containing coccidiostats all can be used in beef calves and dairy replacement heifers to increase the rate of gain and to prevent coccidiosis.

REMARKS

In the oocyst stage, the parasite may resist freezing and certain disinfectants and may remain viable

Fig. 20-1. Individual feeding, strict sanitation, and the use of feed additives can prevent coccidiosis in calves.

outside the body for months, but it is readily destroyed by direct sunlight or complete drying.

Most cattle outbreaks are among dairy calves and feeder calves.

Flies

Several species of flies attack and annoy cattle, with the most economically important species being *Haemotobia irritans*, the horn fly, and *Musca autumnalis*, the face fly. Other species are Culicoides, or biting midge, buffalo fly, and the house fly, *Musca domestica*.

BITING MIDGE

Biting midge (ceratopogonid no-see-um, *Culicoides variipennis*) is common in the United States. It is a small (<0.5 mm) fly that is usually brownish, and has a hump-backed appearance similar to the black fly. Wings are often patterned. It is a blood-feeder which often leaves a bloody lesion at the site of the bite.

Distribution and Losses

The biting midge occurs worldwide, in both warm and temperate climates. Production losses result from blood-feeding of the flies. Biting midges also transmit the virus that causes bluetongue in cattle, sheep, and some wild ruminants.

Preadult stages are found most often in boggy areas near bodies of water—streams or ponds. The larvae prefer conditions where there is a large amount of

Fig. 20-2. Good fly control will keep animals productive and healthy.

organic matter from the feces and urine of farm animals.

Species Affected

All classes of livestock and humans are affected.

Symptoms and Signs

Biting midge flies cause production losses, restless animals, and large, weeping, crusting lesions on the skin where large numbers of flies have fed.

Treatment

Treat areas where larvae are found with insecticide.

Prevention and Control

Stabilize banks of bodies of water to reduce the habitat of the larvae. Spray farm animals with insecticide.

BLACK FLY

The black fly (buffalo gnat, no-see-um) is a small (<0.5 mm), dark-colored fly, with a hump-backed appearance. It is a blood-sucker that leaves a small, bloody lesion at the site of feeding.

Distribution and Losses

Losses occur worldwide, in all climates, and result from irritation of animals, blood loss, and transmission of some animal diseases in the tropics.

Species Affected

All farm animals and humans are affected.

Symptoms and Signs

Restlessness; production losses; and swollen, weeping lesions on the skin are symptoms of black fly infestation.

Treatment

No satisfactory treatment is available.

Prevention and Control

Black flies develop in well-aerated, running water. Flies often emerge in hoards in the spring, and attack both animals and humans.

Prevention consists of treatment of streams with insecticides.

Remarks

Sometimes black flies occur in such large numbers that cattle are smothered by the onslaught.

Larval development often occurs at irrigation structures and downstream from dams.

FACE FLY

The face fly (*Musca autumnalis*) was first found in this country in New York in 1953. It is a close relative of and similar in appearance to the house fly.

Distribution and Losses

Face flies are most prevalent in the eastern United States, along waterways, irrigated pastures, and areas with greater than 30 inches of rainfall.

Fig. 20-3. Face fly. *Musca autumnalis*.

Face flies have been linked to severe outbreaks of infectious keratoconjuncitivitis, or pinkeye, as flies are a vector for transmission of *Moraxella bovis*. Flies are attracted to calves and adults, so dust bags and oilers must be positioned to treat calves as well as adults.

Species Affected

The face fly is primarily a pest of cattle, although it also attacks horses and sheep.

Symptoms and Signs

The face fly does not bite, but its habit of clustering around the eyes, mouth, and nostrils is extremely annoying to animals, interfering with their vision and breathing, and preventing normal grazing. Large populations force animals to leave pastures and seek relief in wooded areas and shelters.

Treatment

Insecticide impregnated ear tags or insecticide boluses may be used.

The following insecticides are effective for face fly control:

Crotoxyphos plus	Fenvalerate (Ectrin)
Dichlorvos (Cio Vap)	Flucythrinate (Guardian)
Dichlorvos (Vapona)	Permethrin

Prevention and Control

Prevention consists in scattering or removing fresh cow manure.

Remarks

When cattle enter a barn or darkened area, the fly leaves the animal's face and rests on fence posts, gates, sides of barns, etc. The adult fly hibernates in attics and other protected places during the winter.

The face fly lays its eggs in fresh manure, where the larvae develop.

Insecticides should be applied in keeping with the directions of the manufacturer.

HORN FLY

The horn fly (*Haematobia irritans*), which is about one-half the size of an ordinary house fly, is one of the most numerous and worst annoyances of cattle.

Distribution and Losses

Stress from horn flies, even at low populations of less than 100 flies per animal, reduces milk yield and average weight gain. Horn flies feed exclusively on blood, and the damage to the United States cattle industry is estimated to be $750 million annually. The cattle will spend much of their time trying to avoid flies. Significant reductions of 0.5 pound daily gain are reported in fly infested beef herds. Milk production can drop in beef and dairy cattle as the cattle seek water and shade and fail to graze.

Species Affected

Cattle and horses are affected.

Symptoms and Signs

Horn flies are always present on the host, and are found on the back, belly, and base of the horns. In extremely hot, sunny, or rainy weather, the flies migrate to the belly. The flies aggravate the cattle and they will stand in water, congregate together, or enter brush and weeds to seek relief.

Tormented cattle often refuse to graze during the day and seek protection by hiding in dark buildings, brush, or tall grass.

Heavily infested cattle may also have rough, sore skin and suffer an inevitable loss in condition.

Treatment and Control

There are several treatment options to control horn flies on cattle. Care must be taken to read the drug/insecticide label to insure that the proper withholding times for meat and milk are observed. For cattle that are handled frequently, frequent applications of fast-acting insecticides, like pyrethrins, are effective. These products may be applied by sprays, dusts, or pour-ons. Avermectins can be used for horn fly control up to 28 days in the pour-on formulation.

Self-treatment devices (like dust bags or back rubbers) and sustained release devices are popular for treatment and control. Insecticide ear tags are the most popular sustained release devices. Formulations in the plastic ear tags are synergized pyrethroid, pyrethroid, or organophosphates. Horn fly resistance to pyrethroids has been noted to occur. Current recommendations to reduce the resistance of horn fly to control products are to remove the tags from the ears of the cattle at the end of the fly season, use an avermectin product to kill insecticide-resistant flies at the end of fly season, and rotation of fly tag types. In areas where pyrethroid resistance is known, using organophosphate tags for two years followed by one year of pyrethroid tags has been helpful.

Insecticide formulations can also be included in feed, minerals, salt blocks or as sustained release boluses that are administered to the animal with a balling gun. These insecticidal formulations kill the developing larvae in the manure. The larvicide program needs to be started before the fly season, as it will have no effect on the adult flies that feed on the cattle.

Horse Fly and Deer Fly

Horse flies and deer flies are biting flies that attack cattle. The two most troublesome genera are *Tabanus* (horse flies) and *Chrysops* (deer flies).

Distribution and Losses

Tabanids are found in all parts of the United States, and large numbers may be expected wherever there are extended areas of permanently wet, undeveloped land and a mild climate. Generally, horse flies are more of a problem to livestock than deer flies, but deer flies are often extremely annoying in the coastal areas of the South and the mountain areas of the West.

In many areas, tabanids are the principal source of *Anaplasma* transmission between cattle.

Species Affected

Cattle and horses are most often affected. These flies are less common on other animals.

Symptoms and Signs

The bite from the slashing mouthparts of these insects is very painful, and animals try to dislodge the fly with their tail or tongue or by stamping their feet. Heavily attacked animals stop grazing and tend to bunch together or seek shelter. Severe outbreaks can seriously affect weight gain.

Treatment

For both beef and dairy animals, the following insecticides are effective for horse flies, deer flies, stable flies, and mosquitoes:

Crotoxyphos plus   Fenvalerate (Ectrin)
Dichlorvos (Cio Vap)   Permethrin

Prevention and Control

If possible, avoid pasturing cattle near swampy wooded areas when these flies are numerous. Also, sheltering animals is often beneficial since tabanids do not ordinarily enter enclosures.

Remarks

Horse flies are also implicated in disease transmission because their habit of feeding on one animal and immediately attacking another can result in the direct mechanical transfer of pathogenic organisms that live in blood.

House Fly

House flies (*Musca domestica*) are nonbiting flies that are common around barns and lots.

Distribution and Losses

House flies become numerous both inside and outside barns and farm buildings. Perhaps they are the most abundant insect pest of feedlots and confinement livestock housing. House flies are annoying to livestock and people, and they can spread human and animal diseases.

Species Affected

Cattle, sheep, swine, horses, and poultry are affected.

Symptoms and Signs

Although house flies are nonbiting, they cause serious economic losses through annoyance of livestock and by disease transmission. Also, they create public health problems.

Treatment

Several insecticides in fogs, mists, surface sprays, or baits may be used. The following are approved:

Fog, mist, or surface spray:

Dichlorvos (Vapona)
Naled (Dibrom)

Fog or mist:

Pyrethrins

Surface sprays (with animals removed):

Dimethorate (Cygon)
Fenvalerate (Ectrin)
Permethrin
Rabon

Prevention and Control

Insecticides alone will not control house flies. Adequate sanitary measures, including proper disposition or handling of manure, are necessary to eliminate fly breeding areas. Spread manure thinly in fields so fly eggs and larvae will be killed by drying and heat.

Remarks

House flies breed in manure, garbage, and decaying vegetable matter. The eggs hatch after an incubation period of 12 to 36 hours. The larvae feed on the organic medium and grow to full size in 6 to 11 days.

STABLE FLY

Stable flies (*Stomoxys calcitrans*) are bloodsucking flies that are similar to horn flies, but they are larger. They are strong fliers and can migrate over 20 miles.

Distribution and Losses

The stable fly is responsible for over $400 million in losses in the United States. The flies breed in decaying organic matter, and spilled feed mixed with dirt and manure are ideal breeding areas in feedlots or

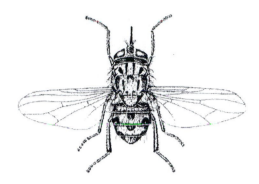

Fig. 20-4.   Stable fly. *Stomoxys calcitrans.*

dairies. The control of stable flies is difficult because they feed on the lower legs and migrate.

Losses occur in all temperate regions of the world and can include (1) decreased gains in beef cattle, (2) lowered milk production in dairy cattle, (3) possible transmission of certain diseases and parasites, and (4) gains and/or milk production being lowered by as much as 50 percent in seasons when the number of flies becomes large.

Species Affected

All farm animals and humans are affected.

Symptoms and Signs

Fly-fighting and restlessness are symptoms of stable fly presence. In seeking natural protection, animals frequently resort to mudholes, brush, etc.

Treatment

With stable flies, insecticides should be used as a supplement to good sanitation, rather than as the principal method of control, because alone they may not do a satisfactory job. Residual sprays are the most effective method of treatment. They should be applied inside and outside barns and other farm structures where stable flies rest. The spray may be applied by spray gun or by fogging or misting devices. Care should be taken to prevent contamination of feed and drinking water, and animals should be removed from the area during the spraying. Application of insecticides to the cattle may afford only temporary relief. The preferred method of application is spraying since the insecticide should be applied to the legs and lower body of the animal.

The recommended insecticides for both beef and dairy animals are the same as approved for horse flies, deer flies, and mosquitoes. (See "Horse Fly and Deer Fly.")

Prevention and Control

Control of stable flies by direct application of insecticides to cattle is usually not satisfactory. They are best controlled by sanitation and by application of insecticides to the resting surfaces. Sanitation is the most effective method of controlling stable flies in such areas as feedlots and barnyards because it breaks the cycle by removing the breeding sites. Barnyards and feedlots should be well drained; manure and decaying organic matter should be removed from inside and outside buildings and disposed of weekly or more often, if possible, by spreading it out to dry (this kills developing larvae). If manure cannot be spread, it should be placed in compact piles where the surface will dry quickly and become unattractive to females.

Remarks

Sprays do not provide the control for stable flies that is achieved with the use of sprays for horn flies, as walking through wet vegetation or water washes the chemical off the animal. Dust bags and oilers are not effective because the chemical does not get on the legs, where the flies feed.

Gastrointestinal Worms (Nematodes)

Ostertagia ostertagi is the most economically significant and pathogenic gastrointestinal parasite in the temperate regions of the world, including most areas of the United States. Other species of gastrointestinal parasites include *Cooperia, Trichostrongylus,* and *Bunostomum,* but they are less likely to be the cause of severe production losses.

LIFE CYCLE

The eggs of *Ostertagi* spp are passed in the feces and may hatch and develop into infective larvae within one week in optimal conditions. The larvae migrate from the feces to plants that are grazed by cattle. The ingested larvae penetrate into the glands of the stomach where they complete the life cycle and emerge as adults in 21 days. In many regions with hot, dry summers, or alternating wet and dry periods during the summer, pasture contamination and infectivity decreases. Areas with mild winters can have very high larvae numbers and high transmission of infection during winter. In northern regions of the United States, major accumulations of infective larvae occur in the summer.

CLINICAL SIGNS

The larvae encysted in the abomasum destroy the cells that produce hydrogen chloride, which raises the abomasal pH from 2 to 7. The elevated pH results in the inability to digest protein, leading to diarrhea and ill-thrift. Appetite suppression is a significant effect of the parasitism, which is another important role in the production losses observed. Decreased weight gain, decreased milk production, immune suppression and susceptibility to respiratory disease are all significant economic effects of parasitism.

Anorexia, weight loss, diarrhea, anemia, submandibular edema, and mortality are additional clinical signs.

Subclinical disease is manifested by reduced weight gains in stocker cattle and replacement heifers. Adult cattle can develop immunity and not show a consistent benefit from deworming.

CONTROL STRATEGIES

"Blanket" recommendations where all herds are treated the same, with the same products, is not applicable due to the seasonal incidence of the disease for the northern and southern climates, the sex-related susceptibility to worms, and the acquired immunity that can develop with age. Bulls are more susceptible to worms than steers, which are more susceptible than heifers. Herd owners are encouraged to consult with a veterinarian to develop deworming programs tailored to the individual farm.

Pasture Conditions

Non-immune cattle grazing lush pastures are at the greatest risk of acquiring infections. Warm, wet conditions favor the development of infective larvae, and drought conditions are conducive to larval mortality. Once a pasture is highly infective, the larvae can survive long periods. In the southern United States, larvae persist through the fall, winter and spring. In the northern United States, larvae commonly survive from one year until the early summer of the next year.

Treatment

Most modern dewormers are highly efficaceous for gastrointestinal worms, and some of them are effective in controlling lungworms. A variety of delivery systems are available for administering dewormers, and the individual animal administration through pour-ons, injectibles, oral formulations, or the rumen injector, are preferred. The advantage of treating every individual animal is to be sure that each animal receives the correct dosage, therefore achieving maximum worm kill. Medicated feed additives, minerals, or feed blocks may have variable consumption of the drug, leading to an increased risk of drug-resistant worm populations. The producer should consult with a veterinarian regarding dewormer products with regards to meat and milk withhold times, and the ability of the dewormer to control larvae that are encysted in the abomasal glands.

Cattle Grubs

Cattle grubs are the maggot stage of honeybee-like insects known as heel flies, warble flies, or gad flies. There are two species of cattle grubs in the United States—the common cattle grub *Hypoderma lineatum*, and the northern heel fly, *Hypoderma bovis*.

Distribution and Losses

Losses occur throughout the United States and include: (1) decreased gains or milk production, mechanical injury, and even death; (2) carcass damage; (3) shock to animals; and (4) downgrading of hides.

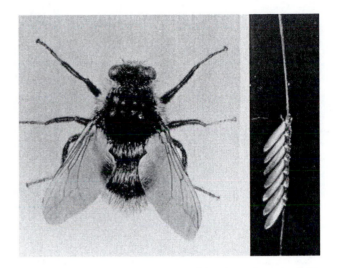

Fig. 20-5. Adult female heel fly, *Hypoderma lineatum* (left), and her eggs (right). The eggs of the northern heel fly, *Hypoderma bovis*, are deposited singly. (Courtesy, U.S. Bureau of Entomology and Plant Quarantine)

Species Affected

Cattle are affected.

Symptoms and Signs

The adult flies are relatively large, similar to honeybees and are called "heel flies." The northern cattle grub lays its eggs one at a time as it flies around the animal, which can cause cattle to run widly with their tail in the air. The eggs hatch in several days and the larvae penetrate the skin and migrate through the tissues of the animal for a period of 4 to 6 months. The larvae of the northern cattle grub spend 2 to 4 months in the epidural fat of the vertebral column. The larvae of the common cattle grub spend 2 to 4 months in the submucosa of the esophagus. Both species migrate to the tissue in the mid-back region, where they cut breathing holes in the skin. Extensive hide and carcass damage results from breathing holes and jellied tissues in the back when the animals are slaughtered.

Treatment

Apply a systemic insecticide to cattle as soon as possible after the activity of the heel fly ceases since

Fig. 20-6. Ox warbles being pressed out of the subcutaneous tissue of a steer. (Courtesy, Oklahoma Department of Agriculture)

these insecticides kill the young larvae in the animal's body.

When the grubs are near the back or located in the back, treatments are less effective, and possible side effects are more likely. Side effects may also occur when there is a concentration of grubs in the gullet or spinal cord of treated cattle. A single treatment with a systemic insecticide should give excellent control of cattle grubs. For the correct timing, each owner is advised to check with the local veterinarian, county agent or consultant. Systemics may be administered as sprays, dips, injectables, or as feed additives. Never use more than one systemic insecticide at a time, and al-

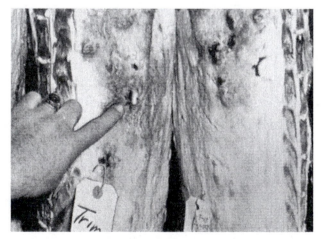

Fig. 20-7. High-priced cuts of meat must be removed because of grub damage to the carcass. (Courtesy, Oklahoma Department of Agriculture)

ways use a systemic in keeping with the manufacturer's directions.

For beef and nonlactating dairy animals, the following insecticides are effective:

Coumaphos (Co-ral)   Ivermectin (Ivomec)
Famphur (Warbex)   Prolate (GX–118)
Fenthion (Tigavon)   Trichlorfon (Neguvon)

PREVENTION AND CONTROL

Cattle grubs can be controlled by the use of systemic insecticides applied as soon as possible after the adult fly season. Organophosphate insecticides are available in spray, pour-on, or injectible formulations. Avermectins in the injectible, sustained release bolus, or pour-on formulations are also effective against cattle grubs.

Larvae that are killed while in the tissues of the esophagus or spinal canal can lead to bloat, salivation, or paralysis. These reactions are most severe around 24 hours post treatment, and subside in 72 hours.

REMARKS

The cattle grub or heel fly is probably the most destructive insect attacking beef and dairy animals.

Follow the label instructions on insecticides carefully, including the minimum intervals to slaughter and freshening (dairy).

Lice

Lice are small, flattened, wingless insect parasites of which there are several species, most of which are specific for a particular class of animal.

DISTRIBUTION AND LOSSES

Losses are widespread. Lice retard growth, lower milk production, and produce unthriftiness.

SPECIES AFFECTED

Lice affect cattle (with other species for other classes of animals).

 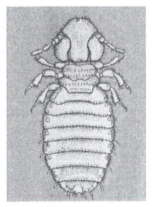

Fig. 20-8.   Blood-sucking louse.   Fig. 20-9.   Biting louse.

Symptoms and Signs

Symptoms include intense irritation, restlessness, and loss of condition. There may be severe itching and the animal may be seen scratching, rubbing, and gnawing at the skin. The hair may be rough, thin, and lack luster; and scabs may be evident. Lice are apt to be most plentiful around the root of the tail, on the inside of the thighs, over the ankle region and along the neck and shoulders. One type sucks blood and may cause the animal to become anemic.

Treatment

For effective control, all members of the herd must be treated simultaneously at intervals, and this is especially necessary during the autumn months about the time they are placed in winter quarters. Cattle

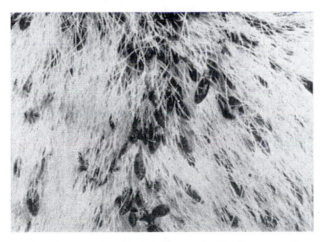

Fig. 20-10.   Bloodsucking lice on cattle. (Courtesy, Cornell University College of Veterinary Medicine, Ithaca, N.Y.)

should be inspected for lice periodically throughout the winter and spring and retreated when necessary.

For beef animals, the following insecticides are effective for lice control.

Coumaphos (Co-ral)	Lindane
Dioxathion (Delnav)	Malathion
Famphur (Warbex)	Rabon
Fenvalerate (Ectrin)	RaVap
Ivermectin (Ivomec)	Trichlorfon (Neguvon)

For both beef and dairy animals, the following insecticides are approved for lice control:

Amatraz (Taktic)	Permethrin
Cio Vap	Rabon
Coumaphos (Co-ral)	

Prevention and Control

Because of the close contact of cattle during the winter months, it is practically impossible to keep them from becoming infested with lice.

Remarks

Lice show up most commonly in winter and on ill-nourished and neglected animals.

Self-applicating devices can aid in louse control.

Liver Fluke

The liver fluke (*Fasciola hepatica*) is a flattened, leaflike brown worm, usually an inch long.

Distribution and Losses

The liver fluke is found wherever there are low-lying wet areas and suitable snails (worldwide).

Lowered gains, decreased milk production, and feed inefficiency are the chief losses attributed to liver flukes. In addition, vast quantities of liver are condemned each year at the time of slaughter.

Species Affected

Cattle, sheep, goats and other animals are affected.

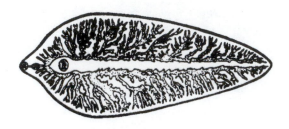

Fig. 20-11. Enlarged drawing of a liver fluke. The actual size is ⅛ to ½ inch wide.

SYMPTOMS AND SIGNS

Anemia, as indicated by pale mucous membranes, digestive disturbances, loss of weight, and general weakness results from liver fluke infestation. Severe liver damage occurs, with liver condemned and not permitted for use as human food. When the liver tissue is damaged, nearby clostridial spores may vegetate and release fatal toxins.

TREATMENT

Albendazole and Ivomec-F are the only drugs in the United States approved and effective for liver fluke treatment.

PREVENTION AND CONTROL

Drainage or avoidance of wet pastures can aid in prevention. Where relatively small snail-infested areas are involved, it may be practical to destroy the snail (carrier of the liver fluke), preferably in the spring season, through: applying 3 to 6 pounds of copper sulfate (bluestone or blue vitrol) per acre of grassland, mixing and applying the small quantity of copper sulfate with a suitable carrier (such as a mixture of one part of the copper sulfate to 4 to 8 parts of either sand or lime); and treating ponds or sloughs with one part of copper sulfate to 500,000 parts of water.

REMARKS

When copper sulfate is used in the dilutions indicated, it is not injurious to grasses and will not poison farm animals, but it may kill fish. Copper sulfate solutions are not curative for infected animals.

Snail-infested pastures should not be used for making hay.

Lungworms

Lungworms (*Dictyocaulus viviparus*) are white, threadlike worms 1½ to 3 inches long, found in the trachea and bronchi of cattle.

DISTRIBUTION AND LOSSES

Lungworms are present throughout the United States, particularly in association with wet pastures, and along coastal areas and around the Great Lakes.

SPECIES AFFECTED

Cattle—especially calves, sheep, goats, swine, and cats are affected.

SYMPTOMS AND SIGNS

Symptoms include coughing, labored breathing, loss of appetite, unthriftiness, and intermittent diarrhea. Death may follow, probably from suffocation or pneumonia.

TREATMENT

See Table 20-1 for approved and effective wormers. Like all wormers, they should be given according to the manufacturer's directions.

PREVENTION AND CONTROL

Practice rigid sanitation. Do not spread infested manure on pastures. Where practical, segregate calves from older animals and keep calves on a good ration. Utilize dry pastures, if possible.

REMARKS

Replacement stock from herds with a history of lungworm infection should be (1) received with great caution, and (2) treated with levamisole.

Adults often gain immunity.

Bovine Measles

Bovine measles is a parasitic disease of cattle, cysticercosis is an invasion of the musculature and viscera by larvae. *Cysticercus bovis*, or *Taenia saginata*, is the beef tapeworm of humans. Cattle are the intermediate host and people the definite host.

DISTRIBUTION AND LOSSES

Beef measles is worldwide; however, the incidence is highest in Africa, the Middle East, Asia, and South America. In the United States, the measles problem is largely confined to the Southwest.

At the time of slaughter, losses result from extensive trimming and prolonged storing of mildly infected carcasses and from condemning heavily infected carcasses.

SPECIES AFFECTED

Bovine measles affects cattle.

SYMPTOMS AND SIGNS

Most cases of beef measles produce few signs in live cattle. In the carcass, the cysticerci (cysts) are readily discernible. Carcasses that are excessively infested are unsatisfactory for food and should be condemned. If only a few cysticerci are found, the entire carcass is frozen sufficiently long to ensure that all the cysticerci are killed.

TREATMENT

No effective treatment for bovine measles is known.

PREVENTION AND CONTROL

Humans are the sole host for the adult tapeworm, and cattle are the only intermediate host. No other animals are involved in the life cycle. Thus, control consists of disposing of human excrement in such manner that it cannot come in contact with cattle.

In endemic areas, workers employed in and around feedlots, cattle pastures, and dairies should be medically examined for beef tapeworm parasitism, and infested individuals should be treated for removal of the worms. Sanitary latrines should be provided for caretakers, and they should be forbidden to defecate in feedlots or pastures where cattle feed. At slaughter, cysticecus-infested meat should be disposed of in a manner which avoids the inclusion of viable cysticerci in human food. Meats should be thoroughly cooked to destroy viable cysticerci.

REMARKS

The name is a misnomer, for it has no relationship and little resemblance, to human measles. In humans, the disease is caused by a virus; in cattle, the disease is caused by a tapeworm cyst which lodges and grows in the muscle tissue.

Mites

Mites are very small parasites that produce mange (scabies, scab, itch). In cattle, there are two types of scabies: psoroptic scabies caused by *Psoroptes ovis* and sarcoptic scabies caused by *Sarcoptes scabiei*. Mange is less severe than scabies in cattle. Mange types are chorioptic mange caused by *Chorioptes bovis*, demodectic mange caused by *Demodex bovis*, and psorergatic mange caused by *Psorergates bos*.

DISTRIBUTION AND LOSSES

Distribution of mites is widespread.

Mites retard growth, lower milk production and gains, and produce unthriftiness. Also, the skin is made less valuable for leather.

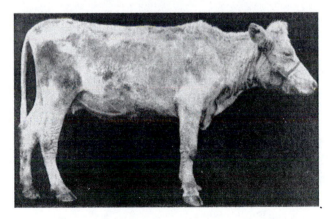

Fig. 20-12. Sarcoptic mange. (Courtesy, Cornell University College of Veterinary Medicine, Ithaca, N.Y.)

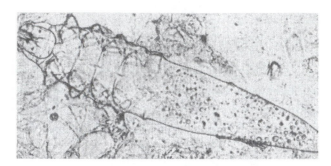

Fig. 20-13. Demodectic mange mite.

Fig. 20-14. Demodectic mange on a cow. Note the nodules over the forelegs and shoulder. (Courtesy, Cornell University College of Veterinary Medicine, Ithaca, N.Y.)

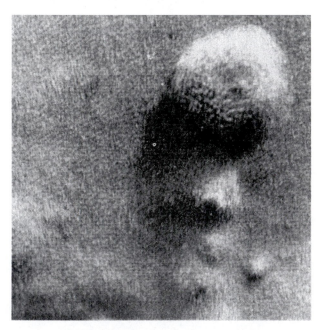

Fig. 20-15. Demodectic mange nodules on the skin of an animal. (Courtesy, Cornell University College of Veterinary Medicine, Ithaca, N.Y.)

SPECIES AFFECTED

Mites affect cattle, although each class of animals has its own species or subspecies of mange mites.

SYMPTOMS AND SIGNS

Marked irritation, itching and scratching, and crusting over of the skin accompanied by formation of thick, tough, wrinkled skin are symptoms of the presence of mites.

TREATMENT

Mites can be controlled by spraying or dipping infested animals with suitable insecticidal solutions, and by quarantine of affected herds.

The following insecticides may be used in keeping with the manufacturer's label:

Coumaphos (Co-ral)	Lime-sulfur
Crotoxyphos	Phosmet
Ivermectin (Ivomec)	Toxaphene

Eprinex is approved for use in lactating cows to control mange.

PREVENTION AND CONTROL

Avoid contact with diseased animals or infested premises.

Scabies is a reportable disease in the United States, so in the case of an outbreak, contact the local veterinarian or livestock sanitary official.

Control by spraying or dipping infested animals with suitable insecticides, and quarantine affected herds.

REMARKS

There are two chief forms of mange: sarcoptic mange (caused by burrowing mites), and psoroptic mange (caused by mites that bite the skin and suck blood but do not burrow.

The disease appears to spread most rapidly during the winter months and among young and poorly nourished animals.

Mosquitoes

Mosquitoes—particularly species of the genera *Aedes, Psorophora,* and *Culex*—are a severe nuisance to cattle in many areas.

DISTRIBUTION AND LOSSES

Mosquitoes are rather widely distributed, but they are most numerous in the southeastern United States, especially in swampy regions that have permanent pools of water or that are exposed to frequent flooding. Sometimes they kill cattle, although this is rare.

SPECIES AFFECTED

All farm animals are affected.

SYMPTOMS AND SIGNS

Mosquitoes may occur in such abundance that cattle refuse to graze. Instead, they bunch together or stand neck deep in water to protect themselves from attack. Moreover, mosquitoes will annoy cattle day and night, so they can cause serious losses in meat production—or even death in extreme cases. Also, they may be disease carriers.

TREATMENT

The recommended insecticides for the control of mosquitoes on both beef and dairy animals are the same as approved for horse flies, deer flies, and stable flies. (See "Horse Fly and Deer Fly.")

PREVENTION AND CONTROL

Mosquitoes can be controlled in several ways: (1) by elimination of breeding places, through providing fills, ditches, impoundments, improved irrigation methods and other means of water manipulation; (2) by chemical destruction of larvae, by treating the relatively restricted breeding areas with proper larvicides; and (3) by chemical destruction of adults. Elimination of breeding sites is by far the most satisfactory and effective method of control. However, either this method or chemical destruction of larvae may not be economically practical if the breeding area is extensive. When the latter is the case, control can best be accomplished through group action, such as mosquito abatement districts. The cattle producer can achieve some control by fogging or spraying the pasture or rangeland, and some relief can be obtained by spraying the animals with insecticides.

REMARKS

Almost all female mosquitoes must take a blood meal before they can lay eggs. (The males do not suck blood, but feed on nectar and other plant juices.) Eggs are laid singly or in rafts on the surface of the water or on the ground in depressions that are flooded by tidal waters, seepage, overflow, or rainwater. The larvae and pupae are aquatic.

Ringworm

Ringworm is a contagious disease of the outer layer of skin caused by certain microscopic molds or fungi.

DISTRIBUTION AND LOSSES

Ringworm is found throughout the United States. It is unsightly and affected animals may experience considerable discomfort, but actual economic losses are not too great.

SPECIES AFFECTED

All animals and humans are affected.

SYMPTOMS AND SIGNS

Signs of ringworm presence include round, scaly areas almost devoid of hair appearing mainly in the vicinity of the eyes, ears, side of the neck, or the root of the tail. Mild itching usually accompanies the disease.

TREATMENT

Ringworm is usually a self-limiting disease. Animals that are on a good plane of nutrition and not heavily parasitized can clear the infection naturally. Thickened ringworm scabs on the eyelids can cause the lashes to turn in against the eye and create corneal ulcers. Betadine solution can be applied through a sprayer to the affected areas every day until new hair growth appears.

PREVENTION AND CONTROL

Prevention of ringworm is difficult, but transmission can be controlled by disinfecting combs and brushes, and insuring proper nutrition and parasite-control programs are followed.

REMARKS

Though ringworm may appear among animals on pasture, it is far more prevalent as a stable disease.

It is usually a winter disease, with recovery the following summer after the animals are turned out to pasture.

Screwworm

Screwworm maggots develop in the open wounds of animals.

DISTRIBUTION AND LOSSES

Screwworms have been eradicated in the United States, although they occasionally reappear.

SPECIES AFFECTED

All farm animals are affected.

SYMPTOMS AND SIGNS

Signs of screwworm presence are loss of appetite and condition, and listlessness due to infested wounds.

TREATMENT

When maggots (larvae) are found in an animal, they should be removed and sent to the proper authorities for identification, and the animal should be treated.

For maggots in wounds, treat with coumaphos (Co Ral) spray bomb.

PREVENTION AND CONTROL

Prevention in infested areas consists mainly of keeping animal wounds to a minimum and of protecting those that do materialize.

In 1958, the USDA initiated an eradication program. Screwworm larvae were reared on artificial media. Two days before fly emergence, the pupae were exposed to gamma irradiation at a dosage which caused sexual sterility but no other deleterious effects. Sterile flies were distributed over the entire screwworm infested region in sufficient quantity to outnumber the native flies, at an average rate of 400 males per square mile per week.

The female mates only once, and therefore, when mated with a sterile male does not reproduce. There was a decline in the native population each generation until the native males were so outnumbered by sterile males that no fertile matings occurred and the native flies were eliminated. This program has virtually eliminated all the losses caused by screwworms in the United States.

REMARKS

Screwworms sometimes penetrate the dimple in front of a cow's udder.

Ticks

The lone star tick (*Amblyomma americanum*), the Gulf Coast tick (*A. maculatum*), the Rocky Mountain wood tick (*Dermacentor andersoni*), the Pacific Coast tick (*D. occidentalis*), and the American dog tick (*D. variabilis*) are three-host species that attack cattle during the summer months. The black-legged tick (*Ixodes scapularus*) is a three-host tick that is common in late winter and early spring. The winter tick (*D. albipictus*) is a one-host species found on cattle and horses in the fall and winter. In addition, larvae and nymphs of the so-called "spinose" ear tick (*Otobius megnini*), a one-host species, attach deep in the ears of cattle and feed there for several months. The cattle tick (*Boophilas annulatus*) and southern cattle tick (*Boophilas microplus*) are one-host ticks that were once very important pests in the southern United States. They are vectors of

Texas cattle fever. They have been eradicated in the United States, but they are still present in Mexico.

DISTRIBUTION AND LOSSES

Ticks are widely distributed, especially throughout the southern part of the United States; but they are usually seasonal in their activities.

Ticks suck blood. They cause economic losses by transmitting diseases; by restlessness, anemia, and inefficient feed utilization; and by necessitating expensive treatments. Among the diseases transmitted to or produced in cattle by ticks are Texas fever, anaplasmosis, Q fever, tick paralysis, and piroplasmosis.

SPECIES AFFECTED

All farm animals are affected.

SYMPTOMS AND SIGNS

Generally speaking, injury to cattle from tick parasitism varies directly with numbers of parasites. Ticks feed exclusively on blood. Thus, when several hundred ticks feed, the host becomes anemic, unthrifty, and loses weight. In addition, some female ticks generate a paralyzing toxin. The spinose ear tick, commonly called the *ear tick*, takes up residence along the inner surface of the ear and in the external ear canals, where it is extremely annoying. Cattle heavily parasitized by spinose ear ticks droop their heads, rub and shake their ears, and turn their heads to one side.

TREATMENT

Effective treatment for ticks varies by the tick species, but several methods include spraying, dipping, and the use of certain insecticidal ear tags. To prevent reinfestation, the sprays need to have residual activity for several days. Pour-ons, dusts, and backrubs may help control some species of ticks. Ticks are difficult to control since some have two and three different hosts during their life cycle. The young ticks may feed on field rodents and other wildlife before feeding upon cattle, thus increasing the control problems.

PREVENTION AND CONTROL

In some areas, it is possible to use habitat modification such as brush control and selective grazing to remove the favorable habitat of ticks.

REMARKS

Apply insecticides in keeping with manufacturer's directions.

Sheep and Goat Parasites

Blow Fly (Wool Maggot)

The blow fly group consists of a number of species of flies, all of which breed in necrotic animal flesh and in exudates, and the wool maggot that attacks soiled fleece areas.

DISTRIBUTION AND LOSSES

Distribution of the blow fly is widespread, but it presents the greatest problem in the Pacific Northwest and in the South and southwestern states. Death losses are not excessive, but production is lowered.

SPECIES AFFECTED

All farm animals are affected.

Fig. 20-16.   A heavily parasitzed ewe. Note the bottlejaw.

SYMPTOMS AND SIGNS

Signs of blow fly presence include: infested wounds and soiled fleece; and maggots spread over the body, feeding on the skin surface, producing severe irritation, and destroying the ability of the skin to function. Infested animals rapidly become weak, fevered, and unthrifty.

TREATMENT

When animals become infested with blow fly maggots, their wounds should be treated directly with coumaphos (Co Ral) spray bomb, used according to manufacturer's directions on the label.

PREVENTION AND CONTROL

Eliminate the blow fly and decrease the susceptibility of animals to infestation by: destroying dead animals by burning or deep burial; using traps, poisoned baits, and electrical screens; using repellents; and docking lambs and tagging sheep at intervals.

REMARKS

Most blow fly damage is limited to the wool maggot fly.

Sheep Gastrointestinal Parasites

Severe production losses and clinical disease is often observed in sheep with worm infections. Ewes and lambs are more susceptible to infection; dry ewes and wethers are more resistant. Lambing ewes commonly have an elevation in fecal egg counts at lambing, contributing to high pasture infectivity. Heavily parasitized sheep on a low plane of nutrition can develop into a situation with high mortality and losses.

MAJOR PARASITES

Haemonchus, Ostertagia, and *Trichostrongylus* are the most harmful genera of worms in the temperate regions of the world. *Haemonchus contortus* is a voracious bloodsucker and is the most damaging species in warm and wet regions. *Ostertagia* and *Haemonchus* have a short life span of about one month, so the number of worms present in the animal depends on the level of infective larvae in the pasture. Sheep worms exhibit a seasonal pattern of infectivity, with the highest survival of worm larvae present on the pastures in the winter in the south, and the spring and summer in the north.

CLINICAL SIGNS

H. contortus infected sheep may die suddenly without clinical signs. Pale membranes, anorexia, submandibular edema, and weakness are signs of the severe blood loss from parasitism. *Trichostronglyus* and *Ostertagia* infections cause anorexia and protein losses into the intestines, as well as diarrhea. Fecal egg counts are a useful aid in diagnosis, but necropsy results and a total worm count is more reliable.

TREATMENT AND PREVENTION

By the time clinical disease is apparent, possibly 95 percent of the total worm population is in the environment, so treatment will remove only 5 percent of the total worm population. To reduce worm burdens, repeated deworming at short intervals of every two weeks, or appropriate grazing management, is used. Pasture regrowth, following mechanical harvesting for hay, straw, or small grain crops, generally have low worm burdens. Rotational grazing is not effective for parasite control because of the length of time that larvae can survive. For the northern United States, four prophylactic treatments at three-week intervals after turnout onto spring grazing are effective. The southern United States is less conducive to prophylactic treatments due to the extended grazing season allowing massive buildup of *H. Contortus* larvae on the pastures. There are only four dewormers approved for sheep in the United States (ivermectin, levamisole, phenothiazine, and thiabendazole). Sheep and goat parasites have developed resistance to all of these drugs, and the future outlook for worm control in sheep and goats is dismal. Farmers need to determine if worm resistance has occurred, and to what dewormers. This procedure is done by checking fecal samples for eggs 10 to 14 days after deworming. Modern dewormers should cause a fecal egg reduction of at least 90 percent if they are working effectively.

Guidelines for appropriate deworming programs are to select the appropriate dewormer, use the full dosage by animal weight, rotate the drugs, treat all new introductions to the herd, and use management strategies to reduce worm exposure to susceptible animals.

Coccidiosis

Coccidiosis is a parasitic disease caused by protozoan organisms known as coccidia.

DISTRIBUTION AND LOSSES

Losses occur worldwide. There are lowered gains and production, and frequently high mortality in feedlot lambs.

SPECIES AFFECTED

Sheep, cattle, goats, pet stock, and poultry are affected.

Each class of animals harbors its own species of coccidia; thus, there is no cross infection between animals.

SYMPTOMS AND SIGNS

Diarrhea and bloody feces, and pronounced unthriftiness and weakness are signs of coccidiosis.

TREATMENT

Lasalocid is approved for prevention of coccidiosis in sheep maintained in confinement. Decoquinate is approved for the prevention of coccidiosis in young goats. Both of the above products should be used in keeping with the manufacturer's instructions on the label.

In the United States, amprolium is permitted in feed only as an aid in the control of coccidiosis of chickens and turkeys.

PREVENTION AND CONTROL

In feedlot lambs, where the disease is most prevalent, good management and natural resistance are important. To these ends, move feeders into feedlots with a minimum of stress and shrink, allow for plenty of space, keep water and feed troughs free from fecal pellets, maintain dry lots and bedding, start animals on grain feed gradually, and segregate affected animals if practical. The same principles and practices apply to other sheep, when trouble is encountered.

Also, when practical, drain or fence low, wet areas; and keep animals in a sunny, dry place.

REMARKS

In the oocyst stage, the parasite resists many disinfectants and may remain viable outside the body for months, but it is readily destroyed by direct sunlight or complete drying.

Some sheep producers protect feeder lambs from coccidiosis by using a feed mixture containing sulfa drugs (such as sulfaguanidine) which check the growth of parasites. Medicated feed may also be fed to lambs when they have to be kept closely confined with ewes.

Cooperias

The four species of parasites classed as *Cooperias* (*C. curtice, C. oncophora, C. punctata, C. pectinata*) are small, hairlike worms less than ⅕ inch in length.

DISTRIBUTION AND LOSSES

Cooperias is widely distributed, but damage is not heavy except in an occasional flock with excessive infection.

SPECIES AFFECTED

Sheep, goats, and cattle are affected.

SYMPTOMS AND SIGNS

No specific symptoms are notable, but affected animals exhibit diarrhea (scours), depression, loss of appetite, loss of weight, and retarded growth.

TREATMENT

See Table 20-1, "Recommended Compounds for Control of Internal Parasites, by Animal Species." Use drug of choice according to manufacturer's directions.

Gid Tapeworm

The gid tapeworm (*Coenurus cerebralis*) is the larval form of one of the four species of bladder worm or tapeworm found in dogs and related carnivores, which also affect sheep and goats.

TREATMENT

No known treatment exists; however, surgery is successful in some cases.

SPECIES AFFECTED

Sheep, goats, and pet stock are affected.

SYMPTOMS AND SIGNS

The disease is known as coenurosis, the symptoms of which are defects in vision and disturbances in movements. Affected animals may stumble, run into objects, walk with the head high or in circles, and there may be at least a partial paralysis of the hindquarters.

DISTRIBUTION AND LOSSES

Losses are spotty and rare over the United States.

PREVENTION AND CONTROL

Prevention is aided by the elimination of stray dogs; and the examination, and proper worm treatment when necessary, of all dogs that may come in contact with sheep and goats.

Proper disposal of all carcasses of infested animals also aids in control.

REMARKS

In afflicted sheep and goats, cysts containing the larvae *(Coenurus cerebralis)* of the tapeworm eggs voided by dogs or other carnivorous animals are found on the brain and spinal cord.

Hydatid

The hydatid (*Echinococcus granulosus*) is the larval form of one of the four species of bladder worm or tapeworm, found in dogs and related carnivores, which also affect sheep and goats.

DISTRIBUTION AND LOSSES

Distribution is sparsely scattered over the United States, and no great numbers of sheep and goats are seriously affected. Few deaths occur.

SPECIES AFFECTED

Sheep, goats, cattle, swine, horses, and humans are affected.

Hydatid disease, infection with *Echinococcus granulosis*, is a serious infection of humans that warrants great care in prevention.

SYMPTOMS AND SIGNS

Ordinarily, no specific or distinctive symptoms are observed, but in heavy infestations there may be shallow respiration, emaciation and weakness.

TREATMENT

No known treatment exists for hydatid.

PREVENTION AND CONTROL

Farm dogs should not be fed portions of carcasses of sheep that are farm slaughtered.

Eliminate stray dogs. Examine—and administer proper worm treatment when necessary—all dogs that may come in contact with sheep and goats.

Properly dispose of, by burying or burning, all carcasses of infested animals.

REMARKS

Correctly speaking, the parasite *Echinococcus granulosus* causes Hydatid disease.

Tapeworm eggs are excreted in the feces of infected dogs and can remain alive for many months.

Lice

Lice are small, flattened, wingless insect parasites of which there are two groups: sucking lice and biting lice. Biting lice, *Damalinia (Bovicola) ovis*, are most common but least harmful. The sucking species are (1) body lice *(Linognahus ovillus* and *L. africanus)*, and (2) the foot louse *(L. pedalis)*. Sucking lice are larger than biting lice.

Body lice are found anywhere on the body where wool is dense, while foot lice are usually found on legs below knees and hock.

SPECIES AFFECTED

Sheep, with other species for other classes of animals, are affected.

SYMPTOMS AND SIGNS

Symptoms include intense irritation, restlessness, and loss of condition. There may be severe itching and the animal may be seen scratching, rubbing, and gnawing at the skin. The wool may be matted and lack luster, and scabs may be evident.

DISTRIBUTION AND LOSSES

Lice are not as common on sheep as on other domestic animals, but they do occur occasionally.

Lice retard growth, lower wool production, and produce unthriftiness.

TREATMENT

For the control of lice, apply one of the following insecticides in keeping with the manufacturer's label:

Coumaphos (Co Ral)	Malathion
Fenvalerate (Ectrin)	Methoxychlor
Lindane	

PREVENTION AND CONTROL

For effective control, all sheep should be treated simultaneously.

REMARKS

Lice show up most commonly in winter and on ill-nourished and neglected animals.

Spray at high pressure (up to 400 psi).

Liver Fluke

The liver fluke (*Fasciola hepatica*) is a flattened, leaflike brown worm, usually about an inch long.

DISTRIBUTION AND LOSSES

Distribution is worldwide. In the United States liver flukes are most prevalent in coastal areas of the Southeast, the Pacific Northwest, and portions of the Rocky Mountains.

Infestation results in lowered meat and wool production and produces many "fluky livers" or "rotten livers," which are condemned in packing houses.

In conjunction with black disease, it is usually fatal in sheep.

SPECIES AFFECTED

Sheep, goats, cattle, and other animals are affected.

SYMPTOMS AND SIGNS

With heavy fluke infestations, death may occur without any definite symptoms being evident. Sometimes there may be a distinct pot-bellied condition caused by the escape of fluids into the body cavity through damage to the liver.

TREATMENT

Under the direction of a veterinarian, administer one of the available compounds. Currently, the drugs of choice are albendazole and carbon tetrachloride.

PREVENTION AND CONTROL

Prevention is aided by drainage or avoidance of wet pastures which may be snail infested.

There is no approved chemical for control of snails in the United States. Copper sulfate has been used in the past, but it is not presently approved.

REMARKS

Snail-infested pastures should not be used for making hay.

Infested sheep are usually treated in November or December, with two or even three treatments at four- to six-week intervals.

The fluke may also carry *Clostridium novyi*, an anaerobic bacterium which causes black disease in sheep.

Lungworms

Thread lungworms (*Dictyocaulus filaria, Muellerius capillaris*) are white and up to 4 inches long, whereas hair lungworms are thinner and much shorter.

DISTRIBUTION AND LOSSES

Thread lungworms are widely distributed throughout the world. Death losses occasionally occur.

SPECIES AFFECTED

Sheep and goats are affected. Thread lungworms also occur in cattle, but with different species of worms.

SYMPTOMS AND SIGNS

No specific symptoms are associated with infestations of hair lungworms.

The thread lungworm is especially damaging to lambs and kids. The first symptom is coughing, usually accompanied by a dirty, pussy discharge from the nostrils. Other symptoms include rapid and difficult breathing, unthriftiness, loss of appetite, and lowered head and extended neck. Death may follow in two to three months, probably from infection or pneumonia.

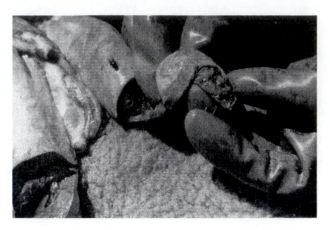

Fig. 20-17. Lungworm damage in sheep.

TREATMENT

See Table 20-1, Recommended Compounds for Control of Internal Parasites, by Animal Species. Treatment should be followed by good feeding and nursing.

PREVENTION AND CONTROL

Remove animals from infested ground and place them on dry pastures where clean water is available. Drain or avoid low, wet pastures. Do not spread fresh manure of infested sheep and goats on pastures used by these animals.

Since old animals may be carriers without showing symptoms, weaned lambs and kids should be kept away from mature animals whenever practical to do so.

Mites (Sheep Scab, Scabies, Mange)

Mites are very small parasites that produce mange of which there are two forms: (1) sarcoptic mange, caused by burrowing mites; and (2) psoroptic mange, caused by mites that bite the skin but do not burrow. Psoroptic or common scab is the most important form of sheep scabies in the United States.

DISTRIBUTION AND LOSSES

Distribution is widespread as mange is easily transmitted by contact from one sheep to another, and it spreads very rapidly after being introduced in a flock.

Mites retard growth and gains, lower wool production, and produce unthriftiness. Also, the pelt is made less valuable.

SPECIES AFFECTED

Sheep are affected; however, each class of animals has its own species or subspecies of mange mites.

SYMPTOMS AND SIGNS

Symptoms include marked irritation, itching, and scratching. Crusting over of the skin, accompanied or followed by the loss of wool and the formation of thick crusts or scabs on the skin, is also seen; hence, the name scabies.

The disease appears to spread most rapidly among young and poorly nourished animals.

TREATMENT

Apply one of the following insecticides in keeping with the manufacturer's label:

Coumaphos (Co Ral)	Phosmet
Ivermectin (Ivomec)	Toxaphene

PREVENTION AND CONTROL

Avoid contact with diseased animals or infested premises.

Scabies is a reportable disease in the United States, so, in case of an outbreak, contact the local veterinarian or livestock sanitary official.

Control by spraying or dipping infested animals with suitable insecticides, and quarantine affected flocks or bands.

REMARKS

The disease appears to spread most rapidly during the winter months.

Nodular Worms

Nodular worms (*Oesophagostonum columbianum*) are white worms about ⅜ inches long, found in the cecum and colon of sheep and goats.

DISTRIBUTION AND LOSSES

Nodular worms are found worldwide. In the United States they occur mostly in central, eastern, and southern areas with only limited damage.

In addition to lowered fleece and mutton yields and death losses, the presence of nodular worm in slaughtered animals causes all or a considerable portion of the large and small intestine to be unfit for surgical sutures (catgut) or for casings. It is estimated that the latter two losses total half a million dollars annually.

SPECIES AFFECTED

Sheep and goats are affected, as are cattle and swine, but with a different species of worms in each.

SYMPTOMS AND SIGNS

Unthriftiness, reduced fleece and mutton yields, and death losses are signs of nodular worm presence. In general, the symptoms accompanying severe nodular worm infestation are not unlike those of general parasitism.

TREATMENT

See Table 20-1, Recommended Compounds for Control of Internal Parasites, by Animal Species. Use drug of choice according to manufacturer's directions.

Ringworm

Ringworm is a contagious disease of the outer layers of skin caused by a microscopic mold or fungus of the genus *Trichophyton*.

DISTRIBUTION AND LOSSES

Ringworm of sheep is of little economic importance in the United States.

Species Affected

All animals and humans are affected, but ringworm in sheep is seldom observed.

Symptoms and Signs

Round, scaly areas almost devoid of hair (or wool) appear mainly in the vicinity of the eyes, ears, side of the neck, or the root of the tail. Mild itching usually accompanies the disease.

Treatment

Clip the hair or wool from the affected areas and paint sores with tincture of iodine every three days until cleared up.

Prevention and Control

Isolate affected animals and disinfect everything that has been in contact with them, including cards and brushes. Practice strict sanitation.

Remarks

Although ringworm may appear among animals on pastures, it is far more prevalent as a stable disease among livestock during winter.

Screwworm

The screwworm fly raises its maggots in the open wounds of animals, where they feed on live flesh.

Distribution and Losses

Screwworm is found mostly in the southern or southwestern states, in which areas it formerly caused 50 percent of the normal annual livestock losses.

Species Affected

All farm animals are affected.

Symptoms and Signs

Symptoms include loss of appetite and condition, and lowered thrift and vigor.

Treatment

Apply the following insecticide in keeping with the manufacturer's label: coumaphos (Co-ral) spray bomb.

Prevention and Control

Keep animal wounds to a minimum. Schedule branding, castrating, docking, and other stock operations that necessarily produce wounds during the winter season or early spring when the flies are least abundant and active. Kill all possible maggots during the winter and spring months. If possible, keep wounded and infested animals in a screened, fly-proof area.

Screwworm infestation is lessened by castrating with Burdizzo pincers, by eliminating sources of mechanical injury, and by having newborn animals arrive at the season of least fly activity.

Remarks

Some producers apply pine-tar oil to recent wounds and to the navels of newborn animals to repel flies.

The federal program of release of sterile male flies has been very successful in screwworm eradication in the southern United States. Screwworms have been eradicated from the United States except for sporadic outbreaks in south Texas near Mexico.

Sheep Bots (Nasal Botflies)

Sheep bots (*Oestrus ovis*), commonly called grubs-in-the-head, are due to a beelike fly about the size of the common horse fly.

Distribution and Losses

Sheep bots occur worldwide.

Although death losses are rare, there is loss in condition, both at the time the fly attacks and while the larvae are in the nasal passage.

Species Affected

Sheep are affected.

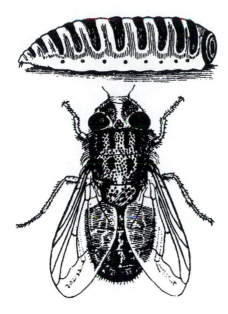

Fig. 20-18.   Sheep botfly. *Oestrus ovis*, and larva.

SYMPTOMS AND SIGNS

When the flies attempt to deposit their larvae around the nostrils of sheep, the animals cease to feed, become restless, press their noses against other sheep, and/or seek shelter.

Grub infestation results in a snotty nose, and there may be difficulty breathing and frequent sneezing.

TREATMENT

No insecticide is registered and approved for sheep nose bot control.

Infestation can be prevented by painting the nostrils of sheep with pine tar weekly during the season when the adult bot fly is active.

PREVENTION AND CONTROL

The following measures, during the sheep nasal botfly season, may lessen, but not eliminate, sheep bots. With a small flock and where practical, keep sheep in a darkened barn during the day; apply pine-tar oil about the nostrils of the sheep every few days, thus repelling some of the flies.

Sheep Keds (Sheep Ticks)

Sheep keds (*Melophagus ovinus*) are hairy, bloodsucking flies without wings, which range up to ¼ inches in length.

DISTRIBUTION AND LOSSES

Sheep keds are found throughout the United States, but are especially prevalent in the northern states.

Losses include retarded growth of young animals, loss in condition of mature animals, and damage to the fleece. Fine wool breeds of sheep and Angora goats are not seriously affected.

SPECIES AFFECTED

Sheep and goats are affected.

SYMPTOMS AND SIGNS

Symptoms include marked reduction in condition, anemia, biting and scratching, and loss of and damage to wool.

TREATMENT

Treat after shearing—as soon as the shear cuts heal (including unshorn lambs)—with any one of the following insecticides, used according to the manufacturer's label:

Coumaphos (Co Ral)	Diazinon
Fenvalerate (Ectrin)	Permethrin

PREVENTION AND CONTROL

Prevention and control involves spraying or dipping all sheep as soon as the cuts heal up following shearing.

REMARKS

Poorly housed and poorly fed animals are most likely to suffer from sheep ticks.

The method of application of an insecticide depends on the product used; it may involve dipping, spraying, pour-ons, or dusting.

Sheep Measles

Sheep measles (*Cysticercus ovis*) are the larval form of one of the four species of bladder worm or tapeworm, found in dogs and related carnivores, which also affect sheep and goats.

DISTRIBUTION AND LOSSES

Distribution is worldwide. In the United States, they are most prevalent in the West.

Few death losses occur due to sheep measles. Chief economic loss occurs at slaughter, because infested carcasses are trimmed or condemned according to the degree of infestation.

SPECIES AFFECTED

Sheep and goats are affected.

SYMPTOMS AND SIGNS

No specific symptoms have been attributed to infestation with sheep measles.

TREATMENT

There is no known treatment for the removal of the parasite.

PREVENTION AND CONTROL

Prevention consists of elimination of stray dogs; the examination, and proper worm treatment when necessary, of all dogs that may come in contact with sheep and goats; and proper disposal of all carcasses of infested animals.

REMARKS

The parasite is not transmissible to humans; the removal or condemnation of affected carcasses is done to assure clean, wholesome meat.

Tapeworms

Tapeworms (common tapeworm—*Moniezia expansa, M. benedeni*; fringed tapeworm—*Thysanosoma*

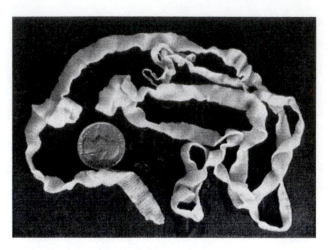

Fig. 20-19. The tapeworm breaks into sections called proglottids which contain eggs.

actinoides) are long, flat, ribbonlike worms which range up to a width of ¾ inch and a length of 20 feet. The fringed tapeworm differs in that it is smaller and it has fringed segments.

DISTRIBUTION AND LOSSES

Tapeworms are found worldwide, with lambs being more susceptible than older sheep.

Common tapeworms are not an important factor in sheep raising, because they are practically harmless. Only under poor nutritional conditions will they increase unthriftiness.

The fringed tapeworm is confined largely to the range bands of the western United States, whereas the other species occur mostly in the central, eastern, and southern states.

The fringed tapeworm may cause death. In the packing plant, livers infected with this parasite (that is, when the worms are found) are condemned as unfit for human consumption.

SPECIES AFFECTED

Sheep and goats are affected.

SYMPTOMS AND SIGNS

Common tapeworms do not produce any marked or specific symptoms. However, the fringed tapeworm may cause death of the host through blocking the cystic duct, gallbladder, and the ducts of the liver and pancreas. Infected animals usually have normal appetites.

Monezia infections are sometimes associated with diarrhea and poor growth, because the adult tapeworms compete with the host for nutrients and can interfere with gut motility.

TREATMENT

Albendazole, fenbendazole, and oxfendazole have excellent activity for tapeworms, and are safe for use in all classes of sheep.

PREVENTION AND CONTROL

The fringed tapeworm has an intermediate host, the psocid louse, so the parasite is limited to the western United States. Periodic treatment of the flock with drugs to control the adult tapeworms will reduce the amount of tapeworm eggs in the environment. The oribatid pasture mite is an intermediate host for tapeworms, and elimination of the mites from pastures is impractical as a means of control.

Thin-Necked Bladder Worm

The thin-necked bladder worm (*Cysticercus tenuicollis*) is the larval form of one of the four species of bladder worm or tapeworm, found in dogs and related carnivores, which also affect sheep and goats.

DISTRIBUTION AND LOSSES

Thin-necked bladder worms are found worldwide.

SPECIES AFFECTED

Sheep and goats are affected.

SYMPTOMS AND SIGNS

Usually there are no external symptoms, and since light infestations are the rule, no attention is called to the parasite.

Thin-necked bladder worms burrow into the liver and/or thin membranes of the abdominal cavity of sheep and goats, producing tissue damage.

TREATMENT

There is no known treatment for infestations of thin-necked bladder worms.

PREVENTION AND CONTROL

Prevention is aided by eliminating stray dogs; examining—and properly treating for worms when necessary—all dogs that must come in contact with sheep and goats; and properly disposing of all carcasses of infested animals.

Whipworm

The whipworm (*Trichuris ovis*) is a white worm 1½ to 2 inches in length, with slender anterior and heavy posterior portions that resemble the lash and handle, respectively, of a whip.

DISTRIBUTION AND LOSSES

Whipworms are found worldwide.

Damage can be severe in specific flocks with numerous death losses.

SPECIES AFFECTED

Sheep and goats are affected.

SYMPTOMS AND SIGNS

There are no well-defined symptoms, though persistent bloody diarrhea and unthriftiness may be evident.

TREATMENT

See Table 20-1, Recommended Compounds for Control of Internal Parasites, by Animal Species. Use drug of choice according to manufacturer's directions.

PREVENTION AND CONTROL

Clean pastures and rotation grazing are the key to prevention and control.

Swine Parasites and Their Control

The advant of confinement operations has caused a decline in the incidence of some swine parasites, although the prevalence of some parasites is as high as in pastured swine. The stomach worms, thorny headed worm, and the lungworms have decreased in incidence with confinement operations. However, roundworms, nodular worms, and whipworms continue to be a serious limiting factor to successful swine production in some herds.

Ascarids (Large Intestinal Roundworm)

Ascarid infections in swine are caused by *Ascaris suum*, which are large worms that can be up to 12 inches long.

DISTRIBUTION AND LOSSES

Ascarids are found worldwide.

Ascarid larvae are very damaging to the host during the migration from the gut, to the lungs, and back to the gut. When the pigs eat infective eggs, the eggs hatch in the intestines and pentrate the intestinal wall and migrate to the liver within 24 hours. The larvae then pass through the bloodstream to the lungs, where they break out into the lungs about one week after ingestion of the eggs. The larvae are coughed up and swallowed, where they mature into adults in the intestines. Female ascarids can produce up to half a million eggs daily.

The larval penetration of the gut provide a way for bacteria, viruses, and other microorganisms to enter the body. The migration of the larvae through the liver cause "milk spots," resulting in the condemnation of the liver for human consumption. As the larvae migrate through the lungs, the pigs become more susceptible to mycoplasma pneumonia and swine influenza.

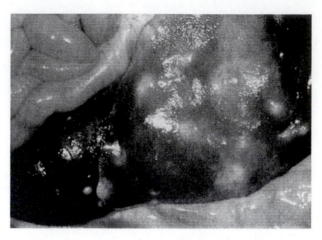

Fig. 20-20. Liver damage in swine due to roundworm migration. (Courtesy, Shell Chemical Company)

SPECIES AFFECTED

Swine are affected.

SYMPTOMS AND SIGNS

Young pigs become unthrifty and stunted, and there is usually coughing, "thumpy" breathing, and there may be a yellow color to the mucous membrane due to blockage of the bile ducts.

Principal damage is produced by migrating larvae which produce liver damage and lung lesions resulting in verminous pneumonia.

TREATMENT

See Table 20-1, Recommended Compounds for Control of Internal Parasites, by Animal Species. Use drug of choice according to manufacturer's directions.

PREVENTION AND CONTROL

Due to the ability of roundworm eggs to survive for periods up to 40 years, sanitation is important in the control program. Deworming sows several days before farrowing will allow the fecal eggs to be expelled before exposing the pigs.

REMARKS

See manufacturer's label for proper drug withdrawal prior to slaughter for food.

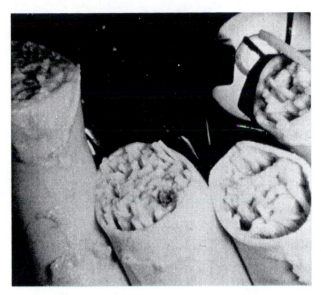

Fig. 20-21. Intestinal cross section showing blockage by roundworms. (Courtesy, Shell Chemical Company)

Coccidiosis

The important species involved in swine coccidiosis is *Isospora suis.*

DISTRIBUTION AND LOSSES

Coccidiosis occurs worldwide.

Isospora suis infections are responsible for up to 20 percent of the cases of piglet diarrhea in the United States. The disease is very common in confinement operations utilizing farrowing crates. The disease occurs in piglets less than two weeks, and is rare in pigs older than three weeks.

SPECIES AFFECTED

Swine, cattle, sheep, goats, pet stock, and poultry are affected.

Each class of animals harbors its own species of coccidia; thus, there is no cross infection between animals.

SYMPTOMS AND SIGNS

Diarrhea and bloody feces, and pronounced unthriftiness and weakness are signs of coccidiosis.

Piglets may appear weak, dehydrated, and undersized; and they may die.

TREATMENT

Drugs effective against coccidia species in other farm animals have very little effectiveness against *Isospora.*

PREVENTION AND CONTROL

Sanitation utilizing high-pressure sprayers, hot water, and disinfectants is the best control measure at this time.

REMARKS

In the oocyst stage, the parasite resists low temperatures and disinfectants and may remain viable outside the body for months, but it is readily destroyed by sunlight or complete drying.

Flies

The common species of flies are fully discussed in the section "Flies" under "Cattle Parasites"; hence, the reader is referred thereto.

TREATMENT

For house fly control, the following insecticides are recommended:

    Femvalerate (Ectrin)
    Rabon oral larvacide

Kidney Worm

The kidney worm (*Stephanurus dentatus*) is a thick-bodied, black-and-white worm up to 2 inches in length.

DISTRIBUTION AND LOSSES

The kidney worm is found mostly in the southern United States where it is one of the most damaging worm parasites affecting swine. Losses incurred include (1) inefficient gains, (2) upon slaughter, severe trimming or even condemnation of the damaged carcass, and (3) liver condemnations.

SPECIES AFFECTED

Swine and cattle are affected, but it is not so damaging in the latter.

SYMPTOMS AND SIGNS

No specific symptoms can be attributed to kidney worm infestation. The growth rate is markedly retarded and frequently pus is discharged in the urine. The parasite may also seriously affect the ability of the sow to produce young.

TREATMENT

See Table 20-1, "Recommended Compounds for Control of Internal Parasites, by Animal Species." Use drug of choice according to the manufacturer's directions.

PREVENTION AND CONTROL

The gilt-only method was first proved at the Coastal Plain Experiment Station, Tifton, Georgia. This method is based on the fact that the kidney worm may take as long as a year to reach the egg-laying stage. With this method, gilts are bred only once; then at weaning time, they are sent to slaughter before mature kidney worms develop. By using only young breeding stock for four farrowing seasons (two years), the parasites are eliminated.

Adult and larval stages of the kidney worm are controlled with fenbendazole. Adult worms are controlled with levamisole.

REMARKS

Carcass and liver losses at slaughter are ultimately borne by the swine producer in the form of lowered market prices.

See manufacturer's label for proper withdrawal prior to slaughter for food.

Lice

Lice are small, flattened, wingless insect parasites of which there are several species, most of which are specific for a particular class of animals. Only one species is found on swine.

DISTRIBUTION AND LOSSES

Lice presence is widespread.

Lice retard growth, lower milk production, and produce unthriftiness.

SPECIES AFFECTED

Swine are affected, with other species for other classes of animals.

SYMPTOMS AND SIGNS

Intense irritation, restlessness, and loss of condition are symptoms of lice infestation. There may be severe itching and the animal may be seen scratching and rubbing. The hair may be rough, thin, and lack luster; and scabs may be evident. Lice are apt to be most plentiful around the root of the tail, on the inside of the thighs, and around the neck and ears.

TREATMENT

For lice control, the presently available and effective insecticides include:

Coumaphos (Co Ral)	Malathion
Fenthion (Liguvon)	Permethrin
Fenvalerate (Ectrin)	Prolate
Ivermectin	Rabon

PREVENTION AND CONTROL

Because of the close contact of swine during the winter months, it is practically impossible to keep them from becoming infected with lice.

For effective control, all swine should be treated simultaneously at intervals. Ivermectin is available as a feed premix formulation that will treat lice infestations.

REMARKS

Lice show up most commonly in winter and on ill-nourished and neglected animals.

Always use insecticides according to manufacturer's directions.

Lungworms

Lungworms (*Metastrongylus* spp) are threadlike in diameter, 1 to 1½ inches long, white or brownish in color, found in the air passages.

DISTRIBUTION AND LOSSES

Lungworms are found throughout the United States, but heaviest infestation of swine occurs in southeastern states.

In addition to the lowered growth and feed efficiency, there is evidence to indicate that lungworms may be instrumental in the spread of swine influenza.

SPECIES AFFECTED

Swine are affected, but lungworms also occur in cattle, sheep, and goats, but with different species of worms in each.

SYMPTOMS AND SIGNS

Pigs heavily infected with lungworms become unthrifty, stunted, and are subject to spasmodic coughing.

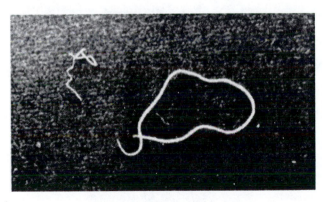

Lungworm (Male & Female)

Fig. 20-22. Swine lungworms.

TREATMENT

See Table 20-1, Recommended Compounds for Control of Internal Parasites, by Animal Species. Use drug of choice according to manufacturer's directions.

PREVENTION AND CONTROL

Keep hogs away from those areas where earthworms are likely to abound. Remove manure piles and trash, and drain low places. Ring the snout. Routinely treat with levamisole.

REMARKS

See manufacturer's label for proper drug withdrawal prior to slaughter for food.

Mites

Mites are very small parasites that produce mange (scabies, scab, itch).

DISTRIBUTION AND LOSSES

Mite presence is widespread, and they retard growth, lower gains, and produce unthriftiness.

Mange causes significant economic damage to the swine herds due to lowered milk production in infected sows, slower growth rate in pigs, carcass damage, and a shortened lifespan of swine equipment due to damage by swine rubbing.

SPECIES AFFECTED

Swine are affected, but each class of animals has its own species or subspecies of mange mites.

SYMPTOMS AND SIGNS

Symptoms are marked irritation, itching and scratching, and crusting over of the skin, accompanied by formation of thick, tough, wrinkled skin.

TREATMENT

The following insecticides are approved for the control of mange mites:

Chlordane   Lindane
Ivermectrin (Ivomec)   Malathion

Ivermectin is available as a feed premix to eliminate the need to inject every animal on the premises at one time.

PREVENTION AND CONTROL

Avoid contact with diseased animals or infested premises.

Control by spraying or dipping infested animals with suitable insecticides, and quarantine affected herds.

REMARKS

The disease appears to spread most rapidly during the winter months and among young and poorly nourished animals.

Follow container label for mixing directions, application, and safety precautions.

Nodular Worms

Nodular worms consist of four species occurring in swine, all of which are slender, whitish to grayish in color, and ⅛ to ½ inches in length.

DISTRIBUTION AND LOSSES

Nodular worms are widely distributed over the United States, but damage is heaviest in southeastern states. In addition to the usual lack of thrift, the intestines of severely infested animals are not suited for either sausage casings or food (chitterlings).

Nodular worms cause calcification of the intestinal tissue forming nodules, which interfere with absorption of nutrients. The worms cause weight loss and diarrhea, but rarely death.

SPECIES AFFECTED

Swine are affected.

Fig. 20-23.   Nodular worms in the intestinal tract of a pig. (Courtesy, Shell Chemical Company)

SYMPTOMS AND SIGNS

No symptoms are specific to nodular worms; however, weakness, anemia, emaciation, diarrhea, and general unthriftiness occur.

TREATMENT

See Table 20-1, Recommended Compounds for Control of Internal Parasites, by Animal Species. Use drug of choice according to manufacturer's directions.

REMARKS

Dichlorvos and levamisole are broad-spectrum wormers; hence, they control ascarids and whipworms, in addition to nodular worms.

Screwworm

Screwworm (*Cochliomyia hominivorax*) fly maggots develop in the open wounds of animals.

DISTRIBUTION AND LOSSES

Screwworms are found mostly in the southern and southwestern states; in which areas it formerly caused 50 percent of the normal annual livestock losses.

SPECIES AFFECTED

All farm animals are affected.

SYMPTOMS AND SIGNS

Screwworm symptoms include loss of appetite, unthriftiness, and lessened activity.

TREATMENT

When maggots (larvae) are found in an animal, they should be removed and sent to the proper authorities for identification, and the animal should be treated. Coumaphos (Co Ral) is the insecticide of choice.

PREVENTION AND CONTROL

Keep animal wounds to a minimum. Schedule stock operations that necessarily produce wounds during the winter season when the files are least abundant and active. Kill all possible maggots during the winter and spring months. If possible, keep wounded and infested animals in a screened, fly-proof area.

REMARKS

The screwworm eradication program in which sterile males were used, has eliminated the screwworm in the United States.

Stomach Worms

Stomach worms consist of three species which infest swine. *Ascarops strongylina* and *Physocephalus sexalatus,* commonly known as "thick stomach worms," are reddish in color and up to an inch long. *Hyostrongylus rubidus,* commonly known as the "red stomach worm," is reddish, small, delicate, slender and about ⅕ of an inch in length.

DISTRIBUTION AND LOSSES

Stomach worms are widespread throughout the United States. Losses are chiefly in stunted growth and waste of feed.

SPECIES AFFECTED

Swine are affected, but stomach worms also occur in cattle, sheep, and goats, but with a different species of worms in each.

SYMPTOMS AND SIGNS

Symptoms include unthriftiness and marked loss of appetite.

TREATMENT

See Table 20-1, Recommended Compounds for Control of Internal Parasites, by Animal Species. Use drug of choice according to manufacturer's directions.

PREVENTION AND CONTROL

Preventive measures for the control of stomach worms are similar to those advocated for the control of ascarids; hence, the reader is referred thereto.

REMARKS

On farms where a good parasite control program is practiced, this parasite is no problem.

Thorn-Headed Worms

Thorn-headed worms (*Macracanthorhynchus hirudinaceus*) are white to bluish worms, cylindrical to flat, up to the size of a lead pencil, with rows of hooks which it uses for attachment purposes.

DISTRIBUTION AND LOSSES

Thorn-headed worms are common in the southern United States.

Losses include slow growth, inefficient feed utilization, death losses, and damaged intestines that are unfit for sausage casings.

SPECIES AFFECTED

Thorn-headed worms affect swine.

SYMPTOMS AND SIGNS

There are no specific symptoms, although swine infested with thorn-headed worms exhibit the general unthriftiness commonly associated with parasites. Digestive disturbance accompanies severe cases.

TREATMENT

No known drug treatment is entirely satisfactory for removing thorn-headed worms.

PREVENTION AND CONTROL

Keep pigs from feeding in areas where they might obtain the white grub of the June bug, the intermediate host. Sanitation, clean ground, and nose-ringing are effective preventive measures.

Threadworms

Threadworms (*Strongyloides ransomi*) in the adult stage burrow in the wall of the small intestine.

DISTRIBUTION AND LOSSES

They are found mainly in the southern and southeastern states.

SPECIES AFFECTED

Swine are affected, but infestation also occurs in cattle with different species of worms.

SYMPTOMS AND SIGNS

In heavy infestations, it is a serious disease causing scours, anemia, and severe weight and death loss, particularly in young pigs. Light infections may show no symptoms.

TREATMENT

See Table 20-1, Recommended Compounds for Control of Internal Parasites, by Animal Species. Use drug of choice according to manufacturer's directions.

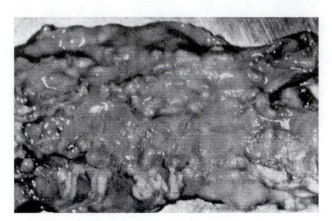

Fig. 20-24. Swine intestine showing severe damage from threadworm infestation. (Courtesy, Shell Chemical Company)

PREVENTION AND CONTROL

Good sanitation, pasture rotation, and frequent worming are effective preventions. Fenbendazole or ivermectin are good products to use pre-farrowing to stop transmission of larvae to the newborn pigs.

REMARKS

See drug manufacturer's label for proper withdrawal prior to slaughter for food.

Larval transmission may occur through sow's milk, frequently resulting in serious infections in young pigs.

Trichinella

Trichinella (*Trichinella spiralis*) is a parasitic disease of humans contracted largely by consuming infested pork, eaten raw, or imperfectly cooked.

DISTRIBUTION AND LOSSES

Trichinella infections in the United States, even though at an extremely low percentage of the swine population, hinders the export of U.S. pork to some countries.

SPECIES AFFECTED

Swine and humans are affected.

SYMPTOMS AND SIGNS

There are no specific symptoms in hogs, even when the parasite is present in the muscle tissue, its usual abode.

TREATMENT

There is no practical treatment for infected hogs. Infected humans should be under care of an M.D.

PREVENTION AND CONTROL

Prevention of trichinosis in humans may be obtained by (1) thoroughly cooking all pork at a temperature of 137°F before it is consumed or (2) freezing pork for a continuous period of not less than 20 days at a temperature not higher than 5°F.

Trichinosis in swine may be lessened by: (1) destruction of all rats on the farm, (2) proper carcass disposal of hogs and other animals that die on the farm, and (3) not feeding garbage.

REMARKS

In humans, the disease is usually accompanied by a fever, digestive disturbances, swelling of infected muscles, and severe muscular pain (in the breathing muscles as well as others).

The presence of trichina can be detected by a microscopic examination of pork, but such a method is regarded as impractical in meat inspection procedure.

The FA and ELISA tests are the most reliable of currently available blood tests to detect infections in animals.

Whipworms

Whipworms (*Trichuris suis*) are 1½ to 2 inches in length, with slender anterior and heavy posterior portions that resemble the lash and handle, respectively, of a whip.

Fig. 20-25.   Whipworms removed from swine intestines.

DISTRIBUTION AND LOSSES

Whipworms are widely scattered throughout the United States, but heaviest infestation is in the southeastern states.

Losses include slow gains, feed inefficiency, and some deaths.

SPECIES AFFECTED

Swine are affected.

SYMPTOMS AND SIGNS

Infected animals may develop a diarrhea, and in heavy infections the diarrhea becomes bloody. In massive infections, growth may be noticeably retarded, and the animal may become weak and finally die. Additional signs of infection are anorexia, weight loss, straining to defecate, and rectal prolapse.

TREATMENT

See Table 20-1, Recommended Compounds for Control of Internal Parasites, by Animal Species. Use drug of choice according to manufacturer's directions.

PREVENTION AND CONTROL

Dichlorvos, fenbendazole, and hygromycin B are effective for whipworms.

REMARKS

The whipworm is increasing in the Corn Belt and in the Southwest.

Horse Parasites

Ascarids (White Worm, Large Roundworm)

Female ascarids (*Parascaris equorum*) vary from 6 to 14 inches in length and males from 5 to 13 inches. When full grown, both are about the diameter of a lead pencil.

DISTRIBUTION AND LOSSES

Ascarids are found throughout the United States.

The presence of ascarids results in loss of feed through feeding worms, lowered work efficiency (including performance on the track and in the show-ring), retarded growth of young animals, lowered breeding efficiency, and death in severe infestations.

SPECIES AFFECTED

Horses, mules, and zebras are affected.

SYMPTOMS AND SIGNS

The injury produced by ascarids covers a wide range from light infections producing moderate effects

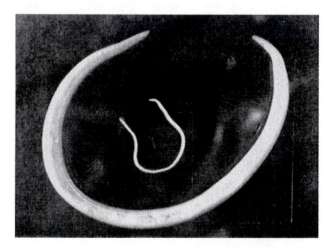

Fig. 20-26. The large roundworm of equines. *Parascaris equorum.*

to heavy infections which may be the essential cause of death. Death is usually due to a ruptured intestine. Serious lung damage caused by migrating ascarid larvae may result in pneumonia. More common, and probably more important, is a retarded or impaired growth and development manifested by potbellies, rough hair coats, and digestive disturbances.

It especially affects foals and young animals, but is rarely important in horses over five years of age; older animals develop acquired immunity from earlier infections.

TREATMENT

See Table 20-1, Recommended Compounds for Control of Internal Parasites, by Animal Species. Use drug of choice according to manufacturer's directions.

In addition to selecting the particular drug(s) for ascarid control, the caretaker should set up a definite treatment schedule, then follow it. The advice of the veterinarian should be sought on both points. Also, to preclude the possibility that worms may become resistant to a drug that is used continuously, the veterinarian may recommend a rotation of drugs.

PREVENTION AND CONTROL

To further ascarid control, keep foaling barn and paddocks clean, store manure in a pit for two to three weeks, provide clean feed and water, place young foals on clean pasture, worm all mares four to six weeks before foaling, and worm foals and yearlings on a regular basis.

REMARKS

Foals usually first acquire ascarid infection from contaminated stalls and paddocks. Foals should be treated early in life, before the ascarids have a chance to mature and become large enough to block the intestine.

Blow Fly

The blow fly group consists of a number of species of flies, all of which breed in animal flesh.

DISTRIBUTION AND LOSSES

Blow fly presence is widespread, but they present the greatest problem in the Pacific Northwest and in the South and southwestern states.

Death losses are not excessive, but production is lowered.

SPECIES AFFECTED

All farm animals are affected.

SYMPTOMS AND SIGNS

Symptoms of blow fly infestation include infested wounds and soiled hair. Maggots spread over the body, feeding on the skin surface, producing severe irritation, and destroying the ability of the skin to function. Infested animals rapidly become weak, fevered, and unthrifty.

TREATMENT

When horses become infested with blow fly maggots their wounds should be treated with coumaphos (Co-ral) spray bomb, used according to the manufacturer's directions on the label.

PREVENTION AND CONTROL

Eliminate the blow fly by: (1) destroying dead animals by burning or deep burial, (2) using traps, poisoned baits, and electrified screens, and (3) using repellants, such as pine-tar oil.

Bots

Bots consist of three species of horse bot flies which are pests of horses in the United States: the common horse bot or nit fly *(Gastrophilus intestinalis)*, the throat bot or chin fly *(G. nasalis)*, and the nose bot or nose fly *(G. hemorrhoidalis)*.

DISTRIBUTION AND LOSSES

Bots are a problem worldwide.

Presence of bots results in loss of feed through feeding worms, itching and loss of tail hair due to rubbing, lowered work efficiency, retarded growth of young animals, lowered breeding efficiency, and death in severe infestations.

SPECIES AFFECTED

Horses, mules, asses, and zebras are affected.

SYMPTOMS AND SIGNS

Animals attacked by the bot fly may toss their heads in the air, strike the ground with their front feet, and rub their noses on each other or on any convenient object.

Infected animals may show frequent digestive upsets and even colic, lowered vitality and emaciation, and reduced work output.

TREATMENT

See Table 20-1, Recommended Compounds for Control of Internal Parasites, by Animal Species. Use drug of choice according to manufacturer's directions.

In the late fall or early winter at least one month after the first killing frost, administer one of the recommended drugs.

PREVENTION AND CONTROL

Control bot flies with frequent grooming, washing, and clipping. Prevention of reinfestation is best assured through community campaigns in which all

Fig. 20-27. Bots, *Gastrophilus intestinalis*, in the stomach of a horse. (Courtesy, Shell Chemical Company)

Fig. 20-28. Three stages in the life cycle of the botfly. (Courtesy, Shell Chemical Company)

horses within the area are treated. Thirty days prior to worming, the eggs of the bot fly which may be clinging to the body should be destroyed by (1) clipping the hair, and/or (2) washing with warm water at 120°F. The insides of the knees and the fetlocks especially should be treated in this manner.

Fly nets and nose covers offer some relief from the attacks of adult bot flies.

REMARKS

As a precaution, mares should not be treated closer than 30 days before foaling or within 14 days after foaling.

Dourine

Dourine is a chronic venereal disease caused by *Trypanosoma equiperdum*, a protozoa, spread mostly through mating.

DISTRIBUTION AND LOSSES

Dourine occurs worldwide, but is now rare in the United States.

SPECIES AFFECTED

Horses and asses are affected.

SYMPTOMS AND SIGNS

Symptoms of dourine include redness and swelling of the reproductive organs of both the mare and

the stallion, frequent urination, and increased sexual excitement in both sexes. A pussy discharge may be noted. Firm, round, flat swellings (dollar plaques) eventually appear on the body and neck. In advanced stages, there may be paralysis of the face, knuckling of the joints of the hind limbs, and dragging of the feet.

TREATMENT

Quinapyrine-sulfate has been used to treat horses infected with dourine.

PREVENTION AND CONTROL

Blood testing to diagnose suspect horses, and isolation and culling of positive-testing horses is the preferred method to control dourine.

REMARKS

An official test of the blood serum may be obtained from the USDA, on request.

Equine Piroplasmosis

Equine piroplasmosis (*Babesiasis*) is caused by either of two protozoa, *Babesia caballi* or *B. equi*, which invade the red blood cells.

DISTRIBUTION AND LOSSES

Equine piroplasmosis occurs worldwide. It was first diagnosed in the United States in 1961, in Florida.

SPECIES AFFECTED

Horses, donkeys, mules, and zebras are affected.

SYMPTOMS AND SIGNS

Equine piroplasmosis is similar to equine infectious anemia (or swamp fever). A positive diagnosis is made by demonstrating the protozoa in the red blood cells or by antigen-antibody serum tests. Clinical signs include fever (103-106°F), anemia, icterus, depression, thirst, lacrimation, and swelling of the eyelids. Constipation and colic may occur. The urine is yellow to reddish in color.

The incubation period is one to three weeks. *B. cabelli* signs may last a few days to two weeks, but horses infected with *B. equi* frequently die within 48 hours of onset of signs.

TREATMENT

Imidocarb propionate can be used to treat horses with piroplasmosis. A veterinarian should be consulted regarding diagnosis and treatment, because treatment regimens vary depending on the organism.

PREVENTION AND CONTROL

Tick control is the most effective approach. The tropical horse tick, *Dermacentor nitens*, is the vector in the United States.

The disease may also be spread by the vampire bat.

REMARKS

Horses can remain carriers of the organism for years, and stress may cause a relapse and appearance of clinical signs.

Flies and Mosquitoes

Flies and mosquitoes are usually classed as follows:

1. Biting flies and mosquitoes—this includes horse flies, deer flies, stable flies, horn flies, black flies, biting midges, and mosquitoes.
2. Nonbiting flies—include the face fly and house fly.

DISTRIBUTION AND LOSSES

Flies and mosquitoes are found wherever there are horses. They are probably the most important insect pests of horses.

SPECIES AFFECTED

Flies and mosquitoes affect horses as well as other farm animals.

SYMPTOMS AND SIGNS

They lower the vitality of horses, mar the hair coat and skin, produce a general unthrifty condition, lower performance, and make for hazards when riding or using horses. Also, they may temporarily or permanently impair the development of foals and young stock.

TREATMENT

For materials and control recommendations for horse flies, deer flies, and mosquitoes, see the section "Flies" under "Cattle Parasites."

For control of horn flies, face flies, house flies, and stable flies, the following materials and controls are recommended, with the admonition to follow label instructions:

Coumaphos (Co-ral)	Permethrin (Ectiban)
Ectrin strips (Farnum),	Permethrin + piperonyl butoxide
use on halter or	Rabon oral larvacide
brow-band	Vapona + pyrethrin +
Fenvalerate (Ectrin)	piperonyl butoxide

Also, the following compounds may be used in automatic spray systems in keeping with the label directions:

Natural pyrethrins + piperonyl	Resmethrin
butoxide	Vapona
Permethrin	

PREVENTION AND CONTROL

Biting flies and mosquitoes: Sanitation—the destruction of the breeding areas of the nests—is the key to the control of biting flies and mosquitoes. Do not allow manure or other breeding areas to accumulate. Spread manure in fields (to dry) every day or two. Control horse flies, deer flies, and mosquitoes by filling low spots in corrals or paddocks and draining all water-holding areas.

Nonbiting flies: Sanitation is the most efficient method of reducing populations of house flies. Sanitation may be additionally important if the horses are located near an urban area, in order to avoid complaints from neighbors. Residual sprays will eliminate many house flies. Also, house flies are attracted to baits (insecticides mixed with sugar or other attractive material), which are effective house fly killers.

REMARKS

Flies and mosquitoes can be the vector (carrier) of serious diseases.

Lice

Lice are small, flattened, wingless insect parasites of which there are several species, most of which are specific for a particular class of animals.

DISTRIBUTION AND LOSSES

Lice presence is widespread.

Lice retard growth, lower work efficiency, and produce unthriftiness.

SPECIES AFFECTED

Horses and mules are affected, with other species for other classes of animals.

SYMPTOMS AND SIGNS

Symptoms include intense irritation, restlessness, and loss of condition. There may be severe itching and the animal may be seen scratching, rubbing, and gnawing at the skin. The hair may be rough, thin, and lack luster; and scabs may be evident. Lice are apt to be most plentiful around the root of the tail, on the inside of the thighs, over the fetlock region, and along the neck and shoulders.

TREATMENT

Treat with periodic application, according to the directions on the label, of one of the following insecticides:

Coumaphos (Co-ral)   Permethrin
Fenvalerate (Ectrin)

PREVENTION AND CONTROL

Because of the close contact of horses during the winter months, it is practically impossible to keep them from becoming infested with lice.

For effective control, all horses should be treated simultaneously at intervals, especially in the fall about the time they are placed in winter quarters.

REMARKS

Lice show up most commonly in winter and on ill-nourished and neglected animals.

Although rarely dipped, horses may be so treated for lice, using any one of the mixtures and procedures recommended for dipping cattle.

Mites

Mites are very small parasites that produce mange (scabies, scab, itch).

DISTRIBUTION AND LOSSES

Mites (which are widespread) retard growth, lower work efficiency, and produce unthriftiness.

SPECIES AFFECTED

Horses and mules are affected. Each species of animals has a species or subspecies of mange mites.

SYMPTOMS AND SIGNS

Symptoms of mites are marked irritation, itching and scratching, and crusting over of the skin, accompanied by formation of a thick, tough, wrinkled skin.

TREATMENT

Any of the following may be used in keeping with the manufacturer's directions on the label:

Ivermectin
Moxidectin

PREVENTION AND CONTROL

Avoid contact with diseased animals or infested premises.

In case of an outbreak, contact the local veterinarian or livestock sanitation official.

Control mites by spraying or dipping infested animals with suitable insecticides, and quarantine affected herds.

REMARKS

There are two chief forms of mange: sarcoptic mange (caused by burrowing mites), and psoroptic mange (caused by mites that bite the skin and suck blood but do not burrow).

The disease appears to spread most rapidly during the winter months and among young and poorly nourished animals.

Although rarely dipped, horses may be so treated if it is practical and convenient to do so.

Pinworms (Rectal Worms)

Two species of pinworms are frequently found in horses. *Oxyuris equi* are whitish worms with long, slender tails, whereas *Probstmyria vivipara* are so small as to be scarcely visible to the naked eye.

DISTRIBUTION AND LOSSES

Pinworms are found throughout the United States.

SPECIES AFFECTED

Horses are affected, but they also occur in humans with different species of worms.

SYMPTOMS AND SIGNS

Symptoms are irritation of the anus and tail rubbing. Heavy infections may also cause digestive disturbances and produce anemia.

The large pinworm is the most damaging to the horse.

TREATMENT

See Table 20-1, Recommended Compounds for Control of Internal Parasites, by Animal Species. Use drug of choice according to manufacturer's directions.

PREVENTION AND CONTROL

Prevention entails sanitation and keeping animals separated from their own excrement.

Ringworm

Ringworm is a contagious disease of the outer layers of skin caused by certain microscopic molds or fungi.

DISTRIBUTION AND LOSSES

Ringworm occurs throughout the United States.

It is unsightly and affected animals may experience considerable discomfort, but actual economic losses are not too great.

SPECIES AFFECTED

All animals and humans are affected.

SYMPTOMS AND SIGNS

Symptoms of ringworm are round, scaly areas almost devoid of hair appearing mainly in the vicinity of the eyes, ears, side of the neck, or the root of the tail. Mild itching usually accompanies the disease.

TREATMENT

Clip the hair from the affected areas, remove scabs with a brush and mild soap. Paint affected areas with tincture of iodine or salicylic acid and alcohol (1 part in 10) every three days until cleared up. Copper napthenate or dichlorphene is also effective.

PREVENTION AND CONTROL

Isolate affected animals. Disinfect everything that has been in contact with infested animals, including curry combs and brushes. Practice strict sanitation.

REMARKS

Though ringworm may appear among animals in pasture, it is far more prevalent as a stable disease.

Screwworm

Screwworm flies raise their maggots in the living flesh of animals.

DISTRIBUTION AND LOSSES

Screwworms are found mostly in the southern and southwestern states, where they once caused 50 percent of the normal annual livestock losses.

SPECIES AFFECTED

All farm animals are affected.

SYMPTOMS AND SIGNS

Loss of appetite, unthriftiness, and lowered activity are signs of screwworm infestation.

TREATMENT

When maggots (larvae) infest the flesh of an animal, a sample of the larvae should be sent to proper authorities for identification, and the animal should be treated with a proper insecticide. For maggots in wounds, treat with coumaphos (Co Ral) spray bomb.

PREVENTION AND CONTROL

Keep animal wounds to a minimum.

The screwworm eradication program, by sterilization, has been very effective. This consists in sterilizing male screwworms, in the pupal stages with gamma rays. Male screwworms mate repeatedly, but females mate only once. Thus, when a female mates with a sterilized male, only infertile eggs are laid. The release of millions of sterilized males has led to the eradication of screwworms from most of the United States.

REMARKS

Screwworms sometimes infest the prepuce of geldings.

Strongyles

Strongyles consist of about 60 species; three are large (up to 2 inches in length) and the rest small (some scarcely visible to the naked eye). The large strongyles are variously referred to as bloodworms (*Strongylus vulgaris*), palisade worms, sclerostomes and red worms.

DISTRIBUTION AND LOSSES

Strongyles are found throughout the United States, wherever horses and mules are pastured.

Parasitism with strongyles, both large and small, is known to cause colic. *S. vulgaris* infection was once considered to cause 90 percent of all colic, because the larvae cause peritonitis, clotting of the arteries supplying blood to the intestines, and intestinal motility disturbances. The larval stages of small strongyles encyst into the large intestine, and can contribute to poor performance, colic, and even death.

SPECIES AFFECTED

Horses and mules are affected.

SYMPTOMS AND SIGNS

Collectively these symptoms indicate the disease known as strongylosis: lack of appetite, anemia, progressive emaciation, a rough hair coat, sunken eyes, digestive disturbances including colic, a tucked-up appearance, and sometimes posterior paralysis and death.

Harmful effects are greatest with younger animals.

One species of large strongyles *(S. vulgaris)* may permanently damage an intestinal blood vessel wall, resulting in death at any age. Also, they may cause severe colic, which may terminate in death.

Small strongyles promote colic, cause general loss of body condition, and stunt the growth and development of horses. Occasionally, sudden onset of

Fig. 20-29. Ulceration of the intestine of a horse due to small strongyles. (Courtesy, Shell Chemical Company)

diarrhea and colic can occur when massive numbers of larvae emerge from the cysts in the large intestine.

TREATMENT

See Table 20-1, "Recommended Compounds for Control of Internal Parasites, by Animal Species." Use drug of choice according to manufacturer's directions.

Small strongyles are hard to diagnose using a fecal exam, as the encysted larvae will not be releasing eggs into the intestine. Often, in consultation with a veterinarian, the horse owner may elect to treat a horse or herd of horses for small strongyles based upon clinical signs alone. Fenbendazole, used at twice the normal dosage for five consecutive days, is the only proven and approved drug to control encysted small strongyles.

PREVENTION AND CONTROL

Approved drugs should be rotated on an annual basis; that is, use one product for about a year, then rotate. Also, it is very important that rotation be between drug classes.

REMARKS

Strongyles are not transmissible to ruminants or swine.

Heavily infected animals may have one or more of the three species of large strongyles along with 10 or 12 species of small strongyles.

Summer Sores

Summer sores (*Cutaneous habronemiasis*) are caused by larval form of *Habronema* spp (large stomach worm) in the skin. Also known as Jack sores or Bursatti.

DISTRIBUTION AND LOSSES

Summer sores occur throughout the United States.

SPECIES AFFECTED

Horses, mules, and jacks are affected.

SYMPTOMS AND SIGNS

Signs of summer sores are unsightly and uncomfortable skin lesions of various sizes.

TREATMENT

Ivomec is commonly used in the treatment of summer sores. Surgical removal or cauterization of the sores may be resorted to.

PREVENTION AND CONTROL

Prevention consists in good fly control, as flies are the vector.

REMARKS

Veterinary diagnosis is sometimes required to differentiate this lesion from "proud flesh" and ringworm.

Tapeworms

Three species of tapeworms are of economic importance in the horse. *Anoplocephala perfoliata* is most common and most damaging.

DISTRIBUTION AND LOSSES

Tapeworms are found throughout the northern part of the United States.

Losses are primarily in wasted feed and retarded growth.

SPECIES AFFECTED

Horses are affected.

SYMPTOMS AND SIGNS

Heavy infections may cause digestive disturbances.

TREATMENT

See Table 20-1, Recommended Compounds for Control of Internal Parasites, by Animal Species. Use drug of choice according to manufacturer's directions.

Pyrantel pamoate is an effective drug for treating tapeworm infections.

PREVENTION AND CONTROL

The process of using drugs effective against tapeworms will reduce the population of eggs on the pasture, and is therefore the most practical method in controlling tapeworms.

REMARKS

Grass mites are the intermedial hosts.

Threadworms

Threadworms (*Strongyloides westeri*) are common in the small intestine of foals.

DISTRIBUTION AND LOSSES

Threadworms are very common in foals.

SPECIES AFFECTED

Threadworms affect horses.

SYMPTOMS AND SIGNS

Diarrhea in foals can be a sign of threadworms.

TREATMENT

See Table 20-1, Recommended Compounds for Control of Internal Parasites, by Animal Species. Use drug of choice according to manufacturer's directions.

REMARKS

Deworming the mare several days before foaling with an avermectin product will prevent threadworms in foals.

Ticks

Tick control may vary, depending on the tick species.

DISTRIBUTION AND LOSSES

Ticks are particularly prevalent on horses in the southern and western parts of the United States.

SPECIES AFFECTED

All farm animals and humans are affected.

SYMPTOMS AND SIGNS

Ticks reduce the vitality of horses through constant irritation and loss of blood.

Massive infestations may cause anemia, loss of weight, and even death. "Head heaviness" is often associated with massive infestations of ear ticks. Other losses may result from the simple presence of the ticks on the animal, a factor called *tick worry*.

TREATMENT

The following insecticides may be used for tick control:

Amatraz	Fenvalerate (Ectrin)
Coumaphos (Co Ral)	Permethrin
Crotoxyphos (Ciodrin)	Rabon

PREVENTION AND CONTROL

Because most species of ticks, except for the ear tick and tropical horse tick, attach to the external surfaces of horses, an application of the recommended insecticide by spray or wipe-on will give effective control. Ear ticks and tropical horse ticks should be treated by applying the chemical into the ears of the horses. Since horses are often confined to rather small areas, treatments of the premises may also help control heavy infestations of ticks.

REMARKS

Ticks are important to those taking care of horses because they may transmit diseases such as equine piroplasmosis (carried by *Anocentor nitens*) or cattle fever (carried by the *Boophilus* species). Also, most of the ticks mentioned may be vectors of anaplasmosis, and several species can cause tick paralysis in hosts.

APPENDIX A

Diseases of Livestock with Designations of the Animal Systems That Are Commonly Affected

Diseases of Livestock with Designations of the Animal Systems That Are Commonly Affected

Disease	Page	Digestive	Genitourinary	Respiratory	Circulatory	Nervous	Epidermal	Musculo-Skeletal	Miscellaneous
Acetonemia (Ketosis)	258				✓	✓			Metabolic
Acidosis.	157	✓				✓			
Actinobacillosis	148	✓						✓	
Actinomycosis	147	✓						✓	
Anaplasmosis.	229			✓	✓	✓			
Anemia	229				✓				
Anthrax.	249				✓		✓	✓	
Aspergillosis (Mycotic diarrhea).	168	✓							
Atrophic rhinitis.	222			✓					
Bacillary hemoglobinuria.	252		✓						Liver
Black disease	253				✓				Liver
Blackleg.	250							✓	
Bloat.	151	✓							
Bluetongue	145	✓						✓	
Bovine thromboembolic meningo encephalitis (Brain fever)	243					✓			
Bovine viral diarrhea (BVD)	161	✓	✓	✓					
Broken penis.	206		✓		✓				
Brucella ovis	192		✓						
Brucellosis.	190		✓						
Campylobacteriosis	194		✓						
Choke.	150	✓							
Coccidiosis	175	✓							
Colic.	166	✓							
Colitis X	177	✓							
Contagious ecthyma.	144	✓							Udder
Cystic ovarian disease	187		✓						
Diarrhea	167	✓							
Displaced abomasum	160	✓							

(Continued)

Diseases (Continued)

Disease	Page	Digestive	Genitourinary	Respiratory	Circulatory	Nervous	Epidermal	Musculo-Skeletal	Miscellaneous
Dysentery									
Lamb.	170								
Bovine winter	171	✓							
Swine	171	✓							
Edema.	233				✓			✓	
Enterotoxemia.	163	✓				✓			
Epididymitis.	192		✓						
Epistaxis	224			✓					
Epizootic bovine abortion.	195		✓						
Equine encephalomyelitis	241					✓			
Equine infectious anemia	231	✓		✓	✓				
Equine rhinopneumonitis	215		✓	✓					
Equine protozoal myeloencephalitis.	246					✓			
Exertional rhabdomyolysis	276							✓	
Fly bite hypersensitivity	280						✓		
Foot-and-mouth disease.	140	✓						✓	
Foot rot	269							✓	
Founder	271	✓						✓	
Freemartin	186		✓						
Gastroenteritis (Scours)	172	✓							
Grass tetany	259					✓			Metabolic
Gut edema	169				✓	✓			
Hair balls.	165	✓							
Heaves	218			✓					
Hematoma	234				✓			✓	
Hemolytic icterus.	230				✓				Liver
Hypoglycemia of pigs	264					✓			
Hog cholera.	254	✓			✓		✓		
Impaction	154	✓							
Infectious bovine rhinotracheitis	214	✓	✓	✓					

(Continued)

Diseases (Continued)

Disease	Page	Digestive	Genitourinary	Respiratory	Circulatory	Nervous	Epidermal	Musculo-Skeletal	Miscellaneous
Infectious keratitis	267								Eye
Influenza									
Equine.	221			✓					
Swine	222		✓	✓	✓			✓	
Johne's disease.	256	✓							
Leptospirosis	192		✓		✓				
Malignant edema	251				✓		✓	✓	
Mastitis	199				✓				Udder
Metritis	196		✓						
Milk fever	257					✓			
Mineral deficiencies	29				✓	✓		✓	
Navel III	278				✓			✓	
Necrotic enteritis.	174	✓							
Oral necrobaccillosis	147	✓		✓					
Orchitis.	190		✓						
Ostertagiasis	320	✓			✓				
Ovaries, cystic.	187		✓						
Parainfluenza₃	213			✓					
Parakeratosis.	265						✓		
Paraphimosis	206		✓						
Pasteurellosis.	210			✓					
Peritonitis	159	✓							
Phimosis	206		✓						
Photosensitization	281						✓		Liver
Pizzle rot.	205		✓				✓		
Poisonings	283	✓		✓		✓			
Polyarthritis.	279				✓			✓	
Potomac horse fever	168	✓							
Porcine respiratory disease complex	219			✓					
Pregnancy disease of ewes	261					✓			

(Continued)

Diseases (Continued)

Disease	Page	Digestive	Genitourinary	Respiratory	Circulatory	Nervous	Epidermal	Musculo-Skeletal	Miscellaneous
Prolapsed rectum	177	✓							
Prolapsed uterus	203		✓						
Prolapsed vagina	203		✓						
Pseudorabies	240					✓			
Pyometra	196		✓						
Rabies	239					✓			
Retained meconium of foals	169	✓							
Retained placenta	188		✓						
Ringworm	275						✓		
Salmonellosis	176	✓	✓						
Scrapie	243					✓			
Shipping fever (BRDC)	211			✓					
Shock	234				✓				
Strangles	217			✓					
Sunstroke	244			✓	✓	✓			
Sweet clover disease	289				✓				
Swine erysipelas	253				✓		✓	✓	
Swine pox	276						✓		
Tetanus	238					✓		✓	
Transmissible gastroenteritis	162	✓							
Traumatic reticulitis (Hardware disease)	153	✓							
Trichomoniasis	194		✓						
Tuberculosis	216			✓	✓				
Ulcerative dermatosis	146	✓	✓				✓		
Urinary calculi	204		✓						
Vaginitis	198		✓						
Vesicular exanthema	143	✓					✓		
Vesicular stomatitis	142	✓							
Vitamin deficiencies	25	✓			✓	✓	✓	✓	
Warts	274						✓		
White muscle disease	262							✓	

APPENDIX B

Mastitis Treatment and Records

Good Dairy Management Practices

Example Herd Plan

System: Mastitis

Personnel: Owner, herdsman, or milker

Process: Clinical mastitis (CM) treatment & evaluation

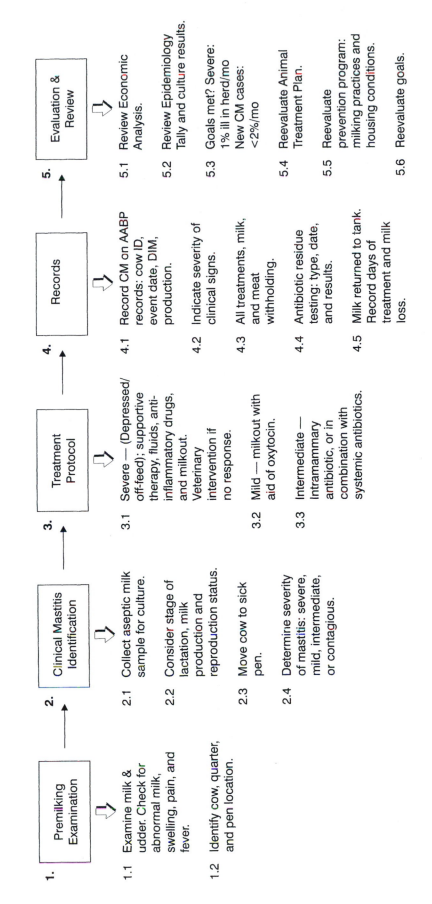

1. Premilking Examination

1.1 Examine milk & udder. Check for abnormal milk, swelling, pain, and fever.

1.2 Identify cow, quarter, and pen location.

2. Clinical Mastitis Identification

2.1 Collect aseptic milk sample for culture.

2.2 Consider stage of lactation, milk production and reproduction status.

2.3 Move cow to sick pen.

2.4 Determine severity of mastitis: severe, mild, intermediate, or contagious.

3. Treatment Protocol

3.1 Severe — (Depressed/off-feed); supportive therapy, fluids, anti-inflammatory drugs, and milkout. Veterinary intervention if no response.

3.2 Mild — milkout with aid of oxytocin.

3.3 Intermediate — Intramammary antibiotic, or in combination with systemic antibiotics.

4. Records

4.1 Record CM on AABP records: cow ID, event date, DIM, production.

4.2 Indicate severity of clinical signs.

4.3 All treatments, milk, and meat withholding.

4.4 Antibiotic residue testing: type, date, and results.

4.5 Milk returned to tank. Record days of treatment and milk loss.

5. Evaluation & Review

5.1 Review Economic Analysis.

5.2 Review Epidemiology Tally and culture results.

5.3 Goals met? Severe: 1% ill in herd/mo New CM cases: <2%/mo

5.4 Reevaluate Animal Treatment Plan.

5.5 Reevaluate prevention program: milking practices and housing conditions.

5.6 Reevaluate goals.

Treatment Records and Treatment Plan

Protocol Number	Diagnosis or Conditions to Treat or Signs	Treatment Plan			Withdrawal Time	
		Antibiotic or Drug Used	Dose and Route	Length of Treatment	Milk (hrs)	Meat (days)

Conversion Table
Withdrawal Time to Number of Milkings

Drug Withdrawal Time (hrs)	Number Milking/Day		
	2X	3X	4X
36	3	5	6
48	4	6	8
72	6	9	12
96	8	12	16

Culture Code

1 Strep ag.	5 Staph species
2 Staph aureus	6 _____
3 Environmental Streps	7 _____
4 Coliforms	8 _____

Route of Administration Code

OR = Oral	IMM = Intramammary
SQ = Subcutaneous	IU = Intrauterine
IM = Intramuscular	TP = Topical
IV = Intravenous	

How to Calculate Withdrawal Times

Each withdrawal day is a full 24 hours starting witht he last time an animal receives the drug.
Here is an example of the pre-slaughter withdrawal time:

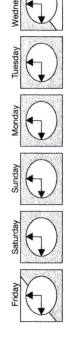

Use of a drug with a five-day pre-slaughter time is discontinued at 9 a.m. on Friday. At 9 a.m. on Saturday, the treated animals have completed their first withdrawal day.
The fifth withdrawal day will end at 9 a.m. on Wednesday.

Here are two illustrations of how milk discard times should be calculated. One shows the milkings and the other milk discard hours:

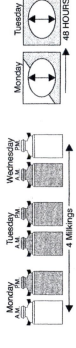

(Courtesy, American Association of Bovine Practitioners and Pharmacia & Upjohn Animal Health)

Daily Treatment Record

HERD: _____   TIME PERIOD: _____

COW ID	TIME OF TREATMENT				PEN	DIAGNOSIS		TREATMENT USED	WITHDRAWAL TIME		CALCULATED WITHDRAWAL PERIOD EXPIRES MILK/MEAT	ACTUAL DATE MILK IN TANK	RESIDUE TEST		REMARKS
	DATE	AM	PM	3X					MILK (hrs)	MEAT (days)			DATE TESTED	TEST RESULTS	(for example: initials of person treating or testing)
						LF	RF								
						LR	RR								
						LF	RF								
						LR	RR								
						LF	RF								
						LR	RR								
						LF	RF								
						LR	RR								
						LF	RF								
						LR	RR								
						LF	RF								
						LR	RR								
						LF	RF								
						LR	RR								
						LF	RF								
						LR	RR								
						LF	RF								
						LR	RR								
						LF	RF								
						LR	RR								
						LF	RF								
						LR	RR								

(Courtesy, American Association of Bovine Practitioners and Pharmacia & Upjohn Animal Health)

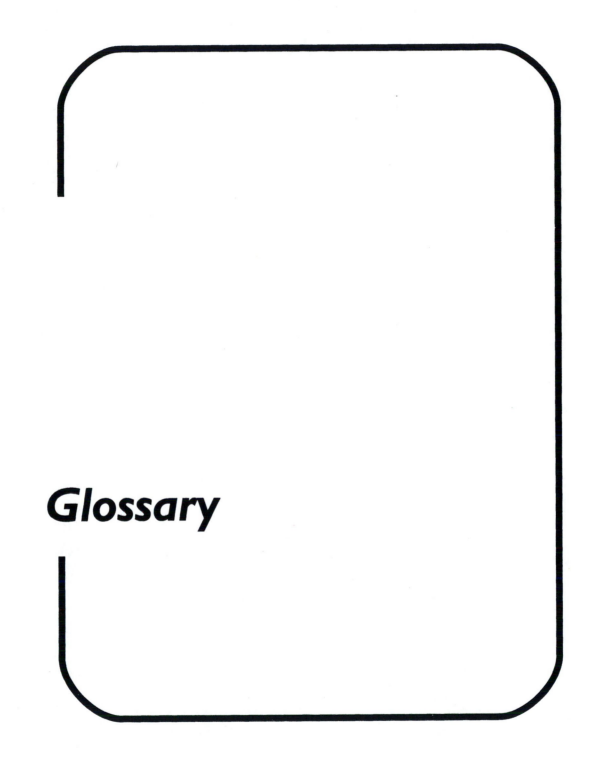

Glossary

Introduction to Veterinary Terminology

This section is divided into two parts. The first part includes selected prefixes, suffixes, and roots of common veterinary terms. The second section is a compilation of words that are often encountered in publications dealing with animal health and disease control.

Words are composed of roots, prefixes, and suffixes. A *root* is the basic part of a word. It may be modified by a *prefix* placed before the root or by a *suffix* which follows it.

PREFIXES

Prefix	Meaning	Example
a	without or lack of	afebrile
an	without	anorexia
ab	from, off, away from	abnormal
ambi	both	ambidextrous
ana	excessive	anadipsia
ante	before	antefebrile
anti	against	antiseptic
bi	twice or two	bilateral
brady	slow	bradycardia
co	together	cohabitation
con	together	concurrent
contra	against, opposite	contraindication
counter	against	counteract
crypto	hidden or undescended	cryptorchid
de	remove from	denude
dis	denoting separation	dislocation
dys	denoting pain, difficulty	dysuria
en	inside	encephalomyelitis

(Continued)

Prefix	Meaning	Example
endo	within, inward	endotoxin
epi	upon, outer	epidermis
eu	normal	eupeptic
ex	outside, from	exudate
hemo	pertaining to blood	hemoglobin
hetero	unlike	heterozygote
homo	likeness	homozygote
hydro	water	hydrocephalic
hyper	above, over, excessive	hyperesthesia
hypo	less, below, insufficient	hypocephalic
im	not	imperforate
in	not	infertile
inter	between	intercellular
intra	inside, within, into	intravenous
intro	into, within	introgastric
macro	large, able to see with the naked eye	macroparasite
micro	small, microscopic	micro-organism
meta	transformation	metamorphosis
multi	many	multiparous
neo	young, new, recent	neonatal
peri	around	pericardium
photo	pertaining to light	photosynthesis
poly	many	polyuria
post	behind or after	post-partum
pre	before	prenatal
pseudo	false	pseudohermaphrodite
pyo	pus, purulent	pyometra
thermo	heat	thermometer
trans	across, through	transfusion

SUFFIXES

Suffix	Meaning	Example
ac	pertaining to	cardiac
al	pertaining to	digital
an	belonging to	ovarian
asis	affected with	acariasis
cide	to kill	insecticide
coccus (cocci)	round-shaped bacterium (bacteria)	streptococcus (cocci)
ectomy	removal of	hysterectomy
fuge	to expel	vermifuge
ity	quality	viscosity
lysis	freeing dissolution	homolysis
mania	pertaining to madness or excess desire	nymphomania
oid	like	mucoid
ology	study of	pathology

(Continued)

Suffix	Meaning		Example
oscopy	inspecting		rumenoscopy
oma	tumor, morbid condition		carcinoma
osis	marked condition		leukocytosis
otomy	opening into		rumenotomy
ous	denoting material		serous
phobia	fear of		hydrophobia
rrhagic	to burst forth		hemorrhagic

ROOTS

Technical Term	Common Meaning
adeno	gland
algia	pertaining to pain
angio	pertaining to blood vessels
arterio	pertaining to arteries
arthro	pertaining to joints
atresia	closed or imperforate
blast	a germ or cell
broncho	pertaining to bronchi
carcino	pertaining to cancer
cardio	pertaining to heart
cephalo	pertaining to head
cervico	pertaining to neck or cervix
chroma	color
cranio	pertaining to skull
cyst	bag filled with fluid
cyto	pertaining to cell
derma	pertaining to skin
encephalo	pertaining to the brain
entero	denoting relation to intestine
esthesia	denoting sensation
fibro	fiber
galacto	milk
gastro	pertaining to stomach
glyco (gluco)	sugar
hemo	pertaining to blood
hepato	pertaining to liver
histo	relating to tissue
hyster	pertaining to uterus
ileo	pertaining to ileum of the intestine
labia	pertaining to lip
lacto	pertaining to milk
leuco (leuko)	pertaining to white
litho	pertaining to stone
mania	madness
mast	pertaining to udder
metro (metri)	uterus

(Continued)

Technical Term	*Common Meaning*
myco	fungus
myo	muscle
necro	pertaining to death
nephro	pertaining to kidney
neuro	pertaining to nerves
odonto	teeth
opthalmo	in relation to the eye
orchido (orchio)	pertaining to testicles
ossa (ossi)	pertaining to bone
osteo	bone
oto	pertaining to ear
ovario	pertaining to ovary
path	pertaining to disease
phagia	to swallow or engulf
phleb	pertaining to vein
pnea	breathing
pneumo	pertaining to lungs or air
polio	pertaining to gray
pyelo	pelvis (kidney)
pyo	pus or purulent
rhino	pertaining to nose
septic	pertaining to poison
thermo	heat
tracheo	windpipe or trachea
trophy	state of nutrition or health
taxis	order
uria	urine
zoo	animal

Compilation of Words

Abacterial—free from bacteria.

Abortion—premature expulsion of the fetus.

Abnormal—deviating from that which is typical.

Abscess—localized collection of pus in a cavity formed by the disintegration of tissue.

Acaracide—drug that destroys mites and ticks.

Acariasis—infestation with mites or ticks.

Acarid—any mite or tick of the family *Acaridae* or order *Acarina*.

Acid-fast—property of not being readily decolorized by acids when stained.

Acute—having a short but relatively severe course.

Adrenal glands—paired, ductless glands lying next to the kidneys.

Aerobic—growing only in air or free oxygen.

Aerogens—gas-producing bacteria.

Afebrile—without fever.

Agalactia—absence of milk or the failure to secrete milk.

Agglutination—the clumping of the cells distributed in a fluid, which is observed when a bacterial culture is treated with serum prepared against that particular organism.

Albuminuria—the presence of albumin in the urine.

Allergy—hypersensitivity to a specific substance.

Alopecia—loss of hair.

Ambidexterous—able to use both hands well.

Amino acids—organic substances from which organisms build protein, or the end product of protein production.

Anabolism—constructive process by which simple substances are convened by living cells into more complex compounds.

Anadipsia—excessive thirst.

Anaerobe—any microorganism that can live without air or free oxygen.

Anaerobic—thriving without air.

Anaphylactic shock—violent attack of symptoms produced by an injection of a serum or protein given to a sensitive animal.

Anaphylaxis—state of increased sensitivity in an animal following an injection of foreign matter.

Anemia—condition in which the blood is deficient either in quality or quantity.

Anesthetics—drugs that produce insensibility to pain.

Anestrus—period of sexual inactivity between two estrus cycles.

Anorexia—lack of or loss of appetite for feed.

Anoxia—oxygen deficiency.

Antefebrile—before the onset of fever.

Anthelmintics—drugs that kill or expel gastrointestinal worms.

Antibiotic—germ-killing substance produced by a bacterium or mold.

Antibody—specific substance produced by and in an animal as a reaction to the presence of an antigen.

Antidote—remedy for counteracting a poison.

Antigen—substance that induces the formation of antibodies in the animal organism under suitable conditions.

Antigenic—having the properties of an antigen.

Antiseptic—substance used to prevent or stop the growth of microorganisms, usually applied to living tissues that have been damaged.

Antiserum—blood serum from an animal that contains antibodies for a special disease.

Antitoxin—antibody capable of combining with and neutralizing a specific toxin.

Apnea—cessation of breathing.

Arthropods—animals with jointed limbs, such as insects, spiders, crustaceans, etc.

Asepsis—freedom from infection.

Aseptic—being free from infectious micro-organisms.

Asphyxiation—suffocation due to a lack of oxygen.

Ataxia—lack of muscular coordination.

Atrophy—wasting of the tissues, emaciation.

Attenuate—to weaken or reduce in virulence.

Attenuation—process of reducing or weakening the virulence of a micro-organism by cultivation on artificial media.

Autogenous vaccine—vaccine that is made from the patient's own bacteria instead of from stock cultures.

Autolysis—spontaneous disintegration of bacteria by their own bacterial enzymes.

Autopsy—post-mortem examination of a body.

Avirulent—without the ability to produce disease.

Bacillus—rod-shaped organism.

Bacteria—minute, one-celled, microscopic plant-like organisms that multiply by fission and have no chlorophyll.

Bacteriocide—substance that kills bacteria.

Bacteriolysis—destruction or dissolution of bacteria inside or outside the animal body.

Bacteriophage—ultramicroscopic virus that produces a transmissible dissolution of specific bacterial cells; regarded by some as a living agent and by others as an enzyme.

Bacteriostatic—preventing the growth or multiplication of bacteria.

Benign—not malignant, favorable for recovery.

Bilateral—having two sides.

Bradycardia—abnormally slow heartbeat.

Bronchiole—one of many subdivisions of the bronchial tubes within a lung.

Bronchitis—inflammation of the mucous membrane of the bronchial tubes.

Carcinoma—malignant cancerous growth.

Cardiac—pertaining to the heart.

Carnivore—flesh-eating animal.

Carrier—animal or person in apparently good health who harbors pathogenic micro-organisms.

Catabolism—process of destroying or breaking down tissues and cells of the body from complex to simpler compounds.

Cathartics—drugs that cause the evacuation of the bowels.

Caudal—of or pertaining to the tail.

Cellulose—carbohydrate substance present in the walls of plant cells.

Cervical—pertaining to the neck of or to any part of the cervix.

Chemotherapy—use of chemicals to treat infectious diseases.

Chromatin—cellular substances containing the genes and forming chromosomes.

Chromosomes—rod-shaped, gene-bearing bodies formed in the cell nucleus during cell division.

Chronic—long continued; not acute.

Coagulant—agent that causes coagulation or clotting.

Cohabitation—two or more organisms living together.

Colostrum—fluid secreted by the mammary gland a few days before or after parturition.

Communicable—readily transferred from one individual to another.

Contagion—spread of disease by direct or indirect contact.

Contagious—easily transmitted from one individual to another, usually by droplet infection.

Contraindication—condition of a disease that renders a particular treatment undesirable.

Convalescence—period of recovery following an illness.

Cryptorchid—condition in which one or both testicles have not descended from the body cavity into the scrotum.

Cumulative—building up of a substance over a long period of time.

Cyanosis—bluish discoloration of the skin.

Cystitis—inflammation of the bladder.

Debilitation—loss of strength, or a weakened condition.

Defecation—act of eliminating fecal material.

Deficient—to be lacking in a substance or quantity.

Dehydration—state of being critically low on body fluids, water.

Deleterious—injurious.

Dermatomycosis—any skin disease caused by a fungus.

Dermatitis—inflammation of the skin.

Dermatosis—any skin disease.

Desiccation—process of drying.

Digestion—process of or act of converting food into materials fit to be absorbed and assimilated.

Dilate—to enlarge, expand; to open.

Disinfectant—any agent that kills bacteria or other micro-organisms.

Disinfection—process of destroying pathogenic organisms.

Dormancy—period of rest or inactivity.

Dose—proper amount of a medicine to be given at one time.

Dysentery—frequent small, watery stools, usually containing blood and mucus, accompanied by pain.

Dyspepsia—impairment of digestion.

Dyspnea—difficult or labored breathing.

Dystocia—painful or slow delivery or birth.

Dysuria—difficult or painful urination.

Edema—presence of abnormally large amounts of fluid in the intercellular tissue spaces of the body.

Egestion—waste material that is excreted from the digestive tract.

Emaciation—abnormally thin and poor condition.

Emetics—drugs that produce vomiting.

Encephalomyelitis—inflammation of the brain and spinal cord.

Encyst—to become enclosed.

Endemic—referring to a disease that is prevalent in a particular district or region at any one time.

Endocarditis—inflammation of the endocardium or epithelial lining membrane of the heart.

Endotoxin—toxin produced within an organism and liberated only when the organism disintegrates or is destroyed.

Enteric—pertaining to the intestines.

Enteritis—inflammation of the intestines.

Environment—external surroundings and influences.

Enzootic—referring to the prevalence of an animal disease in a particular district or region.

Enzyme—catalyst present in digestive fluid that brings about a chemical change in food.

Epidemiology—study of the elements that cause diseases to occur.

Epithelium—surface covering of the body and lining of the inner cavities.

Epizootic—referring to a disease of animals (of the same kind), which spreads rapidly and is widely diffused.

Eradicate—to be rid of; to destroy.

Erythrocyte—red blood corpuscle.

Etiology—study of the causes of diseases.

Eupeptic—having normal digestion.

Eviscerate—to remove the entrails or viscera.

Exotoxin—soluble toxin excreted by specific bacteria and absorbed into the tissues of the host.

Expectorants—drugs that cause expulsion of mucus from the respiratory tract.

Extract—solid preparation obtained by evaporating a solution of a drug, the juice of a plant, etc. (Vitamin extracts are used to supplement a diet.)

Exudate—any substance that is thrown out from the tissues and deposits in or upon them.

Facultative aerobe—micro-organism that lives in the presence of oxygen, but may live without it.

Facultative anaerobe—micro-organism that normally does not grow in the presence of oxygen, but may acquire this property under certain circumstances.

Fats—oily substances that cover the connective tissue of an animal.

Febrile—having fever.

Feces (manure)—waste material of the digestive system.

Fermentation—decomposition of complex molecules through the influence of a substance or an organism.

Fertilization—union of a sperm and an egg.

Filterable—capable of passing through the pores of a filter.

Filterable virus—virus small enough to pass through the pores of a bacteriological filter.

Fomites—substances other than food that may harbor or transmit a disease.

Gamete—male or female reproductive or germ cell.

Gene—unit of the chromosome responsible for the transmission of hereditary traits.

Germ—micro-organism that is usually pathogenic.

Germicide—any agent that kills germs.

Gram-negative—referring to bacteria that lose the initial color stain of the Gram's stain and are decolorized so that they take the final stain and appear red.

Gram-positive—referring to bacteria that take the initial stain of the Gram's stain and are not decolorized so that they appear purple.

Gustatory—pertaining to the sense of taste.

Hematoma—tumor containing effused blood.

Hematopoiesis—formation or production of blood.

Hematuria—the presence of blood in urine.

Hemoglobin—red coloring matter of the red blood corpuscles.

Hemolysis—dissolution or breakdown of red blood corpuscles with the liberation of the hemoglobin from the cells.

Hemorrhagic—characterized by excessive bleeding.

Hemostatics—substances that check internal hemorrhage.

Hepatitis—inflammation of the liver.

Herbivore—plant-eating animal.

Heredity—transmission of traits from parent to off-spring.

Heterologous serum—serum derived from another species or disease.

Heterozygote—plant or animal having a trait or traits formed by unlike genes (one dominant gene and one recessive gene).

Homogenous—of the same kind throughout.

Homologous serum—serum that is derived from the same species or like disease.

Homostasis—state of balance.

Homozygote—plant or animal having a trait or traits formed by like genes (two dominant genes or two recessive genes).

Hormone—chemical substance produced in an organ and transported to another organ to produce a specific effect.

Hybrid—offspring of two animals of different breeds, varieties, or species.

Hydrocephalus—accumulation of fluid on the brain.

Hydrophobia—literally, fear of water; commonly refers to rabies, which is a misnomer. (Rabid animals are not necessarily afraid of water; they often have difficulty drinking water due to paralysis of the tongue.)

Hyperemia—excess blood in any part of the body.

Hyperesthesia—excessive sensitivity of a part of the body to touch or pressure.

Hyperimmunization—process of increasing the immunity of an animal by increasing injection of an antigen, subsequent to the establishment of an initial immunity; enhanced immunity.

Hypocalcemia—low level of blood calcium.

Hypoglycemia—low level of blood sugar.

Hysterectomy—surgical removal of all or part of the uterus.

Icteric—relating to or affected with jaundice.

Immune—having protection against a particular disease as by inoculation or inheritance.

Immunity—level of resistance to a particular infection.

Immunization—process of conferring immunity to an individual.

Immunology—science or study of immunity and its factors.

Imperforate—having no opening; lacking the usual or normal opening.

Incubation period—elapsed time between exposure to infection and the appearance of disease symptoms.

Induration—process of hardening; a hardened spot.

Infection—invasion of pathogenic organisms into body tissues so that the tissues are affected and altered.

Infectious disease—disease that is caused by a living agent, such as bacteria, protozoa, viruses, or fungi, which may or may not be contagious.

Infertile—barren, not productive.

Ingestion—the process of taking food into the digestive tract.

Inorganic—inanimate material not containing a hydrocarbon or its derivative.

Insecticide—chemical used to kill insects.

Intercellular—between cells.

Intracellular—within cells.

Intradermal—within the skin proper.

Intramuscular—within the muscles.

Intraperitoneal—within the peritoneal cavity.

Intravenous—within a vein.

Introgastric—in the stomach.

In utero—within the uterus.

In vitro—in the test tube, outside the animal body.

In vivo—in the living body.

Jaundice—yellowish pigmentation of the skin tissues and body fluids caused by the deposition of bile pigments.

Lactation—secretion of milk by a female mammal.

Lesion—injury or wound.

Lethal—deadly.

Leucocyte—white or colorless blood corpuscle.

Libido—sexual instinct.

Lysin—antibody that dissolves or disintegrates cells: bacteriolysin dissolves bacteria, hemolysin dissolves red blood cells.

Macroparasite—parasite visible to the naked eye.

Malignant—tending to become worse with the passage of time, virulent.

Manifest—to show, to appear.

Masticate—to chew.

Mastitis—inflammation of the mammary gland.

Membrane—layer of tissue covering a surface or dividing a space or organ.

Mesophiles—parasites and often pathogenic bacteria that grow best at a body temperature of 98.6°F.

Metabolism—chemical changes by which the nutritional and functional activities of an organism are maintained.

Metamorphosis—transitional change in structure from one distinct phase to another.

Metastasis—transfer of disease from one organ or part to another not directly connected with it.

Micro-organism—organism so small that magnification is required to observe it.

Migrate—to move from one place to another.

Morbidity—condition of being sick or diseased.

Morphology—science of the form and structure of plants and animals.

Mortality—death rate.

Mucoid—resembling mucus.

Mucous membrane—form of epithelial tissue that secretes mucus and lives in the body openings and digestive tract.

Mucus—viscid, watery secretion of the mucous glands, composed of water, mutin, inorganic salts, epithelial cells, leucocytes, and granular matter.

Multiparous—producing more than one offspring at birth, or experiencing more than one parturition.

Mutant—offspring possessing a characteristic that was not inherited from the parents.

Mutation—permanent change in certain characteristics of an organism, resulting from an alteration or change in the genes.

Myocarditis—inflammation of the muscular walls of the heart or myocardium.

Nausea—tendency to vomit.

Necrosis—local death of tissues characterized by distortion and, in some cases, disintegration of the cells.

Neonatal—pertaining to the first four weeks of life.

Nephritis—inflammation of the kidney.

Non-specific immunity—increase of antibodies or production of immunity, resulting from the injection of some non-specific antigen.

Nucleus—controlling portion of a cell.

Nutrient—substance that can be used as food.

Nymphomania—abnormal sexual desire in a female.

Obligate aerobe—organism that must have free oxygen for its growth.

Opisthotonos—tetanic spasm in which the head is drawn backward and the back is arched.

Oral—pertaining to the mouth.

Orchitis—inflammation of a testis, which is marked by pain and swelling and a feeling of weight.

Organic—pertaining to substances derived from living organisms.

Osteomalacia—softening of the bones.

Oxidation—process of combining any substance with oxygen and the resulting release of energy.

Panzootic—referring to a widespread epidemic among animals.

Paralysis—loss of normal power of motion in some part or organ of the body.

Paraplegia—paralysis of the lower half of the body.

Parasite—organism deriving nourishment from a living plant or host.

Parenteral—administering materials by means other than the alimentary canal.

Paresis—slight or incomplete paralysis.

Parturient—giving birth.

Passive immunity—resistance gained by the injection of serum containing antibodies into an animal or by the ingestion of colostrum by newborn animals.

Pasteurization—process of heating milk or other substances to a moderate temperature for a definite time to kill bacteria.

Pathogenic—capable of producing disease.

Pathogens—disease-producing micro-organisms.

Pathology—study of disease.

Pericarditis—inflammation of the pericardium or membranous sac, which contains the heart.

Pericardium—membrane that encircles the heart.

Peristalsis—(peristaltic) contractions of muscular walls of the alimentary canal to propel contents within its lumen.

Per Os—by the mouth.

Phagocyte—cell capable of ingesting micro-organisms or other foreign bodies.

Phagocytosis—process whereby micro-organisms and injurious cells are engulfed and destroyed by leucocytes.

Phenol coefficient—figure representing the relative killing power of a disinfectant, as compared with phenol acting on the same organism for the same length of time.

Photosynthesis—process by which green plants, using chlorophyll and the energy of sunlight, produce carbohydrates from water and carbon dioxide, and release oxygen.

Plasma—fluid portion of blood before clotting has occurred.

Poison—any substance ingested, inhaled, or developed within the body that causes or may cause damage or disturbance of function.

Polypnea—rapid or panting respiration.

Polyuria—excessive secretion of urine.

Polyvalent—designating a stock vaccine made up of many strains of the same organism or different organisms.

Portal of entry—route that a micro-organism must take in entering the body in order to have any effect.

Post-mortem—examination after death.

Post-partum—period immediately following parturition.

Predator—organism that lives by feeding on another.

Predisposing cause—stress; anything that renders an animal liable to an attack of disease without actually producing it.

Prehension—manner of taking feed into the mouth.

Prenatal—occurring or existing before birth.

Primary cause—principal or original cause of attack.

Prognosis—forecast as to the probable result of an attack of disease, the likelihood of recovery.

Prophylaxis—prevention of disease by various measures.

Pseudohermaphrodite—person or animal having internal genital organs of one sex, while its external genital organs and secondary sex characters resemble in whole or in part those of the opposite sex.

Pruritis—intense itching.

Purulent—containing pus.

Pus—liquid inflammation made up of cells and a thin fluid called *liquor puris*.

Pustule—small elevation of the skin filled with pus or lymph.

Putrefaction—decomposition of animal or vegetable matter, produced by micro-organisms in the absence of oxygen.

Pyemia—general septicemia in which pus is found in the blood.

Pyogenic—pus-producing.

Pyometra—condition in which pus is present in a sealed uterus.

Regurgitate—to return the contents of the paunch to the mouth for further mastication.

Resistance—ability of the individual to ward off infection by bacteria or other micro-organisms.

Rhinitis—inflammation of the mucous membrane of the nose.

Rigor mortis—stiffening of muscles following death.

Rumen—the first of four stomach divisions in ruminants.

Saturated—unable to hold in solution any more of a given solute.

Sensitization—condition of being allergic or sensitive to, for example, a vaccine.

Septicemia—condition in which pathogenic organisms and their associated proteins are present in the blood.

Serous—containing or resembling serum.

Serum—extract of blood that contains antibodies.

Soluble—capable of changing form or changing into a solution.

Somatic—pertaining to the framework of the body.

Spore—reproductive element of one of the lower organisms, such as protozoa.

Sterility—inability to fertilize or produce an egg.

Streptococcus—round-shaped bacterium growing in a chain-like formation.

Stress—borderline condition between health and disease.

Subcutaneous—beneath the skin.

Suppuration—pus formation.

Susceptible—exposed, vulnerable, or non-resistant due to weakness or disease; without defenses.

Symbiosis—living together, or close association of two dissimilar organisms, with mutual benefits for each other.

Symptom—sign or signal, reaction, evidence of disease.

Syndrome—group of signs or symptoms that occur together and characterize a disease.

Synergism—ability of two or more organisms or chemical substances to bring about changes that neither can accomplish alone.

Tachycardia—excessively rapid heartbeat or pulse rate.

Taeniacides—drugs that destroy tapeworms.

Taeniafuges—drugs that expel tapeworms.

Tetany—syndrome manifested by sharp flexion of the joints, muscle twitchings, and convulsions.

Therapy—medical treatment of disease (usually excludes surgery).

Titer—standard strength or degree of concentration of a solution as established titration.

Toxemia—general intoxication due to the absorption of bacterial products, usually toxins.

Toxin—poison.

Toxoid—toxin that has been chemically altered so that it is no longer toxic but is still capable of uniting with antitoxins and/or stimulating antitoxin formation.

Trauma—wound or injury.

Traumatic—resulting from an injury or wound.

Vaccination—injection of a vaccine to induce immunity.

Vaccine—substance composed of attenuated organisms that produces active immunity.

Vector—carrier, especially the animal host that carries protozoal disease germs from one human host to another.

Venereal disease—disease due to or propagated by sexual contact, for example, trichomoniosis, vibriosis.

Ventral—pertaining to or relating to the belly or underside; opposite the dorsal or back.

Vermicides—drugs that destroy intestinal worms.

Vermifuges—drugs that expel intestinal worms.

Verminous—pertaining to or due to worms.

Vesicle—blister-like sac or small bladder containing fluid.

Viable—living.

Viremia—condition where viral organisms are found in an animal's blood system.

Virulence—capacity of a micro-organism to produce disease.

Virulent—exceedingly harmful and capable of producing symptoms of disease.

Virus—submicroscopic infective agent that causes disease.

Viscosity—resistance of molecules to flow over and around each other; a measure of consistency of a substance.

Zoonosis—disease transmissible from animal to human.

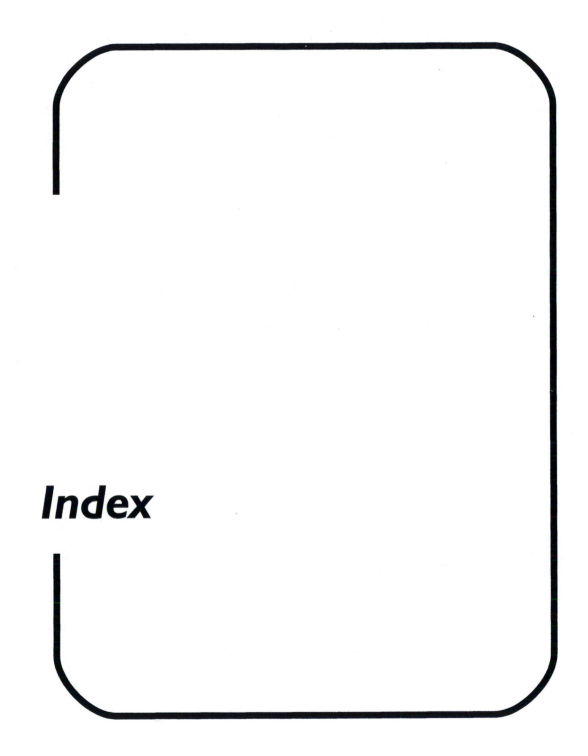

Index

Index